Assistance Robotics and Biosensors

Assistance Robotics and Biosensors

Special Issue Editors

Fernando Torres
Santiago Puente
Andrés Úbeda

MDPI • Basel • Beijing • Wuhan • Barcelona • Belgrade

MDPI

Special Issue Editors
Fernando Torres, Santiago Puente and Andrés Úbeda
University of Alicante
Spain

Editorial Office
MDPI
St. Alban-Anlage 66
4052 Basel, Switzerland

This is a reprint of articles from the Special Issue published online in the open access journal *Sensors* (ISSN 1424-8220) from 2017 to 2018 (available at: https://www.mdpi.com/journal/sensors/special_issues/Assistance_Robotics_Biosensors)

For citation purposes, cite each article independently as indicated on the article page online and as indicated below:

LastName, A.A.; LastName, B.B.; LastName, C.C. Article Title. *Journal Name* **Year**, *Article Number*, Page Range.

ISBN 978-3-03897-394-2 (Pbk)
ISBN 978-3-03897-395-9 (PDF)

Contents

About the Special Issue Editors . . . . **vii**

Fernando Torres, Santiago T. Puente and Andrés Úbeda
Assistance Robotics and Biosensors
Reprinted from: *Sensors* **2018**, *18*, 3502, doi:10.3390/s18103502 . . . . **1**

Dorin Copaci, David Serrano, Luis Moreno and Dolores Blanco
A High-Level Control Algorithm Based on sEMG Signalling for an Elbow Joint SMA
Exoskeleton
Reprinted from: *Sensors* **2018**, *18*, 2522, doi:10.3390/s18082522 . . . . **4**

**Eugenio Ivorra, José M. Catalán, Mario Ortega, Santiago Ezquerro, Luis Daniel Lledó,
Nicolás Garcia-Aracil and Mariano Alcañiz**
Intelligent Multimodal Framework for Human Assistive Robotics Based on Computer
Vision Algorithms
Reprinted from: *Sensors* **2018**, *18*, 2408, doi:10.3390/s18082408 . . . . **22**

**Vu Thi Thu Huong, Felipe Gomez, Pierre Cherelle, Dirk Lefeber, Ann Nowé and
Bram Vanderborght**
ED-FNN: A New Deep Learning Algorithm to Detect Percentage of the Gait Cycle for Powered
Prostheses
Reprinted from: *Sensors* **2018**, *18*, 2389, doi:10.3390/s18072389 . . . . **46**

**Andrés Úbeda, Brayan S. Zapata-Impata, Santiago T. Puente, Pablo Gil, Francisco Candelas
and Fernando Torres**
A Vision-Driven Collaborative Robotic Grasping System Tele-Operated by
Surface Electromyography
Reprinted from: *Sensors* **2018**, *18*, 2366, doi:10.3390/s18072366 . . . . **65**

Karina de O. A. de Moura and Alexandre Balbinot
Virtual Sensor of Surface Electromyography in a New Extensive Fault-Tolerant
Classification System
Reprinted from: *Sensors* **2018**, *18*, 1388, doi:10.3390/s18051388 . . . . **76**

Marisol Rodríguez-Ugarte, Eduardo Iáñez, Mario Ortiz and José M. Azorín
Effects of tDCS on Real-Time BCI Detection of Pedaling Motor Imagery
Reprinted from: *Sensors* **2018**, *18*, 1136, doi:10.3390/s18041136 . . . . **96**

Han Sun, Xiong Zhang, Yacong Zhao, Yu Zhang, Xuefei Zhong and Zhaowen Fan
A Novel Feature Optimization for Wearable Human-Computer Interfaces Using Surface
Electromyography Sensors
Reprinted from: *Sensors* **2018**, *18*, 869, doi:10.3390/s18030869 . . . . **110**

Alan Floriano, Pablo F. Diez and Teodiano Freire Bastos-Filho
Evaluating the Influence of Chromatic and Luminance Stimuli on SSVEPs from Behind-the-Ears
and Occipital Areas
Reprinted from: *Sensors* **2018**, *18*, 615, doi:10.3390/s18020615 . . . . **141**

Qingsong Ai, Chengxiang Zhu, Jie Zuo, Wei Meng, Quan Liu, Sheng Q. Xie and Ming Yang
Disturbance-Estimated Adaptive Backstepping Sliding Mode Control of a Pneumatic
Muscles-Driven Ankle Rehabilitation Robot
Reprinted from: *Sensors* **2018**, *18*, 66, doi:10.3390/s18010066 . . . . **153**

Ana Cecilia Villa-Parra, Denis Delisle-Rodriguez, Jessica Souza Lima, Anselmo Frizera Neto and Teodiano Bastos
Knee Impedance Modulation to Control an Active Orthosis Using Insole Sensors
Reprinted from: *Sensors* **2017**, *17*, 2751, doi:10.3390/s17122751 . . . . **174**

About the Special Issue Editors

Fernando Torres, PhD Industrial Engineering. Fernando Torres was born in Granada, where he attended primary and high school. He moved to Madrid to undertake a degree in Industrial Engineering at the Polytechnic University of Madrid, where he also carried out his PhD thesis. In the last year of his PhD thesis he became a full-time lecturer and researcher at the University of Alicante, and he has worked there ever since. He is currently a professor at this University. He directs the research group, "Automatics, Robotics, and Artificial Vision", was founded in 1996 at the University of Alicante. His research focuses on automation and robotics (intelligent robotic manipulation, visual control of robots, robot perception systems, neurorobotics, field robots and advanced automation for industry 4.0, and artificial vision engineering), and e-learning. In these areas, it currently has more than fifty publications in JCR (ISI) journals and more than a hundred papers in international congresses. In addition, he is a member of TC 5.1 and TC 9.4 of the IFAC, a Senior Member of the IEEE and a member of CEA. He has been deputy director of the EPS, deputy director of the department and director of secretariat at the University of Alicante. He was deputy of the field of Electrical, Electronic, and Automatic Engineering (IEL) at the National Agency of Evaluation and Prospective (ANEP) from 2009 to 2011, and from 2012 to February 2016 he was the coordinator of this IEL area at the National Agency of Evaluation and Prospective (ANEP). Since 2018 he has been the coordinator of the field of Electrical, Electronic, and Automatic Evaluation of the ANECA. Since July 2018, he has been the coordinator of the field of Electrical, Electronic, and Automatic (IEA) of the Spanish Agency of State Research (AEI). At the moment and since its creation, he is the coordinator of the Degree in Robotic Engineering at the University of Alicante, which is the first degree of Robotic Engineering in Spain.

Santiago Puente, PhD Applied Computing. Santiago Puente was born in La Coruña. He moved to Alicante, where he attended primary, high school, and undertook a degree in Computer Engineering at the University of Alicant, where he also carried out his PhD thesis. He obtained a grand to perform his PhD thesis. After one year, he became a full-time lecturer and researcher at the University of Alicante, and he has worked there ever since. He is currently a professor at this University. He is a research member of the research group "Automatics, Robotics, and Artificial Vision", founded in 1996 at the University of Alicante. His research focuses on robotics (intelligent robotic manipulation, neurorobotics, myoelectric control, marine robots, and automatic disassembly) and e-learning. In these fields, he currently has around twenty publications in JCR (ISI) journals and nearly a hundred papers in congresses. In addition, he is a member of CEA. He is deputy director of the EPS and the head of studies of the Degree in Robotic Engineering at the University of Alicante, the first degree in Robotic Engineering in Spain.

Andrés Úbeda, PhD in Bioengineering. Andrés Úbeda is an assistant professor in the Department of Physics, System Engineering, and Signal Theory at the University of Alicante; a member of the research group AUROVA; and a regular collaborator at the Brain–Machine Interface Systems Lab (Miguel Hernández University of Elche). He holds a MSc in Industrial Engineering from UMH (2009), and a PhD in Bioengineering from UMH (2014). He has been a visiting researcher at the Defitech Chair in Non-Invasive Brain–Machine Interfaces (CNBI) at EPFL (École Polytechnique Fédérale de Lausanne, Switzerland) (May–July 2013) and at the Institute of Neurorehabilitation Systems (Universitätsmedizin Göttingen, Germany) (Sep 2015–Sep 2016). His main research is focused on studying the neuromuscular mechanisms of motor control and coordination, by assessing the descending motor pathways during movement execution. His other research topics are centered on the analysis of cortical information in the decoding of motor intentions, and their use in motor neurorehabilitation procedures, human-machine interaction, and assistive technologies in the field of neurorobotics, particularly focused on myoelectric and brain-controlled devices.

sensors

MDPI

Editorial
Assistance Robotics and Biosensors

Fernando Torres [1,2,*], Santiago T. Puente [1,2] and Andrés Úbeda [1,2]

1   Department of Physics, System Engineering and Signal Theory, University of Alicante, 03690 Alicante, Spain; santiago.puente@ua.es (S.T.P.); andres.ubeda@ua.es (A.Ú.)
2   Computer Science Research Institute, University of Alicante, 03690 Alicante, Spain
*   Correspondence: fernando.torres@ua.es; Tel.: +34-965-90-9491

Received: 10 October 2018; Accepted: 15 October 2018; Published: 17 October 2018

Abstract: This Special Issue is focused on breakthrough developments in the field of biosensors and current scientific progress in biomedical signal processing. The papers address innovative solutions in assistance robotics based on bioelectrical signals, including: Affordable biosensor technology, affordable assistive-robotics devices, new techniques in myoelectric control and advances in brain–machine interfacing.

Keywords: electromyographic (EMG) sensors; electroencephalographic (EEG) sensors; assistance robotics applications; robotic exoskeletons; robotic prostheses; advanced biomedical signal processing

1. Introduction

In recent years, the use of bioelectrical information to enhance traditional motor-disability assistance has experienced significant growth, mostly based on the development and improvement of biosensor technology and the increasing interest in solving accessibility limitations in a more natural and effective way. For that purpose, control outputs are directly decoded from the user's biological information. Biomedical signals, recorded from cortical or muscular activity, are used to interact with external devices, such as robotics exoskeletons or assistive robotic arms or hands. However, efforts are still needed to make these technologies affordable for end users, as current biomedical devices are still mostly present in rehabilitation centers, hospitals and research facilities.

2. Contributions

This Special Issue collected ten outstanding papers covering different aspects of assistance robotics and biosensors. In the following, a brief summary of the scope and main contributions of each of these papers is provided as a teaser for the interested reader.

One of the most important issues in assistive robotics is helping people with special needs or disabilities to adequately perform rehabilitation exercises in the friendliest way. In "A High-Level Control Algorithm Based on sEMG Signalling for an Elbow Joint SMA Exoskeleton" [1] the authors designed a high-level control algorithm capable of generating position and torque references from surface electromyography signals (sEMG). They applied this algorithm to a shape memory alloy (SMA)-actuated exoskeleton used in active rehabilitation therapies for elbow joints.

In the same field of assistance, the paper "Intelligent Multimodal Framework for Human Assistive Robotics Based on Computer Vision Algorithms" [2] shows a multimodal interface based on computer vision, which has been integrated into a robotic system together with other sensory systems (electrooculography (EOG) and electroencephalography (EEG)). The results were part of an European project, AIDE, whose purpose is to contribute to the improvement of current assistance technologies.

Undoubtedly, rehabilitation tasks require friendly systems and exoskeletons that are at the same time more precise. In this sense, the improvement of hardware systems is crucial. In the paper

"ED-FNN: A New Deep Learning Algorithm to Detect Percentage of the Gait Cycle for Powered Prostheses" [3] the authors propose a novel gait detection algorithm that can predict a full gait cycle discretized within a 1% interval. In addition, the system provides an opportunity to eliminate detection delays for real-time applications.

In the case of assistive robotics, another field of great interest is creating equipment and friendly environments for people with physical movement disabilities. In this context, the authors of the paper "A Vision-Driven Collaborative Robotic Grasping System Tele-Operated by Surface Electromyography" [4] propose an interface that combines computer vision with electromyography, aiming to allow a person with impeded movement to teleoperate a robotic hand. Experiments were carried out on basic operations of the grasping and shifting of objects.

The problem of more reliable myoelectric systems is addressed in the paper "Virtual Sensor of Surface Electromyography in a New Extensive Fault-Tolerant Classification System" [5]. The authors propose extending the use of virtual sensors used in other research fields to the myoelectric field. With this, they provide a new, extensive, fault-tolerant classification system to maintain the classification accuracy after the occurrence of the following contaminants: ECG interference, electrode displacement, movement artifacts, power line interference, and saturation. The time-varying autoregressive moving average (TVARMA) and time-varying Kalman filter (TVK) models were compared to define the most robust model for the virtual sensor.

In "Effects of tDCS on Real-Time BCI Detection of Pedaling Motor Imagery" [6] the authors sought to strengthen the cortical excitability over the primary motor cortex (M1) and the cerebro-cerebellar pathway by means of a new transcranial direct current stimulation (tDCS) configuration to detect lower limb motor imagery (MI) in real time using two different cognitive neural states: relaxed and pedaling MI. In this case, the use of software or hardware techniques with the purpose of improving the reception of signals was again treated.

The use of assistive robotics is justified when it improves the life of people with certain disabilities. In this sense, the use of appropriate signals for each case is of great importance. In "A Novel Feature Optimization for Wearable Human–Computer Interfaces Using Surface Electromyography Sensors" [7], the authors carried out a study of the signals and selection of optimal-feature selection made according to a modified entropy criteria (EC) and Fisher discrimination (FD) criteria. The feature selection results were evaluated using four different classifiers, and compared with other conventional feature subsets. These experiments validated the feasibility of the proposed real-time wearable HCI system and algorithms, providing a potential assistive device interface for persons with disabilities.

In the same field of achieving improvements for people with certain disabilities, the paper titled "Evaluating the Influence of Chromatic and Luminance Stimuli on SSVEPs from Behind-the-Ears and Occipital Areas" [8] presents a study of chromatic and luminance stimuli in low-, medium-, and high-frequency stimulation to evoke steady-state visual evoked potential (SSVEP) in the behind-the-ears area. These findings will aid in the development of more comfortable, accurate and stable BCI with electrodes positioned in the behind-the-ears (hairless) areas.

The use of exoskeletons in rehabilitation therapies is increasingly widespread and it is one of the most promising and expected future lines in the field of assistive robotics. The authors of "Disturbance-Estimated Adaptive Backstepping Sliding Mode Control of a Pneumatic Muscle-Driven Ankle Rehabilitation Robot" [9] propose the improvement of a therapeutic robot for the rehabilitation of ankle injuries. To do this, they proposed a new method of adaptive backstepping sliding mode control (ABS-SMC) in order to solve the PM's nonlinear characteristics during operation and to tackle the human–robot uncertainties in rehabilitation.

The correct adaptation of an exoskeleton to patients with knee problems is dealt with in "Knee Impedance Modulation to Control an Active Orthosis Using Insole Sensors" [10]. In this case, the authors propose a method for online knee impedance modulation that generates variable gains through the gait cycle according to the users' anthropometric data and gait sub-phases recognized through footswitch signals.

Acknowledgments: The authors of the submissions have expressed their appreciation to the work of the anonymous reviewers and the *Sensors* editorial team for their cooperation, suggestions and advice. Likewise, the special editors of this Special Issue thank the staff of *Sensors* for the trust shown and the good work done.

Conflicts of Interest: The authors declare no conflicts of interest.

References

1. Copaci, D.; Serrano, D.; Moreno, L.; Blanco, D. A High-Level Control Algorithm Based on sEMG Signalling for an Elbow Joint SMA Exoskeleton. *Sensors* **2018**, *18*, 2522. [CrossRef] [PubMed]
2. Ivorra, E.; Ortega, M.; Catalán, J.M.; Ezquerro, S.; Lledó, L.D.; Garcia-Aracil, N.; Alcañiz, M. Intelligent Multimodal Framework for Human Assistive Robotics Based on Computer Vision Algorithms. *Sensors* **2018**, *18*, 2408. [CrossRef] [PubMed]
3. Vu, H.T.T.; Gomez, F.; Cherelle, P.; Lefeber, D.; Nowé, A.; Vanderborght, B. ED-FNN: A New Deep Learning Algorithm to Detect Percentage of the Gait Cycle for Powered Prostheses. *Sensors* **2018**, *18*, 2389. [CrossRef] [PubMed]
4. Úbeda, A.; Zapata-Impata, B.S.; Puente, S.T.; Gil, P.; Candelas, F.; Torres, F. A Vision-Driven Collaborative Robotic Grasping System Tele-Operated by Surface Electromyography. *Sensors* **2018**, *18*, 2366. [CrossRef] [PubMed]
5. De Moura, K.D.O.; Balbinot, A. Virtual Sensor of Surface Electromyography in a New Extensive Fault-Tolerant Classification System. *Sensors* **2018**, *18*, 1388. [CrossRef] [PubMed]
6. Rodriguez-Ugarte, M.D.L.S.; Iáñez, E.; Ortiz-Garcia, M.; Azorín, J.M. Effects of tDCS on Real-Time BCI Detection of Pedaling Motor Imagery. *Sensors* **2018**, *18*, 1136. [CrossRef] [PubMed]
7. Sun, H.; Zhang, X.; Zhao, Y.; Zhang, Y.; Zhong, X.; Fan, Z. A Novel Feature Optimization for Wearable Human-Computer Interfaces Using Surface Electromyography Sensors. *Sensors* **2018**, *18*, 869. [CrossRef] [PubMed]
8. Floriano, A.; Diez, P.F.; Freire Bastos-Filho, T. Evaluating the Influence of Chromatic and Luminance Stimuli on SSVEPs from Behind-the-Ears and Occipital Areas. *Sensors* **2018**, *18*, 615. [CrossRef] [PubMed]
9. Ai, Q.; Zhu, C.; Zuo, J.; Meng, W.; Liu, Q.; Xie, S.Q.; Yang, M. Disturbance-Estimated Adaptive Backstepping Sliding Mode Control of a Pneumatic Muscles-Driven Ankle Rehabilitation Robot. *Sensors* **2017**, *18*, 66. [CrossRef] [PubMed]
10. Villa-Parra, A.C.; Delisle-Rodriguez, D.; Souza Lima, J.; Frizera-Neto, A.; Bastos, T. Knee Impedance Modulation to Control an Active Orthosis Using Insole Sensors. *Sensors* **2017**, *17*, 2751. [CrossRef] [PubMed]

Article

A High-Level Control Algorithm Based on sEMG Signalling for an Elbow Joint SMA Exoskeleton

Dorin Copaci *,†, David Serrano †, Luis Moreno † and Dolores Blanco †

Department of Systems Engineering and Automation, Carlos III University of Madrid, 28911 Leganés, Madrid, Spain; davserra@ing.uc3m.es (D.S.); moreno@ing.uc3m.es (L.M.); dblanco@ing.uc3m.es (D.B.)
* Correspondence: dcopaci@ing.uc3m.es Tel.: +34-91624-8812
† These authors contributed equally to this work.

Received: 19 June 2018 ; Accepted: 30 July 2018; Published: 2 August 2018

Abstract: A high-level control algorithm capable of generating position and torque references from surface electromyography signals (sEMG) was designed. It was applied to a shape memory alloy (SMA)-actuated exoskeleton used in active rehabilitation therapies for elbow joints. The sEMG signals are filtered and normalized according to data collected online during the first seconds of a therapy session. The control algorithm uses the sEMG signals to promote active participation of patients during the therapy session. In order to generate the reference position pattern with good precision, the sEMG normalized signal is compared with a pressure sensor signal to detect the intention of each movement. The algorithm was tested in simulations and with healthy people for control of an elbow exoskeleton in flexion–extension movements. The results indicate that sEMG signals from elbow muscles, in combination with pressure sensors that measure arm–exoskeleton interaction, can be used as inputs for the control algorithm, which adapts the reference for exoskeleton movements according to a patient's intention.

Keywords: exoskeleton; electromyographic (EMG); control systems

1. Introduction

The development of advanced robotic assistive technologies has gained special attention in the scientific community over the last decades. Millions of people worldwide rely on assistive devices to improve their quality of life. For this reason, there is a need to further promote the development of assistive devices by pooling the efforts of engineers and clinicians, together with the feedback and experiences of users, to improve these technologies.

Aging of populations, mainly in developed countries, and the incidence of diseases, such as stroke, spinal cord injuries, and various musculoskeletal injuries, have increased the need for health resources, especially those dedicated to the rehabilitation process. Rehabilitation therapy is the process that assists a person in recovering from serious disorders after an injury, illness, or surgery that causes motor impairments. One of the most common rehabilitation methods consists of musculoskeletal rehabilitation to improve motor functions and the autonomy of patients in typical daily activities. In standard rehabilitation methods, every patient needs one or more therapists, because the therapist must directly manipulate the affected limb. This implies a huge consumption of healthcare and financial resources. The use of robotic devices as rehabilitation tools is proposed as a complement to the traditional rehabilitation sessions effectuated by therapists and can reduce the need for human resources. The main advantage offered by the use of robotic systems in rehabilitation is the capacity to support the work of physiotherapists in simple therapies with repetitive movements, reducing the need for the presence of the therapist. In this way, the costs associated with rehabilitation therapies can be reduced, allowing the same therapies to be carried out for longer, if the patient requires it, and for a larger number of patients to be treated simultaneously. Robotic systems have proven to be as effective as conventional therapy [1,2].

Among the most promising assistive robotic technologies are exoskeletons. An exoskeleton robot is a wearable robot designed to assist limb motions. The ease of use and the intuitive control of the robotic exoskeleton are crucial aspects for acceptance by patients. A step towards a more effective and intuitive control of upper-limb exoskeletons is the use of a myoelectric signal to detect the user's motion intention. Myoelectric signals (MESs) contain information from which data about user movement intention in terms of muscular contractions can be extracted. Control based on MESs provides a more natural interaction with the exoskeleton.

A wearable shape memory alloy (SMA)-actuated exoskeleton with two degrees of freedom (DOF) (for flexion–extension and pronation–supination) was presented in [3]. In that work, the control algorithm made it possible to control the exoskeleton tracking a reference for passive rehabilitation therapy in flexion [4]—only actuating in flexion and recuperating (during the extension movement) with the aid of gravity—and actuating with two SMA-based actuators in flexion and extension [5]. The reference pattern in both cases represents a repetitive movement (for example, a sinusoidal trajectory) defined by the therapist, which makes the rehabilitation passive. In order to activate, in a natural manner, the exoskeleton according the user's intended motion, the control algorithm proposed in this work uses input signals to the controller based on a skin surface electromyogram (sEMG). A key aspect for the success of robotic rehabilitation therapies is to keep the patient involved in carrying out the therapy. This is the objective pursued with the proposed control algorithm. Our new control algorithm analyzes the signal sEMG to detect that the patient is involved in the realization of the movement—that is, the patient intends to move their arm, even if they lack sufficient muscular strength to carry out the movement. The exoskeleton will only receive a reference position to which it will move if the patient is generating an sEMG signal indicating their intention to move.

In order to generate the reference position pattern with good precision, the sEMG normalized signal is compared with a pressure sensor signal to detect the intention to move. The pressure sensor is used to estimate the motion of the user through the force between the user and the robot. The proposed approach has been tested in a single joint for the flexion–extension task.

1.1. Electromyogram Signals

Electromyography (EMG) signals of human muscles are biological signals that record the electrical potential generated by muscle cells to contract. It can be used to detect the user's intention to move, since the amplitude directly correlates with the user's muscle activity. Moreover, according to [6], the EMG signal starts about 20–80 ms before the muscle contraction, so it allows anticipation of the motion intention.

EMG signals can be classified into two types: intramuscular EMG signals, detected from inside of the muscles; and surface EMG signals (sEMGs), detected from the skin surface. The intramuscular EMG signals give a better muscle activation pattern, but their use requires an invasive extraction procedure. Therefore, skin surface EMG signals are used as input for control robotic systems. Although the extraction of sEMG signals is relatively simple, the precise estimation of the motion is difficult because of the variability of EMG signals, which can be affected by multiple factors. EMG signals vary from one person to another, and even between two sessions with the same person making the same movement. In addition, each joint movement involves the activation of many muscles, and one muscle can be involved in various joint movements. Factors such as the changes in limb posture affect the relationship between the EMG signal level and motion estimation. The anatomy and physiological conditions of the user, including any diseases, injuries, fatigue, or pain, also modify EMG signals. Consequently, control strategies that employ sEMG signals require adjusting the controller to the particular user and, in many cases, calibrating the system during each session. Therefore, raw EMG signals are not suitable as input signals to a controller. Data must be filtered and normalized using the maximum voluntary contraction (MVC) level of the user [7].

In the case of an elbow exoskeleton, it must taken into account that the human elbow motion is activated by antagonistic pairs of muscles—biceps (agonist) and triceps (antagonist). According to [8], the biceps brachii, brachioradialis, and brachialis muscles are involved in elbow flexion. Biceps muscles are easily accessible from the skin surface. For this reason, the sEMG electrode circuit used in this work was situated over the bicep muscles to detect the intention of movement in the elbow joint.

1.2. Related Work

Since the 1960s, sEMG signals have been a common way of controlling prostheses [9,10]. More recently, EMG signals have been used for motion control of numerous robotic systems [11,12], prostheses [13], and robotics exoskeletons [14]. A broad review of the related literature can be found in [15].

Prosthesis and exoskeleton movements have frequently been controlled using EMG signals from muscles not involved in the movement. For example, Benjuya and Kenny [14] used the EMG signals from the wrist extensors of the forearm to open/close a pinch action. Also, in [7], the EMG signal from the ipsilateral biceps was used to develop an extremely reliable natural reaching and pinching algorithm. The EMG signals from the residual biceps and triceps of a user with transhumeral amputation have been proposed to control a robotic elbow in a learning from demonstration approach [16].

In the last decades, several research groups have worked on different control algorithms based on EMG signals for use with prostheses and exoskeletons. Many of these works have focused on the use of neural networks and fuzzy algorithms to distinguish the user's intention for movement based on the EMG signals of various muscles. Hudgins [17] proved that artificial neural networks are practical for controlling prostheses by classifying different movements from EMG signals. In [18], the authors evaluated a time-delayed artificial neural network to predict shoulder and elbow motions using only EMG signals from six shoulder and elbow muscles as inputs. Results from both able-bodied subjects and subjects with tetraplegia indicate that the EMG signals contain a significant amount of information about arm movement that could be exploited in advanced control systems.

In [19], a hierarchical neurofuzzy controller based on the EMG signals was presented for real-time control of a shoulder and elbow motion exoskeleton. A wrist force sensor was used when the EMG activity levels were low. In [20,21], an EMG signal-based control method for a seven degrees of freedom (7DOF) upper-limb motion assistive exoskeleton robot (SUEFUL-7) was proposed. In their method, an impedance controller was applied to the muscle-model-oriented control method. Impedance parameters were adjusted in real-time as a function of the upper-limb posture and EMG activity levels. The work presented in [22] proposes a more advanced EMG-based impedance control method for an upper-limb exoskeleton. In that work, a neurofuzzy matrix modifier made the controller adaptable to all upper-limb postures of any user. The neurofuzzy modifier is a neural network with fuzzy reasoning that is trained to adjust its output to each user before operation. The method was applied to the 7DOF exoskeleton for upper-limb joint motions, as presented in [20]. They used 16 channels of EMG signals, with each electrode mainly corresponding to one muscle. Moreover, two force/torque sensors were used to estimate the forces between robot and user. The control algorithm was able to distinguish between different kinds of motion.

As can be seen from the previous studies cited, the EMG-based neurofuzzy control method has proven its effectiveness in controlling exoskeleton robots. However, the rules of control are complicated when increasing the number of degrees of freedom of the exoskeleton.

The amplitude of the EMG signals reflects the muscles' activity levels. Many methods have been developed to estimate human muscular torque from EMG activity levels, using this information to control joint torques in robots. Due to the many factors that modify the EMG signals, this type of control requires a complex calibration process to adapt to the variability of the signals, and depends on the user and the session conditions. In the experimental work presented in [23], the reactions of 10 healthy subjects to the assistance provided through a proportional EMG control applied by an elbow powered exoskeleton were studied. The system did not require calibration. Their results showed that in order to assist movement, an accurate estimate of the muscular torque may be unnecessary and a simpler control algorithm can be more efficient.

The control algorithm presented in this work is similar to the binary control algorithm used in [7,24]. In [7], DiCicco tested binary "on–off" control, and variable and natural control algorithms based on EMG signals. They validated that the EMG signal from the ipsilateral biceps could be used to develop an extremely reliable natural reaching and pinching algorithm. A specific EMG threshold value serves to determinate the output binary value "on" if the EMG signal from the biceps muscle is above the threshold and "off" when it is below.

In our case, the rehabilitation exoskeleton has been designed with the objective of assisting in therapies consisting of performing repetitive movements. This type of therapy is typical of the first phases of rehabilitation, where the patient must repeat defined movements of a certain joint in order to recover muscular strength and increase the range of motion lost. In this context, it is not necessary to discriminate the type of movement that the patient wants to make. The proposed algorithm tries to determine the intention of the patient to initiate a certain movement and its ability to maintain it, even if the patient lacks sufficient muscular strength to carry it out. Consequently, the sEMG signals are detected and analyzed only from muscles directly related to the movement being assisted. In this case, the biceps muscles were targeted to detect voluntary flexion of the elbow joint. In the proposed algorithm, the triceps muscle activity was not considered, as the control algorithm has the limitation that if co-contraction happens and the extension signal is not detected by the pressure sensors, the system needs to be manually turned off.

Our proposed approach fuses sensor data with EMG signals. Force sensors were used to check the interaction between the exoskeleton and the user. In this way, only when the patient actively tries to execute the movement does the control algorithm initiate the movement of the exoskeleton. A similar approach was implemented in [20]. This approach reduces errors caused by low EMG levels or external unexpected forces affecting the patient's arm.

This paper presents an algorithm capable of generating the reference pattern in position and torque based on surface electromyography (sEMG) signals and pressure sensors for high-level control of the SMA exoskeleton. The first part of the paper presents an introduction to the problem. In the second section, materials and methods are explained, including a description of the elbow exoskeleton. The initial assembly of SMA-based actuators is presented, and the elbow exoskeleton design is shown. The electronic hardware is also presented in the second section. The final part of the second section is devoted to explaining the high-level control algorithm in detail. In the third section, the results are presented: first, simulation test results of the high-level control algorithm, followed by performance evaluation of the proposed control method, based on experiments with healthy subjects that were carried out with the SMA elbow exoskeleton. The final part presents brief conclusions of the paper.

2. Materials and Methods

This section presents a brief description of the hardware architecture on which the tests were run: the structure of the exoskeleton, the actuators, and the sensors which are involved in the algorithm, as well as the high-level control algorithm capable of generating the reference patterns for position and torque; the algorithm provides high-level control and is based on sEMG signals and pressures sensors.

2.1. Elbow SMA Exoskeleton

In previous publications, a wearable SMA exoskeleton was presented with two DOF, which permits mobilization of the elbow joint in flexion–extension and pronation–supination movements [3,5]. This device used an SMA actuator for the actuation system and was the first elbow joint rehabilitation device powered by this technology. It has the potential to be a light device, with a weight less than 1 kg (structure, actuators, and electronics), noiseless operation, and low-cost fabrication. The actuator structure is described in Section 2.1.1.

2.1.1. Actuator Design

The simple SMA-based actuator (with only one SMA wire) used in this work was presented in [25]. The SMA wire is made of a metallic alloy—a common mixture of nickel and titanium, called Nitinol [26]. It has the property of recovering its original shape (memorized shape) between two thermic transformation phases: the martensite phase (at low temperature) and austenite phase (at high temperature). The principle on which it works is based on the heating effect (Joule effect), where electrical energy is transformed into thermal energy, after which the thermal energy is transformed into mechanical energy. During this transformation, the SMA wire undergoes a variation of total length, between 3% and 5%. As a function of the diameter and alloy type, the actuator can exert different forces. A 0.51 mm diameter wire of Flexinol® [26] can exert a force of about 35.6 N (with a lifetime of tens of millions of cycles under these force conditions). The SmartFlex® [27] wire with the same diameter can exert a maximum force of 118 N (with a lifetime of hundreds or a few thousand cycles). The activation temperature of the SMA wire depends on the alloy and, in this case, it is 90 °C. In this work, the actuator was composed of multiple SMA wires, a polytetrafluoroethylene (PTFE) tube, a Bowden tube, and the terminal parts (Figure 1).

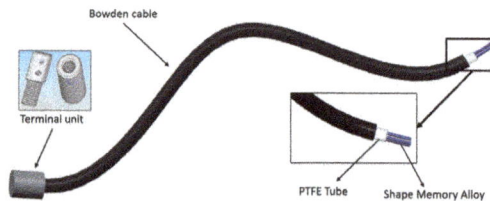

Figure 1. Actuator design. Flexible shape memory alloy (SMA)-based actuator.

- The Bowden cable is a mechanical flexible cable which consists of a flexible inner cable that forms a metal spiral and a flexible outer nylon sheath. This type of wire can guide the SMA actuators and transmit the force. In addition, the metal has the property of dissipating the heat, which is an advantage during the recuperation of the initial position phase.
- The PTFE tube can support high temperatures, more than 250 °C; it is an electrical insulator and does not cause friction.
- The terminal units are used at one end to connect the actuator to the actuated system and at the other to fix the SMA wires to the Bowden cable. They also serve as connectors for the power supply (using the control signal). These units are formed of two pieces that can be screwed to each other to set the tension of the SMA wires. The total SMA wire tension range adjustment is 0.01 m.

There is a relation between the SMA wire diameter, the force, and the cooling time (Table 1). In Table 1, the first column represents the diameter of the wire, the second column is the actuation force which guarantees a lifetime of tens of millions of cycles, and the last two columns represent the cooling time for the two types of wires, with activation at 70 °C and 90 °C, respectively. According to the data shown in the table and the objectives of the exoskeleton, it was decided to work with 0.51 mm wires activated at 90 °C, because the maximum force was obtained with this diameter and the cooling time is lower than when the wire was activated at 70 °C.

If the SMA actuator is designed to operate with the configuration parameters shown in Table 1, the actuator lifetime can be tens of millions of cycles. If the actuator operates with higher forces than those specified, the lifetime drops to only a few thousand cycles.

Table 1. SMA wire characteristics [26].

Diameter Size [mm]	Force [N]	Cooling Time 70 °C [s]	Cooling Time 90 °C [s]
0.025	0.0089	0.18	0.15
0.038	0.02	0.24	0.2
0.050	0.36	0.4	0.3
0.076	0.80	0.8	0.7
0.100	1.43	1.1	0.9
0.130	2.23	1.6	1.4
0.150	3.21	2.0	1.7
0.200	5.70	3.2	2.7
0.250	8.91	5.4	4.5
0.310	12.80	8.1	6.8
0.380	22.50	10.5	8.8
0.510	35.60	16.8	14.0

Regarding the application of the necessary torque to execute defined movements (the necessary torque of each movement was found from a biomechanical simulation [3]), a summary of the system configuration of the actuators can be seen in the Table 2.

Table 2. Exoskeleton actuators.

Movement	SMA Wires	Maximum Actuator Force [N]	Length [m]	Weight [kg]
Flexion	3	354	1.5	0.16
Extension	2	236	1.5	0.15
Pronation	1	118	2	0.1
Supination	1	118	2	0.1

2.1.2. Exoskeleton Design

The exoskeleton was designed according to elbow biomechanics. A biomechanical simulation was performed with the objective of finding the necessary force for various frequencies of movement [3] using the actuator structure presented in Section 2.1.1. The structure of the exoskeleton is displayed in the Figure 2. It was made using simple parts that can be assembled easily, and it permits matching the dimensions of the exoskeleton to those of the user (length of the arm and the forearm), such that the axis of the elbow joint remains aligned with the axis of the exoskeleton. The components of the exoskeleton were a combination of aluminum pieces (such as the Bowden terminals and axis) and other parts made by 3D printer using aluminum with polyamide. The exoskeleton has four points of attachment to the human body, connecting with the arm (two attachments), the forearm, and the hand (Figure 2a). Three force-sensing resistors (FSRs) were placed in the hand piece. These can measure a force between 0.1 and 10 kg. For the safety of the patient, the exoskeleton movement is mechanically limited between 0 and 150 degrees in the elbow flexion–extension direction and between 70 and −70 degrees in the supination–pronation direction. In order to increase comfort, all internal parts in contact with the patient were covered with a soft hypoallergenic material. Compared with current solutions, due to the lack of gears and motors in the mechanism, the proposed rehabilitation device is lightweight. The whole structure with the actuators weighed less than 1 kg. A 960 W DIN rail power supply (24 Vdc/40 A) was used to provide the necessary energy to the actuators. The weight of the power supply unit was 1.9 kg. In addition, it provides noiseless operation, which increases the comfort of the patient during the rehabilitation process. The final version of the exoskeleton installed on the human body can be seen in Figure 2b.

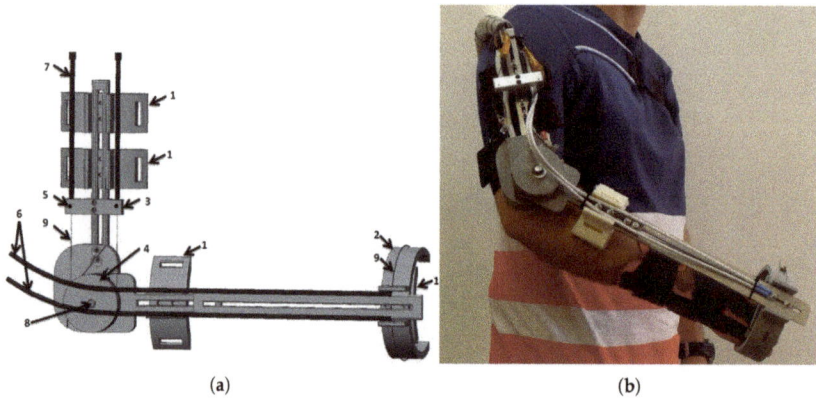

(a)   (b)

Figure 2. SMA exoskeleton design. (**a**) CAD structure: 1—attachment points with the hand and force-sensing resistor (FSR) sensors, 2—fixed structure for supination–pronation, 3—actuator termination for Bowden tube, 4—pulley for linear to rotational transformation, 5—temperature sensors, 6—supination–pronation actuators, 7—flexion–extension actuators, 8—absolute encoder, 9—SMA wires. (**b**) SMA elbow joint exoskeleton on a human body.

2.1.3. Electronic Hardware

The electronic hardware is composed of power electronics, a controller, and sensors placed in the device. The power electronics are capable of supplying the necessary power to four distinct actuators: flexion, extension, supination, and pronation. The system is based on a MOSFET transistor (STMicroelectronics STP310N10F7, STMicroelectronics group, Shanghai, China), which works as a switch circuit and amplifies the control signal (PWM) generated by the controller. The device was connected to the terminal units of the SMA-based actuator.

The controller is a 32-bit microcontroller STM32F4 from STMicroelectronics ®, China, which can be fully programmed with Matlab/Simulink® [28]. It was programmed with four different PWM output ports, which generate the necessary duty cycle for managing the four actuators (each with one or more SMA wires).

The structure of the rehabilitation device includes sensors for position, temperature, force, and sEMG. An absolute angle position sensor with Hall effect (AS5045 made by AMS (Austrian Micro Systems), Premstaetten, Austria) is placed in the shaft of the exoskeleton (pulley for flexion–extension). This sensor has a resolution of 0.0879 degrees and measures the flexion–extension movement. The second position sensor, a membrane potentiometer made by Spectrasymbol, has a length of 0.1 m and is placed on the supination–pronation piece (on the outside) to measure the absolute displacement of this movement. In the same piece, on the inner part which makes the connection between the human forearm and hand and the exoskeleton, three FSR sensors were placed with a 60-degree angular distance between them. These sensors measure the force variation of the elbow during flexion–extension movements—forces that are involved in the high-level control algorithm. Another main sensor involved in this algorithm is the sEMG sensor. The circuit uses three disposable disc electrodes, F-TC1 made by SKINTACT—a low-cost, multi-purpose ECG. It consists of Ag/AgCl electrodes, a conductive gel (Aqua-Tac), an adhesive area with a dimension of 35 × 41 mm, and a snap connection. The gel permits a better connection between the skin and the electrode. This electrode is in the category of non-invasive and wet electrodes.

The sEMG circuit (Figure 3) was made in Carlos III University of Madrid (UC3M), and presents two channels that are connected by two electrodes, which are situated at a distance of 0.03 m from each other over the belly biceps muscle; another channel is used as a reference, which is connected to the last electrode positioned over the shoulder blade. The EMG circuit is composed of various stages, including connectors. There is the differential active feedback stage, the digital stage (where the signal is amplified and filtered), and the stage for the power supply and communication connectors. The communication between the EMG and the microcontroller uses a serial peripheral interface (SPI) bus. For the signal-processing module, we used the same microcontroller STM32F4.

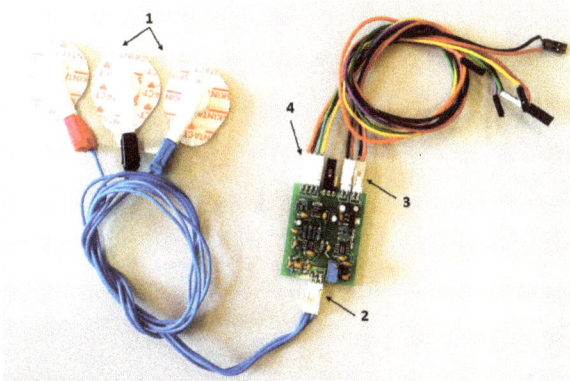

Figure 3. Surface electromyography (sEMG) circuit with two channels and the electrodes: 1—electrodes, 2—electrode connector, 3—connectors for power supply (5 V and GND), 4—connector for serial peripheral interface (SPI) communication.

The temperature sensors are placed in the terminal of the actuator to measure the temperature of the SMA wires, a parameter that is required in the control loop. All the electronics used in this project were based on low-cost components.

The position of the EMG electrodes and FSR sensors over the human body can been seen in Figure 4. A auxiliary piece was built to form the connection between the human hand and forearm-sensor-exoskeleton. This piece (made with a 3D printer using PLA (polylactic acid)) was bent (by introducing it to hot water before the sensors were mounted), taking the form of the patient's forearm, and formed the connection between the forearm and hand with the exoskeleton.

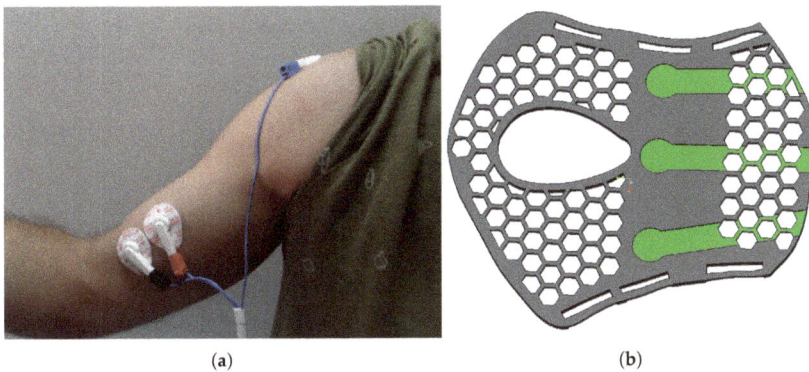

(a)   (b)

Figure 4. (**a**) Surface electromyography (sEMG) electrodes over the subject's arm and shoulder blade. (**b**) The auxiliary piece where are placed the FSR sensors (green parts).

2.2. The High-Level Control Algorithm

Previous publications [3,5] presented a low-level control algorithm based on a BPID (bilinear proportional integral derivative) controller, which governs the SMA-based exoskeleton in position. Their algorithm, involving position and temperature sensors, is capable of acquiring data from the sensors or controlling the exoskeleton in flexion, extension, or in flexion–extension using an antagonistic controller (two BPID controllers in a parallel configuration [5]). With the data acquisition configuration, the SMA-based exoskeleton only offers the possibility to diagnose and evaluate the patient. In the passive mode, the actuators offer all the necessary force to reach and follow the reference position without taking into account the patient force. Through the introduction of sensors for pressure/force and sEMG, the SMA-based exoskeleton offers the possibility of rehabilitation therapies in active mode, where the reference position is generated by the patient's movement intention. In this way, passive reference position (habitually sinusoidal movements) is changed to active reference in a case where the patient presents activity in the motor function (the motor function has been partially affected). Active reference involves the patient undergoing rehabilitation therapy, leading to a faster recovery. The high-level control algorithm, which generates the active rehabilitation therapy (active reference position), uses the sEMG sensors and force-sensing resistor (FSR) sensors, together with position sensors. This is currently available (due to the SMA-based exoskeleton configuration—in fact, the sensors) only for the elbow flexion movement.

The sEMG signals are captured at a sampling frequency of 1 kHz using the circuit presented in Section 2.1.3. The signals are preprocessed: firstly, the raw sEMG data is filtered with a band-pass Butterworth filter, order 8, with a cut-off frequency at 6 dB point below the band-pass value of 20 Hz, and the second cut-off frequency with a value of 480 Hz. This filter was proposed in order to remove the movement artifact [29]. After that, the absolute value of the response of the filter is calculated, and this value is provided to the second filter. This is a low-pass Butterworth filter, order 10, with a cut-off frequency of 20 Hz. The 20 Hz cut-off frequency of the low-pass filter was decided upon according to [29], wherein the authors claim that in the last three decades, various recommendations and standards have been put forth for a cut-off frequency between 5 and 20 Hz. In his publication, he chose such an adequate cut-off frequency of 20 Hz. Both filters were configured at a frequency of 1 kHz. After the filtering process, the EMG signal proceeds to the normalization stage. This consists of an online calibration, where the first 2 s are ignored (in the first 2 s, the circuit experiences some perturbation), and the next 18 s are used to detect the maximum and minimum signals for the normalization process. In this time, the patient is required to flex the forearm as much as possible at least once, followed by an extension movement to return to the original position. During these 18 s, maximum and minimum values are stored to be used in the normalization process, where the normalized signal, E_{norm}, is calculated by Equation (1):

$$E_{norm} = \frac{E_{act} - E_{min}}{E_{max} - E_{min}};$$ (1)

where E_{act} is the actual EMG signal, and E_{min} and E_{max} are the minimum and maximum values of the EMG signal during the 18 s used for normalization.

The entire process of filtering and normalizing of the sEMG signals can be seen in Figure 5.

The normalized signal is compared with a threshold value between 0 and 1. This threshold value is fixed experimentally according to the patient and the desired sensitivity of the algorithm. Lower threshold values imply that the algorithm will be more sensitive to the EMG signal and detect motion intention with less signal intensity, but may be more affected by unexpected external forces. The effect of the threshold, using the same sEMG signals with different thresholds, can be seen in Figure 6. The result of this comparison represents the intention of movement detected through the sEMG signal from the biceps muscle—more precisely, the elbow flexion.

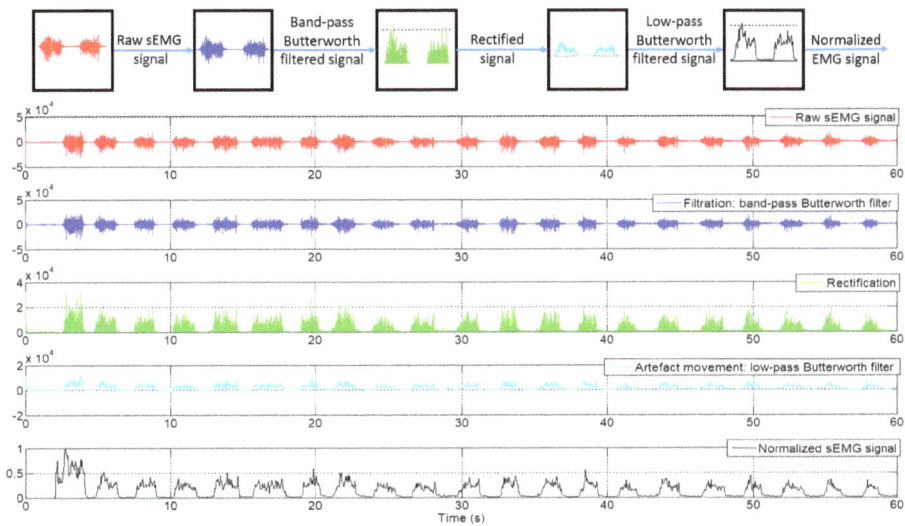

Figure 5. Surface electromyography (sEMG) signals after each processed step.

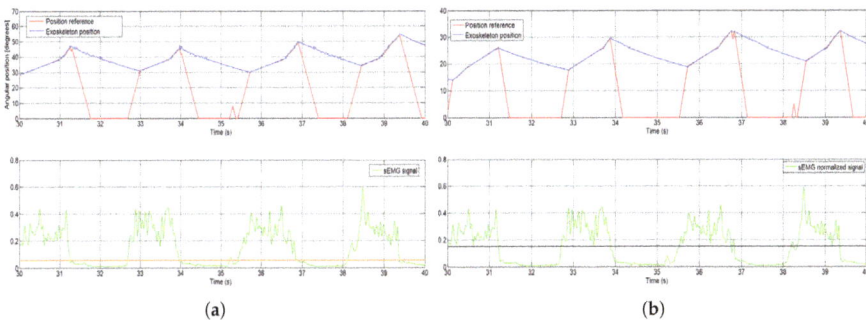

(a) (b)

Figure 6. (a) the orange line shows data using a 0.05 threshold. (b) the black line shows data using a 0.15 threshold.

If the sEMG signal decreases over time, then the following should be taken into account. If E_{act} becomes less than E_{min}, the result according to Equation (1) is a negative signal, which causes the actuators to turn off (the intention of movement is not detected). If the patient's intent is to move the forearm and the result of Equation (1) is less than the threshold, the value of this can be modified online to change the sensibility of the algorithm. If E_{act} becomes less than E_{min} when the patient's intent is to move the forearm, the algorithm needs to be recalibrated.

The proposed control algorithm generates the reference position as an increment of the current joint angle. That is, if movement intention is detected in the sEMG signal, the control algorithm provides a reference to increase the elbow angle of flexion. If no movement intention is detected, the reference position will be null and the actuator is disabled.

According to the actual elbow position and the final movement intention, the system works between two types of increments: one for fast reference position generation and another used to generate a slow reference position. The first increment is used when the actual position of the elbow joint is different from the position of the actuator reference. This case occurs when motion intention is detected, that is, the signal sEMG exceeds the established threshold after a period of deactivation of the actuators caused by the non-detection of intention to move. The exoskeleton used in the flexion movement leaves the joint free to move, as long as the actuator is not activated because of the loss of patient motivation and engagement that results in loss of the EMG signal. At that moment, the reference position is zero, but the actual joint position is not null. This situation is shown in the descending part of the sawtooth-shaped graph in Figure 6. The loss of intention to move produces a null reference that causes deactivation of the actuator, and the recovery of the intention causes a rapid increase of the reference position. If the algorithm is activated and detects an intention to move, the generated reference uses a fast increment until it reaches the elbow position, after which it uses a slow increment to generate the reference that will be followed by the exoskeleton, as long as there exists an intention to move. When intention to move is no longer detected, the high increment is used to decrease the reference position; the actuators are no longer activated and the extension movement is carried out by actuator recuperation (dissipation of the heat).

In order to address the situation caused by small EMG levels and generate the reference position pattern with better precision, the high-level control algorithm uses the sEMG normalized signal together with the FSR sensor signal. Similar to the EMG signals, the signal from the FSR sensors is filtered and normalized. The filter for this signal is a low-pass filter at a frequency of 100 Hz. Filtered signals are normalized in the same way as the sEMG signal, using an equation analogous to (1). After that, it is compared with the threshold defined to detect the intention to move through the force interaction between the patient and exoskeleton. For flexion movement detection, only the signal provided by the FSR sensor placed over the radius bone is taken into account. The patient's movement intention causes the forearm to exert pressure over the rigid part of the exoskeleton, which can be detected with this sensor. The two signals, from sEMG and FSR, are logically compared in order to detect the final intention to move, a binary result that is used later. The logical comparison consists of an AND function, to ensure a higher accuracy of the algorithm, having a minimum of two active signals (above the threshold), or with an OR condition if the reference is generated, where at least one of the signals is above the threshold.

The scheme of the high-level control algorithm capable of generating the reference position pattern can be seen in Figure 7, where $E_{act(k)}$ and $P_{act(k)}$ are the actual EMG and pressure or force signals in the discrete domain, $E_{filt(k)}$ and $P_{filt(k)}$ are filtered EMG and pressure or force signals, $E_{norm(k)}$ and $P_{norm(k)}$ are normalized EMG and pressure or force signals, $\theta_{(k)}$ is the generated angle reference, $V_{(k)}$ is the control signal, and $Y_{(k)}$ is the angular position of the SMA-based exoskeleton.

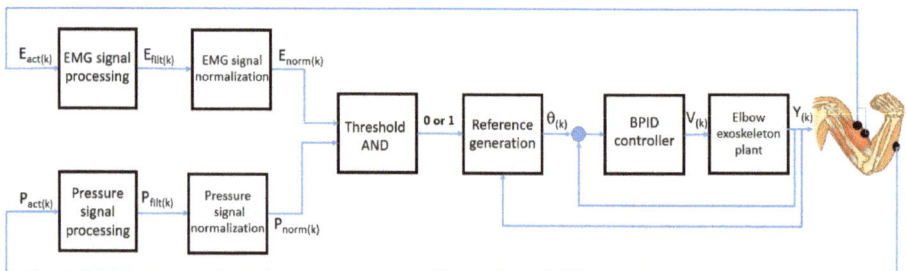

Figure 7. High-level control algorithm based on EMG and pressure signals for reference position generation.

In parallel to the algorithm that generates the reference position, the normalized EMG signal is used to generate a torque assistive reference for rehabilitation therapy. According to the total height and weight of the patient, the weight of the forearm and hand is approximately calculated, as well as the length from the joints to the center of gravity of each. As a function of these parameters and the actual angle, torque on the elbow joint is estimated. Using this torque and the sEMG signal, a percentage of assistance in torque reference can be generated. This percentage can be set by the user.

Torque assistive reference is directly proportional to the sEMG signal. A similar idea is presented in [30], but they did not take the biomechanical structure of the human body into account.

3. Results

In order to highlight the algorithm performance, feasibility, and adaptability to various hardware configurations, a series of tests were done. Firstly, simulation with EMG signals from different circuits, together with an actuator model, was conducted to simulate the behavior of the actuator in the exoskeleton; secondly, the real hardware over the exoskeleton was tested with healthy subjects.

3.1. Results of Simulation

In [31], the model of an SMA-based actuator with a variable charge was presented. This permits the simulation of the actuator with different SMA diameters (0.51 mm and 0.1 mm); in this case, the 0.51 mm diameter was used. According to the simulation results presented in [31], which were compared with the real behavior of an SMA actuator, it can be concluded that the behavior of the model is highly similar to a real actuator. To use this model in the simulation with the high-level control algorithm based on sEMG, a number of settings of the SMA-based actuator were used. Firstly, the charge of the actuator was set according to the forearm and hand weight, and the linear position was converted to an angular position as a function of the exoskeleton characteristics, such as the pulley radius. It is worth noting that the SMA-based actuator model includes the same low-level control algorithm ([3,5]), as well as the exoskeleton.

For the sEMG data acquisition, the electrodes were placed along the biceps muscle fibers and on the midline of the belly of the muscle, taking into consideration that this is where the sEMG signals have the greatest amplitude (Figure 4). The subject was asked to perform some elbow extension–flexion movements, and data was saved to be used in offline simulation. This process was accomplished with two types of sEMG circuits: firstly with the circuit realized in UC3M, presented in Section 2.1.3, and secondly with an sEMG measurement device (DKH Co., Ltd., Tokyo, Japan) with a sampling frequency of 1 kHz. This latter circuit was successfully used in other works, such as for the control of a prosthetic hand [32], and in a rehabilitation finger system [33]. The sEMG signal acquired with the UC3M circuit was similar to the sEMG signal acquired with the DKH circuit.

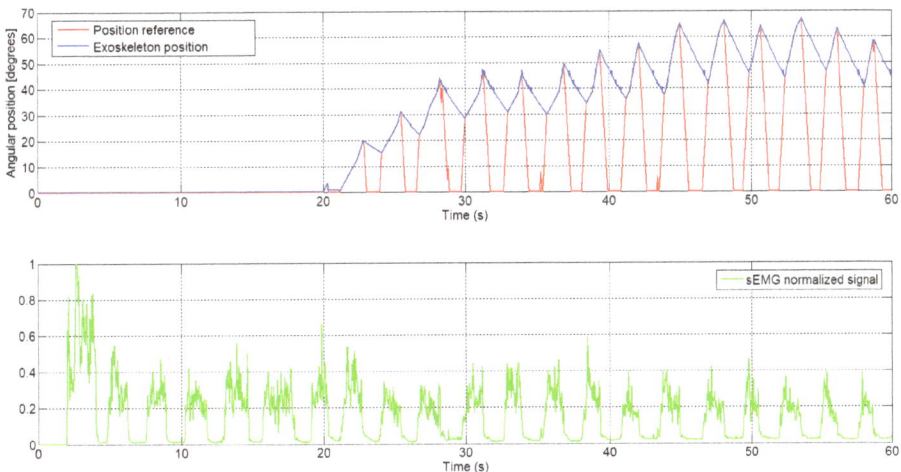

Figure 8. The generated angular reference position from the sEMG signal with the UC3M circuit: first subject (male, 24 years old, 1.73 m height, and 70 kg weight).

Figure 8 shows the normalized sEMG signal acquired from the UC3M circuit, the generated reference, and the angular position of the exoskeleton. This first test was realized offline in a simulation, where the signal of FSR sensor was set to 1 (this means that the signal of the FSR sensor is ignored), and the increment was set empirically to 0.1 for fast increment and 0.01 for slowly increment.

As can be seen, at $t = 20$ s, the reference position is 0 degrees, since this signal from the sEMG was used for calibration, whereas the first 2 s were ignored for perturbation. Next, $t = 18$ s were used to detect the maximum and minimum sEMG signal. After this process of calibration, starting at $t = 20$ s, once muscle activity has been detected in the biceps muscle, the algorithm starts to generate the reference.

We take, as an example, the sEMG signal at $t = 29$ s (Figure 9). From this moment, the normalized sEMG signal changes the amplitude, which means that the circuit detects muscular activity in the bicep muscles, and the algorithm begins to increment the reference position. Because the actual angular position of the exoskeleton is different to the actual reference, by approximately 30 degrees, the algorithm increases the angular reference position with a high increment. Once the angular reference position coincides with the exoskeleton position, the algorithm increases the angular reference position with a slow increment and the exoskeleton begins to follow the voluntary movement intention. In $t = 32.5$ s, the amplitude of the normalized sEMG signal decreases, the high-level control algorithm interprets that there is no intention to move by the user and, therefore, the algorithm decreases the angular reference position. In this case, though the reference decreases very quickly, the angular position of the actuator is limited by the actuator behavior (shows a slow recovery due to heat accumulation). The sEMG threshold can easily be set from the user interface and, in this case, it was set to 0.05.

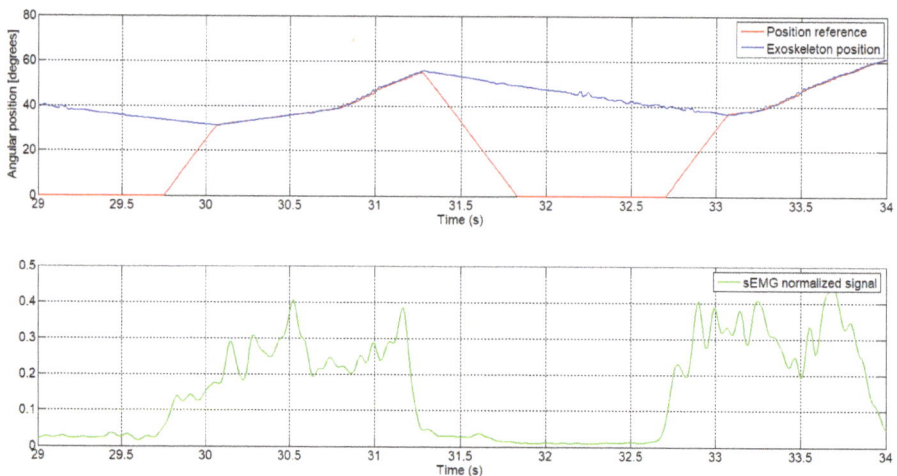

Figure 9. The angular reference position generated by the sEMG signal, first subject (enlarged area).

The second test was performed with a different sEMG circuit and a different person. Similar to the first case, the person was asked to execute some repetitions of flexion–extension of the elbow and the sEMG signal was recorded. The signal can be seen in Figure 10, from which can be observed a higher frequency of movement of the elbow joint. Between $t = 40$ s and $t = 45$ s, we can see a muscle relaxation; the amplitude of the sEMG signal decreases, and in this case the angular reference position goes to 0 degrees. The exoskeleton behavior can be seen when the extension actuator is not active: it represents a slow extension movement.

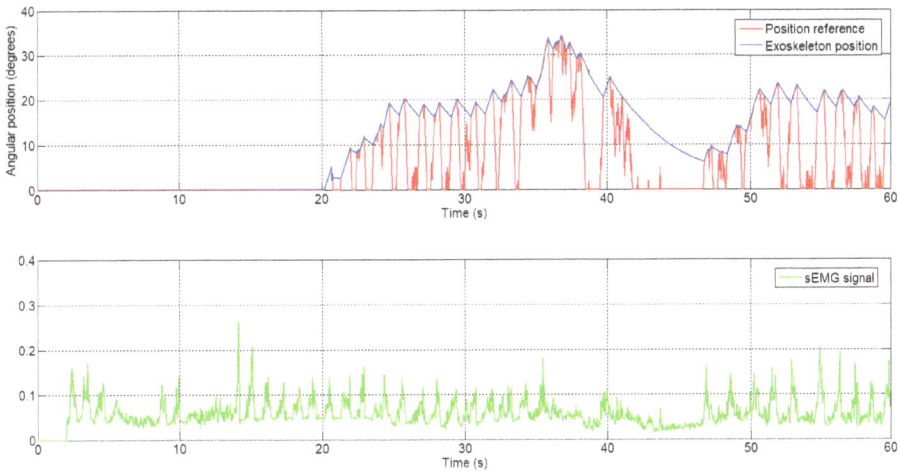

Figure 10. The generated angular reference position by the sEMG signal, second subject.

In parallel, the algorithm offers the possibility to generate the torque reference to assist the movement. This reference is generated according to the biomechanical model of the human body, taking into account that rehabilitation is executed standing or sitting, and that the sEMG signal is detected over the biceps muscles. In Figure 11, the pattern reference in torque assistance is presented for one patient with a weight of 70 kg and height of 1.73 m for two cases: the exoskeleton assists the patient with the total torque, 100% (blue signal), and the exoskeleton assists with 50% of total torque (red signal). The sEMG signal used to generate this reference in torque assistance is the same as the sEMG signal presented in Figure 8.

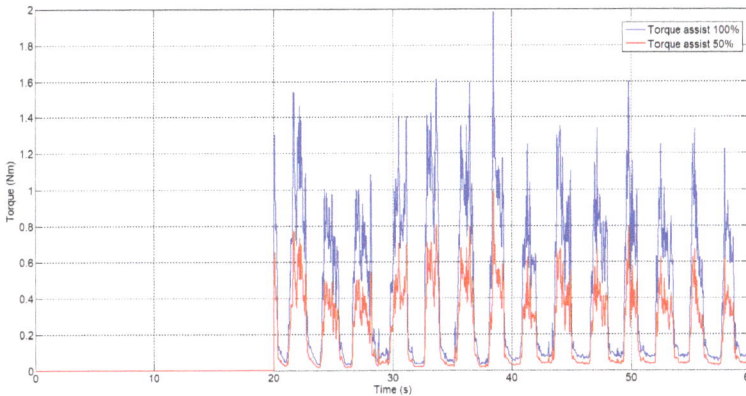

Figure 11. The generated torque reference from the sEMG signal.

3.2. Results with the Real SMA Exoskeleton

The sEMG-based control algorithm was tested in the real exoskeleton presented in Section 2.1. This was tested with healthy people from RoboticsLab laboratory, Carlos III University of Madrid. The characteristics of the subject were: male, 1.73 m height, and 70 kg weight. Firstly, sEMG electrodes were fixed over the biceps (over the belly of the biceps, a good positioning is essential) and the shoulder (reference electrode), and then the exoskeleton was fitted over the body. The exoskeleton was configured on the subject's body so that the elbow axis was aligned with the exoskeleton rotation axis and the FSR sensors were in contact with the hand and forearm. The results of this test can be seen in Figure 12, showing the reference position signal (the blue signal), which is generated by the sEMG signal (purple) and the FSR signal (green), and the real position exoskeleton (red).

According to the high-level control algorithm, in the first 20 s, the exoskeleton user calibrates the algorithm through movements of flexion–extension of the elbow joint. In Figure 12, two movements of flexion–extension can be observed during the first 20 s. In these first seconds, the output reference is 0 degrees.

In the second graphic, the sEMG signals can be seen, where the amplitude is changing during the flexion–extension movement. In the third graphic is the FSR sensor signal variation corresponding to the flexion–extension movement. After the process of calibration, when the algorithm detects the movement intention (from the sEMG signal and FSR sensor), it starts to generate the reference position and the exoskeleton begins to move following the reference. We take as a reference example the interval $t = 23$–40 s. At $t = 23$ s, the FSR sensor presents a signal with a high amplitude which exceeds the value of the threshold, and the sEMG signal also begins to increase in amplitude. Starting from this point, the algorithm begins to generate the angular reference, incrementing slowly, as the angular reference is near the exoskeleton elbow position. Until $t = 30$ s, the amplitude of the sEMG signal remains high, with the angular reference reaching the maximum 120 degrees. Due to the elbow movement, the FSR sensor signal amplitude may have decreased and, for this reason, the weight of this signal (during this period) on the algorithm is lower. After time $t = 30$ s to $t = 40$ s, the sEMG signal has decreased its amplitude and the algorithm starts to decrease the angular reference, finally to 0 degrees.

To successfully use the exoskeleton in this mode of rehabilitation therapy (active mode), the patient needs to present a minimum level of activity in the motor function, otherwise the algorithm is not capable of detecting the movement intention based on the sEMG and force/pressure signals. If this mode of therapy cannot be used by the patient, passive mode rehabilitation therapy can be used, where the exoskeleton follows a passive reference (habitually a sinusoidal reference).

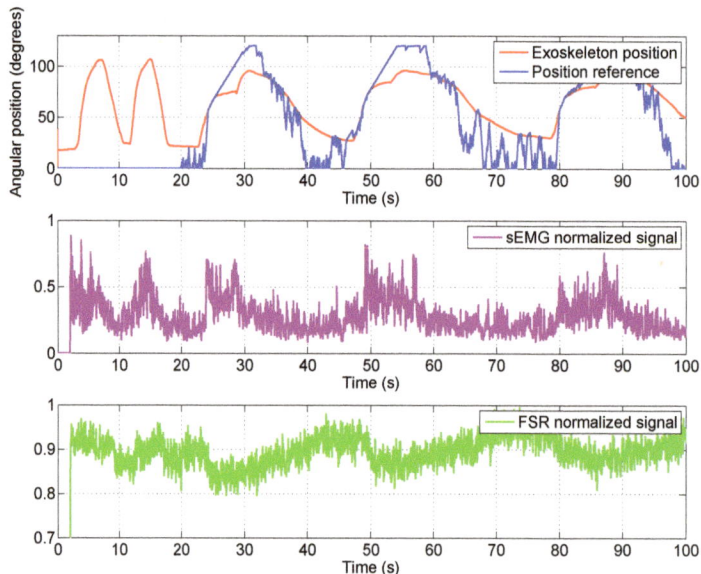

Figure 12. Reference position and response generated by the sEMG signal.

4. Conclusions

In this work, a new high-level control algorithm based on sEMG signals and pressure/force signals capable of generating the angular and torque reference for an active rehabilitation was presented. An algorithm capable of generating the angular and torque reference was successfully tested: in simulations (with the EMG signals provided by the circuit made by the research group and a commercial circuit) and in real applications over the SMA elbow exoskeleton with healthy people. In the latter case, in a real device, the sEMG signal was used together with the force/pressure signals from an FSR sensor.

The SMA-based exoskeleton for an elbow joint presented in this work, together with the low and high-level control algorithm and sensors, is based on low-cost components and offers three modes of operation:

- Data acquisition mode: to evaluate and diagnose the patient. Also, in this mode of operation, the angular limits of elbow movement are saved to set the angular reference limits for the control algorithm.
- Passive rehabilitation mode: The exoskeleton follows a defined angular reference, the most common being a sinusoidal type. In this case, the patient executes repetitive movements, not taking into account the movement intention of the patient. The exoskeleton can support all the movement in flexion, extension, or flexion–extension.
- Active rehabilitation mode: The angular reference for the elbow exoskeleton is generated as a function of the patient's intention for movement, detected by the sEMG signals and force/pressure signals. In this case, the patient is actively involved in the rehabilitation therapy, and if movement intention is not detected, the angular reference goes to 0 degrees. This type of rehabilitation can only be used with patients who present a minimum activity level in their motor function; otherwise, a passive rehabilitation can be used.

The performance of the high-level control algorithm considered the biceps muscle activity and did not take into account the triceps muscle activity. If this co-contraction appears, and the FSR sensor is not capable of detecting it, the user needs to manually stop the system.

The control algorithm presented in this paper permits the adjustment of the parameters of the generated reference position, such as the increments (which modify the angular velocity response of the exoskeleton) and the thresholds of the sEMG and FSR signals, which change the sensibility of the algorithm (from which signal value the algorithm starts to increment to the reference position).

The main advantage provided by the proposed high-level controller is that it forces the patient to be involved in the therapy task on a constant basis. If the patient loses attention, the exoskeleton is deactivated. In this way, the controller promotes active rehabilitation.

Author Contributions: L.M. was in charge of project administration and funding acquisition. D.C. and L.M. designed the exoskeleton. D.C. developed the control method and carried out the experiments. D.S. collaborated in experiments. D.B. supervised the research. D.C. and D.B. wrote the manuscript.

Funding: The research was funded by RoboHealth (DPI2013-47944-C4-3-R) and the EDAM (DPI2016-75346-R) Spanish research projects.

Acknowledgments: The authors are grateful for the collaboration of the LAMBECOM research group, of Rey Juan Carlos University of Madrid, Spain, in defining the design requirements of the rehabilitation device and their participation in the evaluation of preliminary designs.

Conflicts of Interest: The authors declare no conflict of interest.

Abbreviations

The following abbreviations are used in this manuscript:

SMA	Shape Memory Alloy
UC3M	Carlos III University of Madrid
FSR	Force Sensing Resistor
PWM	Pulse-Width Modulation
sEMG	Surface electromyography
PTFE	Polytetrafluoroethylene
DOF	Degrees of freedom
SPI	Serial Peripheral Interface
PLA	Polylactic Acid
MES	Myoelectric signals

References

1. Harwin, W.S.; Murgia, A.; Stokes, E.K. Assessing the effectiveness of robot facilitates neurorehabilitation for relearning motor skills. *Med. Biol. Eng. Comput.* **2011**, *49*, 1093–1102. [CrossRef] [PubMed]
2. Pons, J.L. *Wearable Robots*; John Wiley & Sons: Chichester, UK, 2008.
3. Copaci, D.; Flores, A.; Rueda, F.; Alguacil, I.; Blanco, D.; Moreno, L. Wearable Elbow Exoskeleton Actuated with Shape Memory Alloy. In *Converging Clinical and Engineering Research on Neurorehabilitation II, Proceedings of the 3rd International Conference on NeuroRehabilitation (ICNR2016) Segovia, Spain, 18–21 October 2016*; Springer: Basel, Switzerland, 2016; pp. 477–481.
4. Copaci, D. Non-Linear Actuators and Simulation Tools for Rehabilitation Devices. Ph.D. Thesis, Carlos III University, Getafe, Spain, 2017.
5. Copaci, D.; Blanco, D.; Moreno, L. Wearable elbow exoskeleton actuated with Shape Memory Alloy in antagonist movement. In Proceedings of the Joint Workshop on Wearable Robotics and Assistive Devices, International Conference on Intelligent Robots and Systems (IROS 2016), Daejeon, Korea, 9–14 October 2016.
6. Norman, R.W.; Komi, P.V. Electromechanical delay in skeletal muscle under normal movement conditions. *Acta Physiol. Scand.* **1979**, *106*, 241–248. [CrossRef] [PubMed]
7. DiCicco, M.; Lucas, L.; Matsuoka, Y. Comparison of Two Control Strategies for a Muscle Controlled Orthotic Exoskeleton for the Hand. In Proceedings of the IEEE International Conference on Robotics and Automation, New Orleans, LA, USA, 26 April–1 May 2004; pp. 1622–1627.
8. Martini, F.H.; Timmons, M.J.; Tallitsch, R.B. *Human Anatomy*; Pearson Education Inc.: Old Tappan, NJ, USA, 1997; ISBN 0-13-049178-0.
9. Battye, C.K.; Nightingale, A.; Whillis, J. The use of myo-electric currents in the operation of prostheses. *Bone Jt. J.* **1955**, *37*, 506–510. [CrossRef]
10. Bottomley, A.H. Myo-electric control of powered prostheses. *Bone Jt. J.* **1965**, *47*, 411–415. [CrossRef]
11. Farry, K.A.; Walker, L.D.; Baraniuk, R.B. Myoelectric Teleoperation of a Complex Robotic Hand. *IEEE Trans. Rob. Autom.* **1996**, *12*, 775–788. [CrossRef]
12. Fukuda, O.; Tsuji, T.; Ohtsuka, A.; Kaneko, M. EMG-based Human-Robot Interface for Rehabilitation Aid. In Proceedings of the IEEE International Conference on Robotics and Automation, Leuven, Belgium, 16–20 May 1998; pp. 3942–3947.
13. Kuribayashi, K.; Shimizu, S.; Okimura, K.; Taniguchi, T. A discrimination system using neural netwok for EMG-control prostheses-Integral type of emg signal processing. In Proceedings of the 1993 IEEERSJ International Conference on Intelligent Robots and Systems, Yokohama, Japan, 26–30 July 1993; pp. 1750–1755.
14. Benjnya, N.; Kenney, S.B. Myoelectric Hand Orthosis. *J. Prosthet. Orthot.* **1990**, *2*, 149–154.
15. Singh, R.M.; Chatterji, S. Trends and Callenges in EMG Based Control Scheme of Exoskeleton Robots—A Review. *Int. J. Sci. Eng. Res.* **2012**, *3*, 506–510.
16. Vasan, G.; Pilarski, P. Learning from Demonstration: Teaching a Myoelectric Prosthesis with an Intact Limb via Reinforcement Learning. In Proceedings of the 15th International Conference on Rehabilitation Robotics (ICORR2017), London, UK, 17–20 July 2017.

17. Hudgins, B.; Parker, P.; Scott, R. A new strategy for multifunction myoelectric control. *IEEE Trans. Biomed. Eng.* **1993**, *40*, 82–94. [CrossRef] [PubMed]
18. Au, A.T.C.; Kirsch, R.F. EMG-based prediction of shoulder and elbow kinematics in able-bodied and spinal cord injured individuals. *IEEE Trans. Rehabil. Eng.* **2000**, *8*, 471–480. [CrossRef] [PubMed]
19. Kiguchi, K.; Tanaka, T.; Fukuda, T. Neuro-Fuzzy Control of a Robotic Exoskeleton with EMG Signals. *IEEE Trans. Fuzzy Syst.* **2004**, *12*, 481–490. [CrossRef]
20. Gopura, R.; Kiguchi, K. An Exoskeleton Robot for Human Forearm and Wrist Motion Assist-Hardware Design and EMG-Based Controller. *Int. J. Adv. Mech. Des. Syst. Manuf.* **2008**, *2*, 1067–1083.
21. Gopura, R.; Kiguchi, K. Application of Surface Electromyographic Signals to Control Exoskeleton Robots. In *Applications of EMG in Clinical and Sports Medicine Catriona Steele*; IntechOpen: Rijeka, Croatia, 2012.
22. Kiguchi, K.; Hayashi, Y. An EMG-Based Control for an Upper-Limb Power-Assist Exoskeleton Robot. *IEEE Trans. Syst. Man Cybern. Part B Cybern.* **2012**, *42*, 1064–1071. [CrossRef] [PubMed]
23. Lenzi, T.; de Rossi, S.M.M.; Vitiello, N.; Carroza, M.C. Intention-Based EMG Control of Powered Exoskeletons. *IEEE Trans. Biomed. Eng.* **2012**, *58*, 2180–2190. [CrossRef] [PubMed]
24. Lucas, L.; DiCicco, M.; Matsuoka, Y. An EMG-Controlled Hand Exoskeleton for Natural Pinching. *J. Robot. Mechatron.* **2004**, *16*, 482–488. [CrossRef]
25. Villoslada, A.; Flores, A.; Copaci, D.; Blanco, D.; Moreno, L. High displacement flexible shape memory alloy actuator for soft wearable robots. *Robot. Auton. Syst.* **2015**, *73*, 91–101. [CrossRef]
26. Technical Characteristics of Flexinol, Dynalloy, Inc. Makers of Dynamic Alloys. Available online: http://www.dynalloy.com/ (accessed on 18 June 2018).
27. Saes Group. Available online: https://www.saesgetters.com/ (accessed on 18 June 2018).
28. Flores, A.; Copaci, D.; Villoslada, A.; Blanco, D.; Moreno, L. Sistema Avanzado de Protipado Rápido para Control en la Educación en Ingeniería para grupos Multidisciplinares. *Rev. Iberoam. Autom. Inf. Ind.* **2016**, *13*, 350–362.
29. De Luca, C.J.; Gilmore, L.D.; Kuznetsov, M.; Roy, S.H. Filtering the surface EMG signal: Movement artifact and baseline noise contamination. *J. Biomech.* **2010**, *43*, 1573–1579. [CrossRef] [PubMed]
30. Song, R.; Tong, K.Y.; Hu, X.; Li, L. Assistive Control System Using Continuous Myoelectric Signal in Robot-Aided Arm Training for Patients after Stroke. *IEEE Trans. Neural Syst. Rehabil. Eng.* **2008**, *16*, 371–379. [CrossRef] [PubMed]
31. Copaci, D.; Flores, A.; Villoslada, A.; Blanco, D. Modelado y Simulación de Actuadores SMA con Carga Variable. In Proceedings of the XXXVI Jornadas de Automática, Bilbao, Spain, 2–4 September 2015.
32. Hioki, M.; Ebisawa, S.; Sakaeda, H.; Mouri, T.; Nakagawa, S.; Uchida, Y.; Kawasaki, H. Design and control of electromyogram prosthetic hand with high grasping force. In Proceedings of the 2011 IEEE International Conference on Robotics and Biomimetics, Karon Beach, Phuket, Thailand, 7–11 December 2011; pp. 1128–1133.
33. Hioki, M.; Kawasaki, H.; Sakaeda, H.; Nishimoto, Y.; Mouri, T. Finger Rehabilitation Support System Using a Multifingered Haptic Interface Controlled by a Surface Electromyogram. *J. Robot.* **2011**, *167516*. [CrossRef] [PubMed]

    MDPI

Article

Intelligent Multimodal Framework for Human Assistive Robotics Based on Computer Vision Algorithms

Eugenio Ivorra [1,*], Mario Ortega [1], José M. Catalán [2,*], Santiago Ezquerro [2], Luis Daniel Lledó [2], Nicolás Garcia-Aracil [2] and Mariano Alcañiz [1]

1   Institute for Research and Innovation in Bioengineering, Universitat Politècnica de València, 46022 Valencia, Spain; mortega@i3b.upv.es (M.O.); malcaniz@i3b.upv.es (M.A.)
2   Biomedical Neuroengineering Group, Universidad Miguel Hernández de Elche, 03202 Elche, Spain; sezquerro@umh.es (S.E.); llledo@umh.es (L.D.L.); nicolas.garcia@umh.es (N.G.-A.)
*   Correspondence: euivmar@i3b.upv.es or euivmar@upvnet.upv.es (E.I.); jcatalan@umh.es (J.M.C.)

Received: 4 July 2018; Accepted: 23 July 2018; Published: 24 July 2018

Abstract: Assistive technologies help all persons with disabilities to improve their accessibility in all aspects of their life. The AIDE European project contributes to the improvement of current assistive technologies by developing and testing a modular and adaptive multimodal interface customizable to the individual needs of people with disabilities. This paper describes the computer vision algorithms part of the multimodal interface developed inside the AIDE European project. The main contribution of this computer vision part is the integration with the robotic system and with the other sensory systems (electrooculography (EOG) and electroencephalography (EEG)). The technical achievements solved herein are the algorithm for the selection of objects using the gaze, and especially the state-of-the-art algorithm for the efficient detection and pose estimation of textureless objects. These algorithms were tested in real conditions, and were thoroughly evaluated both qualitatively and quantitatively. The experimental results of the object selection algorithm were excellent (object selection over 90%) in less than 12 s. The detection and pose estimation algorithms evaluated using the LINEMOD database were similar to the state-of-the-art method, and were the most computationally efficient.

Keywords: 3D object detection and pose estimation; assistive robotics; eye-tracking; human–computer interface

---

1. Introduction

Approximately 80 million people in the European Union (one-sixth of its population) have a disability. The percentage of people with disabilities is set to rise as the EU population ages [1].

Accessibility is a basic right for all persons with disabilities according to Article 9 of the United Nations Convention on the Rights of Persons with Disabilities. The purpose of accessibility is to enable persons with disabilities to live independently and to participate in all aspects of life. Assistive technologies help all persons with disabilities to improve their accessibility in all aspects of their life. Current trends in assistive technology for supporting activities of daily living (ADL), mobility, communication, and so on are based on the integration of the capabilities of the user and the assistive technologies. The improvement of the interaction and cooperation between the user and the assistive technologies can be split into three main areas: (1) improvements of the assistive devices, such as mechanical parts, electronic parts, etc.; (2) improvements of the user–technology interface; and (3) improved shared control between the user and the assistive technology. The AIDE European project contributes to improving current assistive technologies by developing and testing a modular and adaptive multimodal interface customizable to the individual needs of people with disabilities as well as a totally new shared control paradigm for assistive devices that integrates information from the identification of the residual abilities, behaviours, emotional state, and intentions of the user on one hand and the analysis of the environment through the use of cameras and context factors on the other. This paper describes the computer vision algorithms part of the multimodal interface developed within the AIDE European project.

There are some examples of multi-modal architectures for the interaction and control of assistive robotic devices. Specifically, Meng et al. presented a non-invasive brain–computer interface (BCI) for controlling a robotic arm to complete reach-and-grasp tasks [2]. They used a Microsoft Kinect Motion Sensor to locate and send the position of a target to the robot control system. Another interesting recent paper presented an assisted feeding strategy that uses a Kinect camera and a modular robotic arm to implement a closed-form system that performs assisted feeding [3]. In contrast to these works, our approach uses two cameras (one looking at the environment in front of the user and the other looking at the user) to locate real objects and the user's mouth position, respectively. In addition, the user wears a pair of eye-tracking glasses to recognise the object at which the user is looking. There are also some works using invasive BCI systems in people with long-standing tetraplegia to control a robotic arm to perform three-dimensional reach and grasp movements [4,5]. Some works can be found in the literature reporting the control of an arm exoskeleton using multimodal interfaces. Specifically, Pedrocchi et al. developed a novel system composed of a passive arm exoskeleton, a robotic hand orthesis, and a neuromuscular electrical stimulation system driven by residual voluntary muscular activation, head/eye motion, and brain signals in the framework of the MUNDUS project [6]. In addition, Frisoli et al. presented a robotic-assisted rehabilitation training with an upper limb robotic exoskeleton for the restoration of motor function in spatial reaching movements [7]. Then, they presented the multimodal control of an arm–hand robotic exoskeleton to perform activities of daily living. The presented system was driven by a shared control architecture using BCI and eye gaze tracking for the control of an arm exoskeleton and a hand exoskeleton for reaching and grasping/releasing cylindrical objects of different size in the framework of the BRAVO project [8]. Most recently, Clemente et al. presented a motion planning system based on learning by demonstration for upper-limb exoskeletons that allow the successful assistance of patients during activities of daily living (ADL) in an unstructured environment using a multimodal interface, while ensuring that anthropomorphic criteria are satisfied in the whole human–robot workspace [9]. In contrast to the previous works, the AIDE multimodal control interface predicts the activity that the user wants to perform and allows the user to trigger the execution of different sub-actions that compose the predicted activity, and to interrupt the task at any time by means of the hybrid control interface based on a system combining gaze tracking, electroencephalography (EEG), and electrooculography (EOG).

Most activities of daily living require complete reach-and-grasp tasks. The grasping task is a common operation for fixed manipulators in a controlled environment, but assistive robotics have the complexity that this environment is not fixed. Moreover, it should be solved in real-time in order to be comfortable for humans and sufficiently precise to perform successful grasps of a variety of objects. To sum up, a grasping task in multimodal assistive robotics requires the processing of a precise location and orientation of common textureless objects in real-time. Some authors have solved it using commercial tracking systems like Optitrack© or ART Track© [10–12], but these solutions require the modification of the objects by adding specific marks. Our proposal employs a computer vision approach that does not have that limitation. There are multiple technical approaches to solving this problem, and despite the great advances made recently in the field of computer vision (especially with the new deep learning techniques), it is still a difficult problem to solve effectively—specifically when the 3D object is textureless. For well-textured objects, several methods based on appearance descriptors like SURF or SIFT [13] can be employed to solve this problem. However, most common objects in our context (home) are textureless.

Considering the technical features required, these methods should be efficient, accurate, scalable, and robust to changes in the environment (no controlled light conditions or occlusions).

The main lines of investigation in the field of 3D textureless object pose estimation are methods based on geometric 3D descriptors, template matching, deep learning techniques, and random forests.

Methods based on geometric 3D descriptors employ the information extracted from the geometry of the 3D models of the objects. There are two kinds of 3D descriptors: local descriptors and global descriptors.

On one hand, local descriptors are obtained from characteristic local geometric points from the model. Once descriptors are calculated from the model and from the depth image from the RGB-D camera, a matching correspondence can be obtained. The last stage is usually a refinement of the pose using an iterative closest point algorithm (ICP [14] Among these methods [15] stand out FPFH [16], PPF [17], and SHOT [18]. These methods are very computationally expensive (need several seconds for estimating the object pose), but are robust to occlusions. One of the most employed methods is the point pair features (PPFs). This method was developed by Drost et al. [17] and employs the depth image for estimating local descriptors using normals of the object. Later, several authors have optimised the original implementation, greatly reducing the computational cost of the algorithm (Stefan et al. [19] still requires between 0.1 and 0.8 s for processing an image). However, the algorithm is still too heavy for real-time use.

On the other hand, global descriptors encode the shape of the 3D model in a single vector. The main global descriptor classes are VFH [20], OUR-CVFH [21], and ESF [22]. In contrast to the local descriptors, and as the main disadvantages, it is necessary to first have the 3D reconstruction of the object, as well as to segment the scene before estimating the pose of the object. In addition, these methods are very sensitive to the occlusion of the object. On the other hand and as a main advantage, they are computationally efficient. These types of descriptors are usually used for their efficiency in problems of classification or 3D model retrieval. In addition, note that these geometric methods (both local and global descriptors) can also use the colour information of the object (if it is available), increasing the robustness of the method (e.g., the local descriptor CSHOT [23]).

Methods based on template matching efficiently search through the generated set of templates of a 3D model employing a sliding window approach to find the most similar template within an image, as well as its 2D location using a similarity criterion. Once the most similar template is determined within the image, the initial pose of the 3D object is inferred from the one associated with the template. Within these methods, the algorithm LINEMOD++ [24] stands out. This algorithm is one of the most-employed algorithms by the scientific community for estimating the pose due to its high efficiency and robustness. Specifically, Hinterstoisser et al. [24] were the first to use this LINEMOD detection method to estimate the pose of objects. The LINEMOD method uses the information extracted from the gradients of a colour image and the surface normals. This information is subsequently quantified so that the search for the most similar template is carried out efficiently. Then, Hinterstoisser et al. [24] added a post-process stage in order to strengthen the detection method by eliminating some of the possible false positives. The last stage is a refinement of the pose using an ICP algorithm. This implementation was enhanced by Kehl et al. [25] to increase its scalability.

Another similar template method was proposed by Hodaň et al. [26]. Unlike the original LINEMOD++ method, they initially limit the search to certain areas of the image, by means of a simple filtering technique. The matching between the templates and the remaining possible locations within the image is done with a voting procedure based on hashing. To refine the pose, they use particle swarm optimization (PSO) [26] in place of the ICP algorithm. Following the same line, Zhang et al. [27] proposes that in addition the detection method be invariant to scale, consequently reducing the number of templates on which to perform the search. Despite all these modifications, these methods [25–27] are less precise and more computationally expensive than the LINEMOD++ method. The main limitation of methods based on templates is that they are very sensitive to object occlusion. On the contrary, they are usually computationally efficient methods when compared with methods based on 3D descriptors or deep learning techniques.

Recently, multiple methods have appeared that address the problem of estimating the pose of a 3D object through the use of deep learning techniques. Among the most popular methods are the SSD-6D [28], BB8 [29], Pose-CNN [30], and [31,32] methods. The main advantages of these methods are that they allow estimation of the pose using only the RGB sensor information, the scalability, and the robustness against occlusion. However, and as one of the main disadvantages, most of these methods need a large amount of training images to detect and estimate the pose of an object. Moreover, there is the added difficulty of estimating the actual pose associated with the training images. This supposes a great effort and time of work on the part of the user, as much the compilation of images of the 3D model as the estimation of the pose in each of them. To solve this type of limitation, methods such as the SSD-6D [28] method have been used to estimate the pose of objects using deep learning techniques using only synthetic images extracted from the original 3D model. However, these methods can have problems when there are substantial differences between the appearance of the synthetic images of the 3D model and the appearance of the images captured by the camera [28] (e.g., local changes in the illumination due to specular reflections). In turn, simply the change in the specifications of the capture sensor in the test phase can substantially influence the results [33]. To mitigate this problem, it is necessary to obtain 3D models of 3D objects with photorealistic quality. Although these methods present promising results, the 3D models of the objects must have colour information so that the pose can be detected and estimated correctly. This is a problem because it is common to only have access to a CAD model of the object without colour, or models are obtained through the use of a depth camera/RGB-D and KinectFusion technology [34], resulting in non-photorealistic models. It is importand to remark that training these models requires high-end equipment and/or a lot of time once the training information is ready. On the other hand, except for the SSD-6D [28] method that works at 10 fps, these are very computationally expensive, preventing their use in real-time. In addition, all these methods need a high-performance GPU.

Finally, the last types of methods are those based on forest classifiers. Some examples of these methods are those proposed by Brachmann et al. [35,36] in which they predict the 3D coordinates of an object as well as the labels assigned to each class by means of a random forest. Then, they use the RANSAC algorithm to estimate the initial pose. This method is very robust to the occlusion problem. Another outstanding work is the method of Tejani et al. [37]. They use the "latent-class Hough forest" method with the extracted information (features) of the LINEMOD algorithm on RGB-D patches to estimate the pose of the object. This method is invariant to scale and also allows estimation of the pose of multiple instances of the same 3D object.

In summary, this paper presents the computer vision algorithms developed in the AIDE multi-modal architecture for human assistive robotics that is able to give accessibility to persons with disabilities. The main contribution of this computer vision component is the integration with the robotic system and with the other sensory systems (EOG and EEG). The technical achievements solved are the algorithm for the selection of objects using the gaze and especially the state-of-the-art algorithm for the efficient detection and pose estimation of textureless objects. These algorithms were tested in real conditions with patients, and were also thoroughly evaluated both qualitatively and quantitatively. This paper is organised as follows. Section 2 presents the experimental setup with the multi-modal interface composition, the integration with the robotic system, and the developed computer vision algorithms. Section 3 shows quantitative and qualitative experimental results to evaluate the computer vision algorithm, and finally, Section 4 presents the conclusions and the future work planned.

2. Materials and Methods

2.1. Experimental Section

All participants were sitting in an electric wheelchair in front of a desk. Moreover, a Jaco2 robot is attached to the wheelchair (see Figure 1). In addition, the multimodal interface is composed of: (1) a pair of gaze-tracking glasses and a hybrid brain–computer interface (BCI) based on electroencephalography (EEG) and electrooculography (EOG); (2) context recognition sensors: two cameras to locate the object's position and the user's mouth position; (3) sensors for the monitoring of physiological parameters (breathing rate, heart rate, heart rate variability, galvanic skin response); and (4) a central server (YARP) for the communication. YARP stands for Yet Another Robot Platform. The experiments and results presented in this paper focused on the algorithms used for gaze-tracking and context recognition.

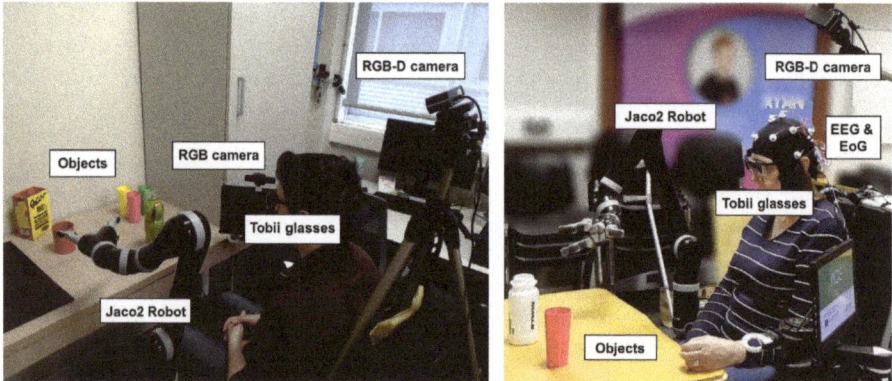

Figure 1. AIDE system integrates three different hardware modules: (i) a full-arm robotic exoskeleton or a Jaco2 robot; (ii) multimodal interfaces, consisting of a pair of gaze-tracking glasses (Tobii glasses) and a hybrid brain–computer interface (BCI) based on electroencephalography (EEG) and electrooculography (EOG); and (iii) context recognition sensors: a RGB-D camera to locate the object's pose and a camera to compute the user's head and mouth pose.

2.2. Calibration Methods Robot <-> RGB-D Camera

The objective of the calibration between the RGB-D camera and the robot is to make it possible to transform the coordinates system from the camera to the coordination system of the robot. This problem in robotics is known as *hand–eye calibration*. Specifically, it consists of estimating the homogeneous rigid transformation between the robot hand, or end-effector, to the camera as well as to the world coordinate system (see Figure 2). In the developed platform, the world coordinate system coincides with the robot base and the camera is not in the final effector of the robot but in a fixed position outside the robot. Let the rigid transformation of the robot-base to the end-effector be $^bB_{ee}$, and ^cA_m be the transformation of the camera to an augmented reality mark system. This system is an Aruco [38] mark mounted on a known pose on the robot thanks to a printed piece as can be seen in Figure 3. The transformation $^{ee}U_m$ between the mark and the robot end-effector is calculated using the CAD schematics of the robot and the printed piece. Thanks to this, the position and orientation of the end-effector can be expressed regarding the robot base and the camera system as shown in Equation (1). From this equation (**Direct Calibration**), the direct transformation ^bT_c can be easily extracted .

However, due to inaccuracies in the measurements and transformations obtained from the robot kinematics, Aruco detection, and U transformation, the following four optimisation methods were employed to increase the accuracy.

1. **Standard Calibration:** The implementation of the shape registration method in C++ [14].
2. **XS Calibration:** The c1 method of Tabb et al. [39].
3. **XS2 Calibration:** The c2 method of Tabb et al. [39].
4. **Ransac Calibration:** The OPENCV library implementation in C++ of the random sample consensus method (RANSAC optimization).

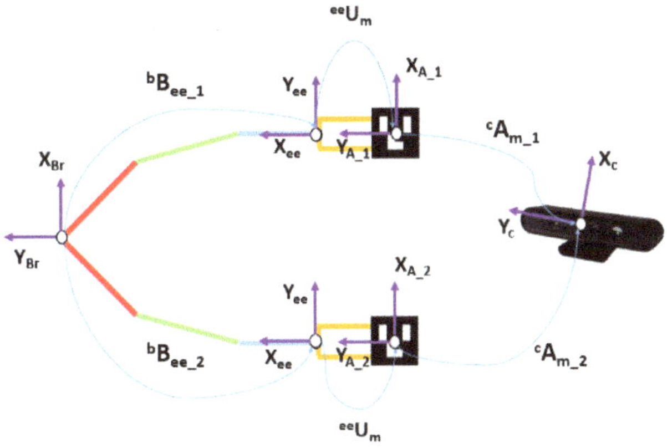

Figure 2. Schematic of the robot—camera problem.

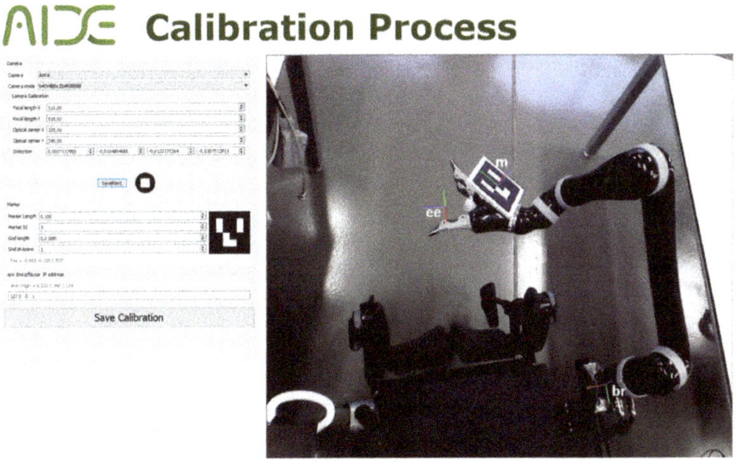

Figure 3. Calibration program.

Methods 1 and 4 employ the strategy of estimating the transformation between a cloud of 3D points expressed in the robot base and a cloud of the same points expressed in the camera system. Methods 2 and 3 were developed by Tabb et al. [39], and are based on the homogeneous matrix equation $AX = ZB$ where Z is the transformation from camera to robot base and X is the transformation from robot base to world coordinate. The difference between both methods is the cost function employed for the optimisation of transformations, as shown in Equations (2) and (3), respectively.

$$\left(^b B_{ee} \longleftrightarrow^c A_m *^m U_{ee}\right) \tag{1}$$

$$c1 = \sum_{n-1}^{i=0} ||A_i X - ZB_i||_F^2 \tag{2}$$

$$c2 = \sum_{n-1}^{i=0} \left|\left|A_i - ZB_i X^{-1}\right|\right|_F^2 \tag{3}$$

2.3. Eye-Tracking Detection

The hardware selected for this task was the Tobii© Pro Glasses 2. This hardware is a mobile lightweight gaze tracker recording both point-of-regard and scene in front of the subject. The gaze point data are sampled at 100 Hz with a theoretical accuracy and Root Mean Square (RMS) precision of 0.5° and 0.3°, respectively [40]. This device has two main components: head unit and recording unit. The head unit is a glasses-shaped device with a full-HD RGB camera with a frame rate of 25 fps. The Tobii© Pro Glasses recording unit can record to a Micro-SD (not used in this project), and has battery support and two network interfaces (wireless and Ethernet). A C++ library was developed that receives the video streaming of the glasses and the synchronized gaze point. No Tobii© SDK or proprietary software was employed for this project. The developed software can configure the glasses to work at different image resolutions, set-up frame rates (until 25 fps), and transmits via wireless or Ethernet connection. For this application, only the wireless connection was employed due to some issues detected during the integration phase of the project. Specifically, Tobii© Glasses internally implement a UDP broadcast and an IP6 discovering devices protocol which is incompatible with the YARP system. Gaze information is received in datagram ASCII. code via UDP protocol, and the streaming video is encoded in H264 (also received using a UDP protocol).

The gaze position obtained from the Tobii© Glasses is enhanced using a median filter, obtaining a more stable gaze point. In addition, due the higher acquisition rate of the gaze position than the RGB camera (100 Hz vs. 25 fps), the median filter allows the filtered gaze position to be synchronized with the RGB image.

A deep learning method called YOLOV2 [41] in combination with the gaze point gives us the initial detection of the desired object. There are other deep learning methods to detect objects, such as Faster-RCNN [42] or SDD [43]. However, YOLOV2 was chosen due its great efficiency and robustness in real-time. Specifically, YOLOV2 was trained with the COCO image database [44], which has 91 classes from the YOLOV2 . These classes cover most of the desired objects to manipulate (e.g., glasses, cutlery, microwave, etc.) in this project. Moreover, in the event that a desired object was not in the dataset, it could been trained. Finally, as a result of this stage, the class of the user-selected object is sent to the object detection and pose estimation stage.

2.4. Detection and Pose Estimation

The method developed for the detection and pose estimation was derived from the detection method of Stefan et al. [24], known as LINEMOD. The eye-tracking stage gives the ID of the object to track, so the Hinterstoisser et al. algorithm [24] only has to search one class of model. Consequently, it is more efficient, has a lower rate of false positives, and removes the scalability problem of different classes of objects that the Hinterstoisser et al. algorithm experiences.

The LINEMOD method starts with 2D images (colour and depth) synthetically rendered from different points of view and scales of the object 3D model. Viewpoints are uniformly sampled around the object, like going over a virtual sphere with the object in its center. For each of the viewpoints, a set of a RGB-D images and the virtual camera pose $\{R, t\}$ are saved. Then, a vector of distinctive points, as well as their associated descriptors, are calculated using the RGB-D information, as described in Hinterstoisser et al. [24]. This method defines a template as $V = (\{O_m\}_{m \in M}, \rho)$. O is the template feature (surface normal orientation or gradient orientation). M is the image information (RGB or depth). ρ is a vector of features locations r in the template image. Then, the generated templates are compared in the region of interest (ROI) of the scene image I at location c based on a similarity measurement over its neighbours ω:

$$(I, V, c) = \sum_{r \in \rho} \max_{v \in c + \omega} f_m(O_m(\omega), I_m(v)))). \tag{4}$$

This function $f_m(O_m(r), I_m(v))$ measures the cosine similarity of the features. Then, an empirical threshold is defined based on the score similarity score in order to decide if it is a match. The template matching stage (Equation (4)) was efficiently implemented taking advantage of the SSE instructions of modern CPUs. Furthermore, the detected templates could contain duplicate object instances, so a template clustering algorithm is performed aggregating templates with similar spatial location. However, this detection method can still throw false positives, so as in the original work, the colour information (in the HSV colour space) and the depth information (using an iterative closest point algorithm) were employed to filter these errors. Finally, the pose associated $\{R, t\}$ with the most similar template was refined with an iterative closest point algorithm—specifically with the *point-to-plane* version.

The main contribution of this part is the optimisation of the LINEMOD detection method [45]. This method was redesigned in order to be multi-processing, so it was split into two independent parts: one process is responsible for extracting the colour information from the RGB image (gradients), while another process is responsible for extracting the depth information (normals from the surface). These processes do not share memory between them so they can be executed in independent physical cores for an optimum performance. Moreover, the post-processing part [24] was also optimised with a multi-threading approach, responsible for eliminating false positives and refining the initial pose obtained. This post-processing part is performed by an ICP algorithm and checking the colour for each of the possible templates in different threads. These threads share memory in order to finish early the execution when one thread finds a valid template. Our method is summarised in Figure 4.

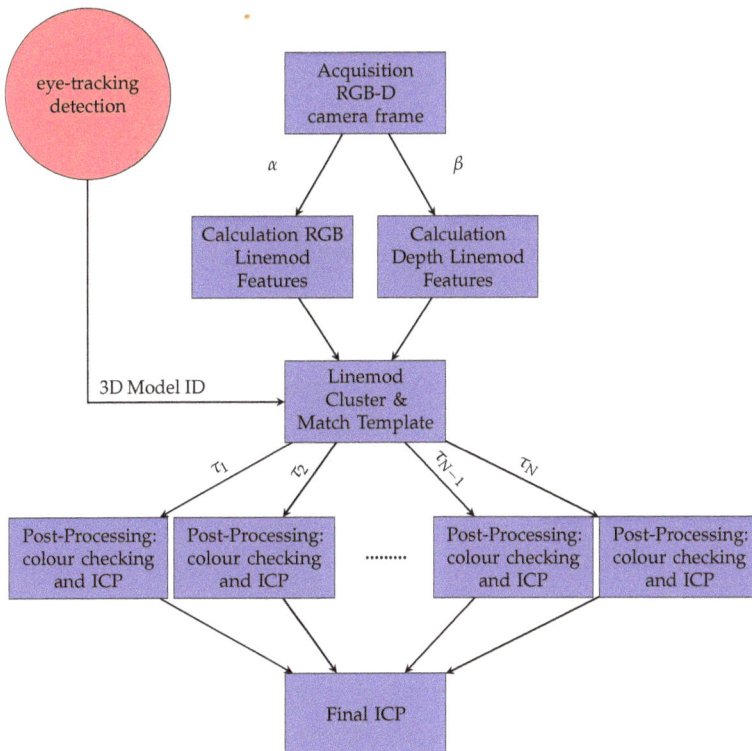

Figure 4. Flow diagram of the LINEMOD++ implemented algorithm. α and β are two independent processes, and τ represents a thread. ICP: iterative closest point.

2.5. Mouth Pose

The first step in the pose estimation of the mouth is to detect and recognise the user's face. Face recognition algorithms generally use 2D images for feature extraction and matching [46]. In order to achieve better performance and accuracy, 3D images captured via RGB-D cameras can be employed [47]. Therefore, we decided to employ one for this project. Specifically, we chose the Intel® RealSense™ SR300 RGB D camera. This camera implements a short-range (maximum 1.20 m) coded light 3D imaging system. The small size of the SR300 provides system flexibility to allow design into a wide range of products.

The mouth pose is obtained using the landmark detection API of the RealSense™ SDK. The algorithm employed returns 78 facial landmark points. For this project, we selected two pupil landmarks and two extreme points of the mouth (left and right). Using these two last 3D points (L and R) we estimate the mouth pose as follows:

$$Mp = \frac{\vec{LR}}{2} + L. \tag{5}$$

The point Mp is the center point of the mouth and the origin of the mouth pose. We set the axis so that the x-axis is in the ($\vec{LR}$) direction, z-axis is in the direction from Mp to the camera, and the y-axis is calculated to be a right-handed coordinate system. In addition, based on the colour information of the detected pupil landmarks, we can estimate if the user is blinking their eyes and know if it is the left or the right eye. For this work, all mouth landmarks were employed to detect when the mouth is open using the area of a convex hull calculated from all of the mouth points provided by the SDK.

3. Results

In this section, the results of different experimental sessions to evaluate the methods and/or algorithms reported in this paper are described.

3.1. Calibration between Camera and Robot

The position and orientation errors measured using different calibration approaches are shown in Figure 5. The most accurate method regarding position error was the standard method. In the case of orientation error, the most accurate methods were XS and XS2 followed by the standard method. After the evaluation of all the methods, we selected the standard one, which had the best results regarding position error and an admissible accuracy regarding orientation error. Moreover, the comparison of the influence of using different number of calibration points can be found in Appendix A.

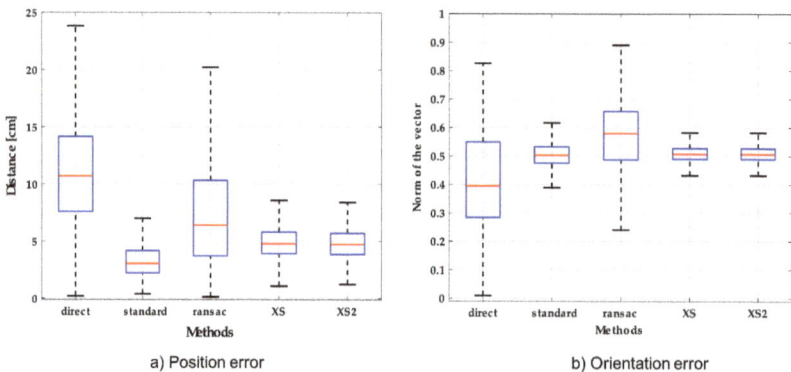

Figure 5. Position and orientation errors using different calibration methods: (**a**) Position error measured as a distance from the correct position; (**b**) Norm of the orientation error vector computed by Rodrigues' expression.

3.2. Detection and Pose Estimation

In this section, the quantitative results of the detection and pose estimation method are described. These results are compared with the works of other authors. Finally, a detailed analysis of the computational performance of the algorithm was carried out, and as in the quantitative analysis, it is compared with the work of other authors.

3.2.1. Quantitative Validation of the Detection and Pose Estimation of 3D Objects

To quantitatively evaluate the developed method, the LINEMOD dataset [24] was used. Although there are other datasets (e.g., T-LESS [48], Tejani [37], among others), the LINEMOD dataset is undoubtedly the most used by the scientific community to quantitatively evaluate detection and pose estimation methods. The LINEMOD dataset is formed by 15 3D non-textured objects, of which 13 colour 3D models are available (see Figure 6). Each model has a sequence of RGB-D images (around 1200 images in each), in which multiple objects appear from different points of view and distances (in a cluttered environment). Each image has the associated real pose ("ground truth") of the object and the intrinsic parameters of the RGB-D camera employed for acquiring the image.

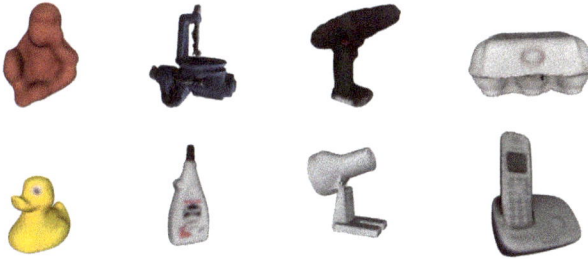

Figure 6. Some 3D models of the LINEMOD dataset.

The most common metrics employed for comparing the different methods are:

- Average distance (AD): This metric was introduced by Hinterstoisser et al. [24] and is the most employed to quantitatively evaluate the accuracy of pose estimation [19,26–29,49]. Given a set of vertices of a 3D model, M, the actual rotation and translation $[R, t]$ ("ground truth") and their estimations $[\hat{R}, \hat{t}]$:

$$m_{AD} = \frac{1}{|M|} \sum_{x \in M} ||(Rx + T) - (\hat{R}x - \hat{T})||_2. \tag{6}$$

when the 3D object is symmetrical, like some of the LINEMOD models ("cup", "bowl", "box", and "glue"):

$$m_{AD} = \frac{1}{|M|} \sum_{x_1 \in M} \min_{x_2 \in M} ||(Rx_1 + T) - (\hat{R}x_2 - \hat{T})||_2. \tag{7}$$

Traditionally, it is considered that the pose is correct if $m_{AD} \leq k_m d$, d being the diameter of the object, and k_m a coefficient ≥ 0. Generally a $k_m = 0.1$ is used (i.e., 10% of the diameter of the object).
- Shotton criteria (5 cm 5°): Using this criteria [24] a pose is considered correct if the rotational error is less than five degrees and the translational error is less than 5 cm. Please note that this metric does not take the size of the object into account.
- 2D Bounding Box: This metric calculates the intersection over union (IoU) [50] between the 2D bounding box obtained by projecting all the vertices of the 3D object with the real pose "ground truth " in the image and the 2D bounding obtained by projecting all the vertices of the object with the estimated pose. A pose is correct if IoU > 0.5.

- 2D Projections: This metric [36] sets a pose as valid if:

$$m_{Proj} = \frac{1}{|M|} \sum_{x \in M} ||K(Rx + T) - K(\hat{R}x - \hat{T})||_2 \tag{8}$$

is less than 5 pixels. M is the set of vertices of the 3D model, K is the matrix of intrinsic parameters, $[\hat{R}, \hat{T}]$ is the estimated pose and $[R, T]$ is the true pose. It should be noted that this metric is the most appropriate when you want to estimate the pose of 3D objects in an Augmented Reality system, and so was not used in this work.

- F1-Score. Given PR as true positive, TPR the recall ratio (true positive rate), and PPV the precision ratio (positive predictive value), $F1 = 2\frac{PR}{PPV+TPR}$ is defined. This metric has been used in References [28,37].

3.2.2. Comparison of the Results with State-Of-The-Art Methods

Firstly, it is worth noting that unlike other authors, quantitative evaluation using the main metrics was carried out in this work. This is of vital importance, since the results obtained often vary substantially depending on the metric employed.

When comparing the results of our method with some of the most popular methods (Table 1) on the LINEMOD dataset and with the AD metric, the results obtained are similar to those of the LINEMOD++ method proposed by Hinterstoisser et al. [24] (95.7% versus 96.6%). This is reasonable since the method developed in this work is based mainly on the LINEMOD++ algorithm. Specifically, the proposed method presents a series of modifications of LINEMOD++ in order to optimise the performance in real scenarios.

Looking in detail at the obtained results (see Tables 1 and 2) and comparing with the results obtained from other similar works, our method improved upon the results of [17,26,27,35]. It also exceeded by a wide margin the method SSD-6D [28], since it got 76.3% using the RGB information and 90.9% with the RGB-D information compared to the 95.7% obtained in our method. Note also that it improved the accuracy of the method of Brachmann et al. [36] when it only used the colour information (50.2%). In addition, the described method improved the BB8 [29] method with or without refinement of the pose (62.7%).

On the other hand, it matched the results obtained by the method of Zhang et al. [51]. In contrast, the method of Brachmann et al. [36] was more accurate when the depth information was employed in addition to the colour information; specifically, it achieved 99.0% with the AD metric in the LINEMOD dataset.

When comparing with the work of [19], it is worth remarking that they only show the best results of 8 of the 13 3D objects available in the LINEMOD database. Consequently, if we calculate the average obtained using the AD metric of our method for these models, we obtained 96.5% versus 97.8% of the method of Hinterstoisser [19].

Finally, it can be concluded that although more precise methods have appeared in recent years [28] (all of them based on deep learning techniques), especially when the objects are partially visible, our method was not only accurate enough compared to many of the methods in the scientific literature (see Tables 1 and 2), but it was also (as will be seen in the next section) the fastest of all the methods analysed in this work, allowing pose estimation in real-time with only the requirement of a 3D model (not necessarily with colour) of the 3D object.

Additionally, Figure 7 shows some qualitative results of the estimated pose using our method in the LINEMOD++ database. Specifically, a projection was done of a bounding box calculated using the estimated pose (in green) and the ground truth pose (in red).

Table 1. Comparison of the results between different detection and pose estimation methods on the LINEMOD dataset [24] using the AD metric and $k_m = 0.1$, given as the percentage of objects in which the pose was estimated with an error smaller than 10% of the object diameter.

Sequence	Our Method	LINEMOD++ [24]	Drost [17]	Hodaň et al. [26]	Brachmann et al. [35]	Hinterstoisser et al. [19]
Ape	97.3%	95.8%	86.5%	93.9%	85.4%	98.5%
Benchwise	95.4%	98.7%	70.7%	99.8%	98.9%	99.8%
Driller	93.0%	93.6%	87.3%	94.1%	99.7%	93.4%
Cam	95.0%	97.5%	78.6%	95.5%	92.1%	99.3%
Can	97.0%	95.9%	80.2%	95.9%	84.4%	98.7%
Iron	98.7%	97.5%	84.9%	97.0%	98.8%	98.3%
Lamp	99.2%	97.7%	93.3%	88.8%	97.6%	96.0%
Phone	97.1%	93.3%	80.7%	89.4%	86.1%	98.6%
Cat	98.8%	99.3%	85.4%	98.2%	90.6%	
Hole punch	92.8%	95.9%	77.4%	88.0%	97.9%	
Duck	99.1%	95.9%	46.0%	94.3%	92.7%	
Cup	97.7%	97.1%	68.4%	99.5%		
Bowl	97.8%	99.9%	95.7%	98.8%		
Box	99.2%	99.8%	97.0%	100.0%	91.1%	
Glue	96.9%	91.8%	57.2%	98.0%	87.9%	
Mean	**95.7%**	**96.6%**	**79.3%**	**95.4%**	**92.5%**	**97.8%**

Sequence	Zhang et al. [27]	Kehl et al. [32]	Zhang et al. [51]	BB8 [29]	SSD-6D with RGB-D [28]	
Ape	96.3%	96.9%	93.9%			
Benchwise	90.4%	94.1%	99.8%			
Driller	95.2%	96.2%	94.1%			
Cam	91.3%	97.7%	95.5%			
Can	98.2%	95.2%	95.9%			
Iron	98.8%	98.7%	97.0%			
Lamp	91.4%	96.2%	88.8%			
Phone		92.7%	92.8%			
Cat	91.8%	97.4%	98.2%			
Hole punch	97.8%	96.8%	88.0%			
Duck	91.8%	97.3%	94.3%			
Cup		99.6%	99.6%			
Bowl		99.9%	99.9%			
Box	99.8%	99.9%	100.0%			
Glue	94.6%	78.6%	98.0%			
Mean	**94.7%**	**95.8%**	**95.7%**	**62.7%**	**90.9%**	

Table 2. Results of our detection and pose estimation system on the LINEMOD dataset [24] using different metrics. The percentage is calculated as the number of times that the pose was estimated correctly with respect to the total number of images for each of the sequences. AD: average distance; IoU: intersection over union.

Model	6D Pose (5 cm 5°)	6D Pose (AD)	2D Bounding Box (IoU)	F1-Score (AD)
Ape (1235)	98.94%	97.33%	98.86%	0.9864
Bench Vise (1214)	95.46%	95.46%	95.46%	0.9768
Driller (1187)	93.09%	91.24%	93.85%	0.9542
Cam (1200)	95.08%	94.50%	95.17%	0.9717
Can (1195)	97.07%	91.88%	97.07%	0.9577
Iron (1151)	98.70%	98.00%	98.87%	0.9899
Lamp (1226)	99.26%	98.04%	99.26%	0.9901
Phone (1224)	97.11%	97.11%	97.11%	0.9853
Cat (1178)	98.89%	98.89%	98.89%	0.9944
Hole punch (1236)	92.80%	91.35%	92.72%	0.9547
Duck (1253)	99.12%	96.96%	99.12%	0.9846
Cup (1239)	97.74%	97.74%	97.66%	0.9881
Bowl (1232)	97.81%	97.81%	97.81%	0.9889
Box (1252)	99.28%	99.28%	99.28%	0.9963
Glue (1219)	96.97%	90.26%	96.97%	0.9495
Mean	**97.15%**	**95.72%**	**97.20%**	**0.9779**

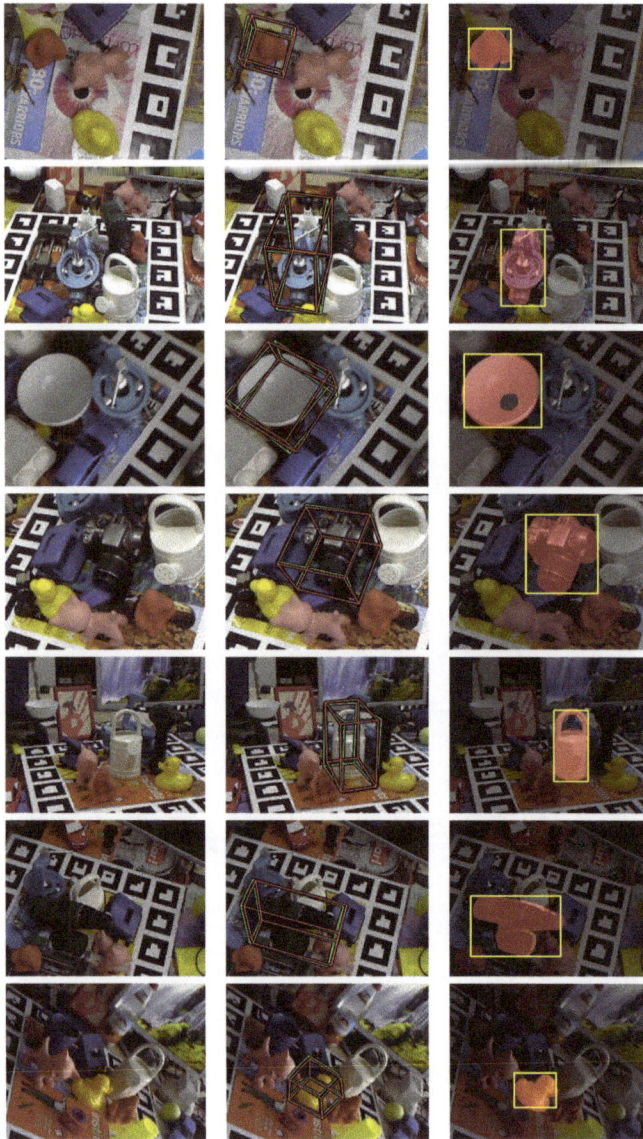

Figure 7. Results visualisation of the estimated pose using the LINEMOD dataset. The bounding box projection of the actual pose is shown in red, and the estimated pose by our method is shown in green (best viewed in colour).

Figure 8 depicts two objects employed in the AIDE project and its pose estimation. Please note that the spoon CAD model was obtained by scanning the actual object using a KinectFusion algorithm [34] and a RGB-D camera, while the plate CAD model was modelled using Autocad 3ds Max. This example sustains the affirmation that our algorithm works with models with poor and high quality.

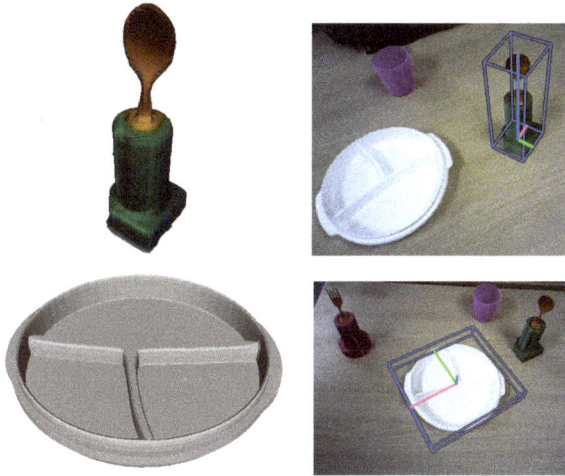

Figure 8. Qualitative results visualisation of the estimated pose of two project objects. The spoon model was scanned and the plate was modelled.

3.2.3. Computational Cost

One of the objectives of this work was to design a detection and pose estimation system that worked in real-time so that the grasping system could correct the position and orientation of the object. This feature is very interesting in our use case because the user selects the object of interest through an eye-tracking system that works in real time so they can change the objective quickly. In addition, it allows the grasping of moving objects. It is important to note that the performance analysis was done with the limitation that only one object is detected and estimated simultaneously (common in grasping systems).

The developed method had a total computational cost of 0.032 s (31.72 fps) obtained using a battery test of sequences of the LINEMOD dataset, as can be seen in Table 3.

Table 3. Breakdowns of time in seconds of the detection and pose estimation algorithm on each of the sequences of the LINEMOD dataset. The algorithm implemented in the project (multi-core version) improved upon the performance of the LINEMOD++ algorithm by a factor of three [24].

Sequence	Total Time (One-Core)	Total Time (Multi-Core)
Ape (1235)	0.1070	0.0401
Bench Vise (1214)	0.0581	0.0289
Bowl (1231)	0.0748	0.0316
Cam (1200)	0.0646	0.0319
Can (1195)	0.0597	0.0288
Cat (1178)	0.0698	0.0308
Cup (1239)	0.0896	0.0367
Driller (1187)	0.0582	0.0291
Duck (1253)	0.0836	0.0333
Box (1252)	0.0830	0.0344
Glue (1219)	0.0837	0.0335
Hole punch (1236)	0.0831	0.0343
Iron (1151)	0.0621	0.0300
Lamp (1226)	0.0577	0.0287
Phone (1224)	0.0624	0.0288
Mean	0.0731	**0.0320**

The developed method considerably improved the computational cost in comparison with other similar works. Specifically, it exceeded by a factor of $3\times$ the method on which it is based (LINEMOD++) [24] (see Table 3), obtaining very similar results in precision (see the previous Section 2). This is due to the optimisation performed in the original method. Table 3 shows the times obtained with our parallelized algorithm and without parallelizing. Please note that the tests were performed using the same number of templates as those cited by Hintertoisser et al. [24]. Specifically, 1235 templates were used for each of the 3D models.

In addition, compared with other works (Table 4), our method considerably decreased the computational cost using only the CPU. In more detail, when compared to detection and pose estimation methods based on deep learning techniques [29,31] (with the exception of the SSD-6D method [28] that works at 10 fps), these are very computationally expensive, preventing their use in real-time. Furthermore, all of these methods require a high-performance GPU. In fact, the implemented method exceeded in efficiency the following template matching methods [24,26,27], methods based on geometric descriptors [17,19], and methods based on random forest [36,52–54].

Another outstanding aspect is that the implemented method estimates the pose independently in each frame; that is, it does not use tracking techniques such as the one proposed by Tan et al. [55]. This makes it possible in future work to further reduce the computational cost if combined with any of these tracking techniques. It is also important to remark that the obtained time results were calculated after the analysis of the complete image captured by the camera (in this case of size 640 × 480). However, in our case it was only necessary to analyse the regions of the image obtained from the eye-tracking phase, and consequently the times obtained were further reduced.

The equipment employed for testing our algorithm was a computer with Intel Core i7-7700 (3.60 GHz) with 16 GB of RAM and an Orbbec Astra S RGB-D camera. Our method was implemented in C++ with OpenMP. The optimisation in the detection algorithm was performed in the LINEMOD algorithm implementation of the OPENCV library.

Table 4. Time comparison (seconds) of different methods for detection and pose estimation.

Method	Time (seconds)	Use GPU
LINEMOD++ [24]	0.12 s	x
Hodaň et al. [26]	0.75 to 2.08 s	✓
Brachmann et al. [36]	0.45 s	x
Drost et al. [17]	6.30 s	x
Hinterstoisser et al. [19]	0.1 to 0.8 s	x
Doumanaglou et al. [53]	4 to 7 s	x
Tejani et al. [52]	0.67 s	x
BB8 [29]	0.30 s	✓
Zhang et al. [51]	0.80 s	–
Zhang et al. [27]	0.70 s	x
Michel et al. [54]	1 to 3 s	x
Do et al. [31]	0.10 s	✓
SSD-6D [28]	0.10 s	✓
Ours	**0.03 s**	x

3.3. Mouth Pose System

The mouth pose algorithm was tested with different users during experimental sessions as can be seen in Figure 9. In this figure, pupil landmarks are coloured yellow while mouth landmarks are red. 3D coordinates are written on the top of the images, and on the top-left corner there are three circle indicators. These indicators change colour to green when the user has their mouth open or if the user is blinking their eyes. These events are also communicated as numerical values and written in blue text on the image. As can be seen, the algorithm worked well with/without facial hair, with glasses, and with different genders.

Figure 9. Qualitative results visualisation of the estimated mouth pose with five users. Red points are the mouth landmarks and yellow points the pupil landmarks. The top-right circles indicate if the user has their mouth or the eyes open (green) or closed (red) (best viewed in colour).

To assess the stability of the developed method, some extreme positions, partial face occlusion, and wearing an eye-tracking system were tested (shown in the second row of Figure 9).

3.4. Eye-Tracking System

To evaluate the performance of object selection using the estimation of gaze point and detection of the type of object already selected, an experiment was conducted with 10 healthy subjects. The experiment consisted of the selection and detection of three kinds of objects (a glass, a bottle, and a fork) wearing the Tobii Glasses. The user had to select the object whose name is shown on a screen in front of the user. The name of the objects appeared randomly, and when the object was selected an audio feedback was provided to the user. The performance of the system was near-excellent since the percentage of average success was 90% and seven out of eleven users only had two or less fails in 20 trials (see Table 5). Regarding the average selection and detection time, the average selection time of all users was around 10 s and the average detection time of all users was around 1 s (see Table 5). Therefore, the users required around 11 s on average to select and detect the object with which they want to interact. Moreover, we measured the angular movements of the neck during the experimental session. The maximum range of motion of each joint was: flexion 15.27°, extension 7.5°, lateral rotation (right) 68.08°, lateral rotation (left) 41.01°, lateral flexion (right) 14.54°, and lateral flexion (left) 35.86°.

Table 5. Object selection using the estimation of gaze point and detection of the type of object.

Users	Average Selection Time (s)	Standard Deviation	Average Detection Time (s)	Standard Deviation	Number of Trials	Successes	Failures
user 1	10.00	13.68	1.02	0.05	20	20	0
user 2	6.38	5.64	1.00	0.02	20	20	0
user 3	18.81	32.52	0.98	0.04	20	20	0
user 4	4.97	2.15	0.96	0.05	20	16	4
user 5	24.63	46.31	0.96	0.05	20	15	5
user 6	6.39	6.98	1.08	0.69	20	18	2
user 7	4.04	1.02	0.96	0.04	20	19	1
user 8	6.05	5.30	1.03	0.03	20	15	5
user 9	14.75	17.32	0.97	0.02	20	18	2
user 10	5.151	1.90	1.06	0.05	20	19	1

3.5. Experimental Results

The algorithms and methods presented in this paper were tested in real environments with healthy subjects and subjects with different neurological conditions. The subjects used the hybrid BCI system to trigger the movements of the Jaco2 robot: EEG to control the open/close movement of the gripper and EOG to trigger the movement to grasp the selected object. In Figure 10, some images of the experiments are shown. The performance of the system was very good, and it is out of the scope of this paper to report on the results regarding the use of the hybrid BCI system.

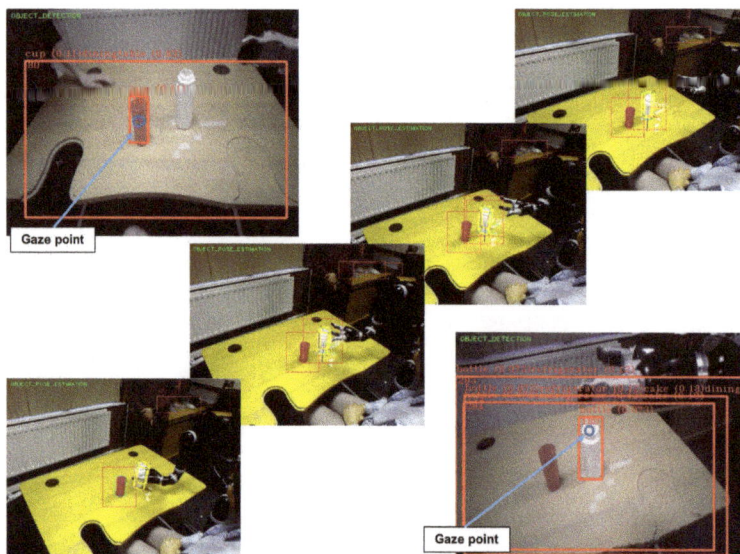

Figure 10. Images of the context recognition and eye-tracking systems in experimental tests. Examples of object detection (cup and bottle) with the estimated gaze position (blue circle). Snapshots of the experiments: grasping a bottle using the Jaco2 robot.

4. Conclusions

The AIDE project has developed a complete assistive robotic solution with a novel and revolutionary modular and adaptive multimodal human–machine interface. The computer vision algorithms have an essential role, as has been shown in this manuscript. On one hand, the object selection algorithm is a very useful and natural robot–human interface because the user only needs to stare at the desired object. Furthermore, the complete selection made by the users only costs around 11 s, with an average success of 90% in the test performed. On the other hand, a major contribution presented here is the real-time detection and pose estimation method of textureless objects that allows for precise grasping tasks. As shown in the results, this algorithm outperformed the state-of-the-art in terms of computational cost, with similar precision results to the top methods. A thorough evaluation was made against the popular LINEMOD so that the results can be compared with future methods. Finally, a mouth pose algorithm was employed with the objective of safely operating the robot system. Moreover, the complete assistive robotic system and sensing solution is mounted on a wheelchair, giving a great deal of independence and accessibility to motion-disabled people.

As a future work regarding the detection and pose estimation topic, it is planned to explore a deep learning approach. The main problems of the deep learning methods of pose estimation is that they are not in real-time and it is difficult to obtain the ground-truth data for the training. It is planned to design convolutional neural network that can be computed very quickly, like the YOLO or SSD methods. Furthermore, this model should be able to learn from synthetic generated images. This approach could improve the weaknesses of our method and maintain its strengths.

Sensors **2018**, *18*, 2408

Author Contributions: M.A. and N.G.-A. conceived and designed the experiments; M.O. developed the detection and pose estimation algorithm and tested it, he also contributed in writing the paper. J.M.C. performed the camera–robot calibration validation, design the eye-tracking experimental validation and analysed the data of the experimental sessions. S.E. and L.D.L. performed the eye-tracking experiments. N.G.-A. contributed in writing the paper. E.I. contributed in writing the paper, developed the eye-tracking and mouth algorithms, implemented the camera–robot methods, and made the integration of the computer vision algorithms.

Funding: The research leading to these results received funding from the European Community's Horizon 2020 programme, AIDE project: "Adaptive Multimodal Interfaces to Assist Disabled People in Daily Activities" (grant agreement No: 645322).

Acknowledgments: The aid for predoctoral contracts for the training of PHD of the Miguel Hernandez University of Elche, Spain.

Conflicts of Interest: The authors declare no conflict of interest.

Appendix A. Experimental Validation: Detailed Figures

Appendix A.1. Comparing the Influence of Using Different Numbers of Calibration Points for Each Method

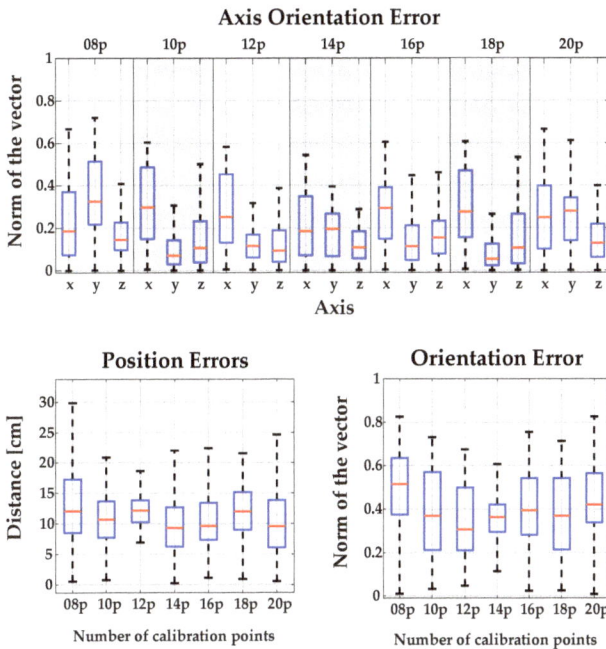

Figure A1. Results obtained by the direct calibration method using different numbers of calibration points.

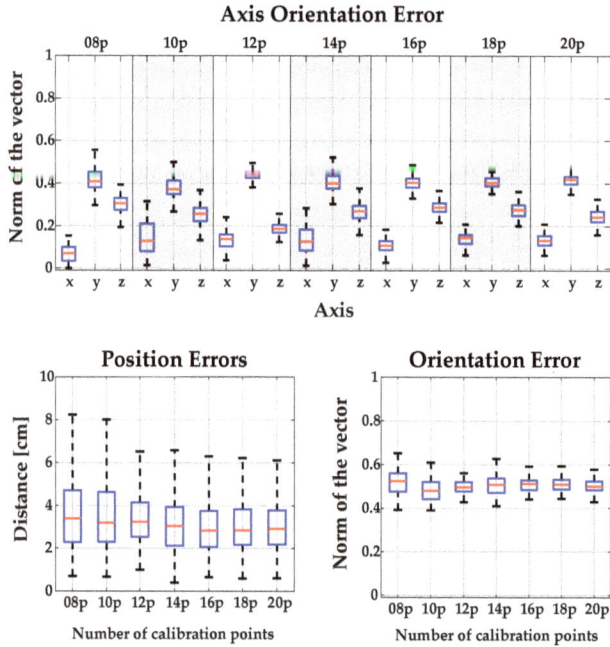

Figure A2. Results obtained by the standard calibration method using different numbers of calibration points.

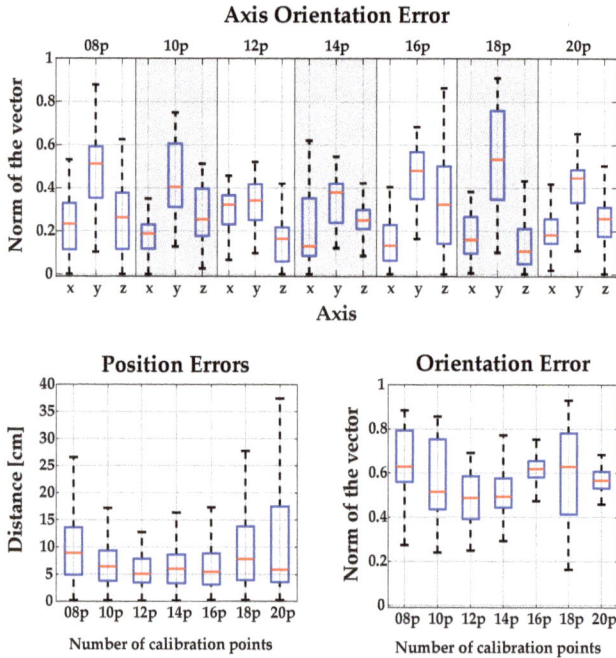

Figure A3. Results obtained by the RANSAC calibration method using different numbers of calibration points.

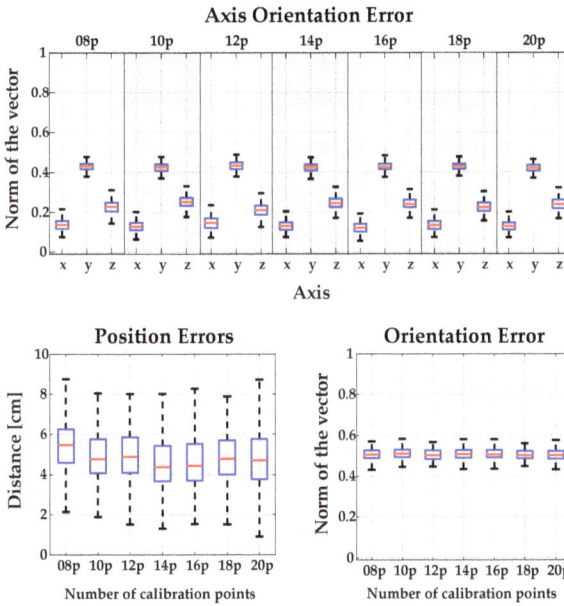

Figure A4. Results obtained by the XS calibration method using different numbers of calibration points.

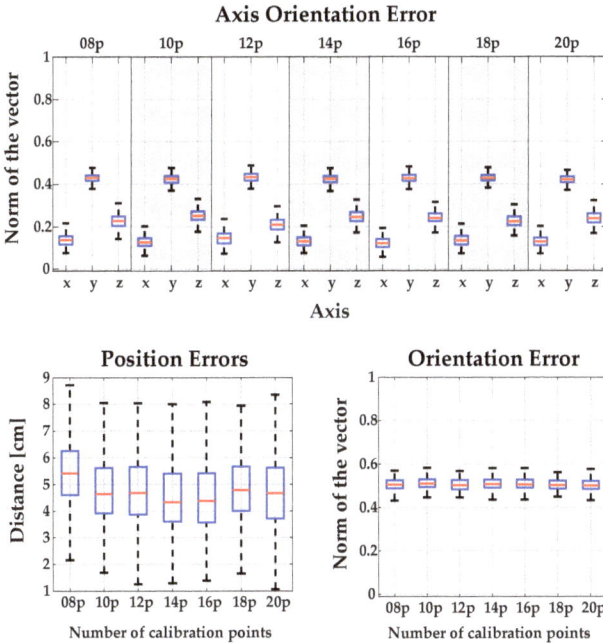

Figure A5. Results obtained by the XS2 calibration method using different numbers of calibration points.

References

1. European Commision. People with disabilities have equal rights. In *The European Disability Strategy 2010–2020*; Technical Report; European Commision: Washington, DC, USA, 2010.

2. Meng, J.; Zhang, S.; Bekyo, A.; Olsoe, J.; Baxter, B.; He, B. Noninvasive Electroencephalogram Based Control of a Robotic Arm for Reach and Grasp Tasks. *Sci. Rep.* **2016**, *6*, 38565. [CrossRef] [PubMed]

3. Silva, C.; Vonghulbhisal, J.; Marques, M.; Costeira, J.P.; Veloso, M. FeedBot—A Robotic Arm for Autonomous Assisted Feeding. In Proceedings of the Portuguese Conference on Artificial Intelligence, Porto, Portugal, 5–8 September 2017; Oliveira, E., Gama, J., Vale, Z., Lopes Cardoso, H., Eds.; Springer International Publishing: Cham, Switzerland, 2017; pp. 486–497.

4. Hochberg, L.R.; Bacher, D.; Jarosiewicz, B.; Masse, N.Y.; Simeral, J.D.; Vogel, J.; Haddadin, S.; Liu, J.; Cash, S.S.; van der Smagt, P.; et al. Reach and grasp by people with tetraplegia using a neurally controlled robotic arm. *Nature* **2012**, *485*, 372. [CrossRef] [PubMed]

5. Collinger, J.L.; Wodlinger, B.; Downey, J.E.; Wang, W.; Tyler-Kabara, E.C.; Weber, D.J.; McMorland, A.J.; Velliste, M.; Boninger, M.L.; Schwartz, A.B. High-performance neuroprosthetic control by an individual with tetraplegia. *Lancet* **2013**, *381*, 557–564. [CrossRef]

6. Pedrocchi, A.; Ferrante, S.; Ambrosini, E.; Gandolla, M.; Casellato, C.; Schauer, T.; Klauer, C.; Pascual, J.; Vidaurre, C.; Gföhler, M.; et al. MUNDUS project: MUltimodal Neuroprosthesis for daily Upper limb Support. *J. Neuroeng. Rehabil.* **2013**, *10*, 66. [CrossRef] [PubMed]

7. Frisoli, A.; Sotgiu, E.; Procopio, C.; Chisari, C.; Rossi, B.; Bergamasco, M. Positive effects of rehabilitation training with the L-EXOS in chronic stroke. In Proceedings of the SKILLS 2011, Montpellier, France, 15–16 December 2011; Volume 1, p. 27.

8. Barsotti, M.; Leonardis, D.; Loconsole, C.; Solazzi, M.; Sotgiu, E.; Procopio, C.; Chisari, C.; Bergamasco, M.; Frisoli, A. A full upper limb robotic exoskeleton for reaching and grasping rehabilitation triggered by MI-BCI. In Proceedings of the 2015 IEEE International Conference on Rehabilitation Robotics (ICORR), Singapore, 11–14 August 2015; pp. 49–54.

9. Lauretti, C.; Cordella, F.; Ciancio, A.L.; Trigili, E.; Catalan, J.M.; Badesa, F.J.; Crea, S.; Pagliara, S.M.; Sterzi, S.; Vitiello, N.; et al. Learning by Demonstration for Motion Planning of Upper-Limb Exoskeletons. *Front. Neurorobot.* **2018**, *12*. [CrossRef] [PubMed]

10. Onose, G.; Grozea, C.; Anghelescu, A.; Daia, C.; Sinescu, C.; Ciurea, A.; Spircu, T.; Mirea, A.; Andone, I.; Spânu, A. On the Feasibility of Using Motor Imagery EEG-Based Brain–Computer Interface in Chronic Tetraplegics for Assistive Robotic Arm Control: A Clinical Test and Long-Term Post-Trial Follow-Up. *Spinal Cord* **2012**, *50*, 599. [CrossRef] [PubMed]

11. Li, M.; Yin, H.; Tahara, K.; Billard, A. Learning Object-Level Impedance Control for Robust Grasping and Dexterous Manipulation. In Proceedings of the IEEE International Conference on Robotics and Automation, Hong Kong, China, 31 May–7 June 2014; pp. 6784–6791.

12. Ahmadzadeh, S.R.; Kormushev, P.; Caldwell, D.G. Autonomous Robotic Valve Turning: A Hierarchical Learning Approach. In Proceedings of the IEEE International Conference on Robotics and Automation, Karlsruhe, Germany, 6–10 May 2013; pp. 4629–4634.

13. Lowe, D.G. Object recognition from local scale-invariant features. In Proceedings of the 7th IEEE International Conference on Computer Vision, Kerkyra, Greece, 20–27 September 1999; Volume 2, pp. 1150–1157.

14. Besl, P.J.; McKay, N.D. A Method for Registration of 3-D Shapes. *IEEE Trans. Pattern Anal. Mach. Intell.* **1992**, *14*, 239–256, doi:10.1109/34.121791. [CrossRef]

15. Hana, X.F.; Jin, J.S.; Xie, J.; Wang, M.J.; Jiang, W. A comprehensive review of 3D point cloud descriptors. *arXiv* **2018**, arXiv:1802.02297.

16. Rusu, R.B.; Blodow, N.; Beetz, M. Fast point feature histograms (FPFH) for 3D registration. In Proceedings of the IEEE International Conference on Robotics and Automation (ICRA'09), Kobe, Japan, 12–17 May 2009; pp. 3212–3217.

17. Drost, B.; Ulrich, M.; Navab, N.; Ilic, S. Model globally, match locally: Efficient and robust 3D object recognition. In Proceedings of the 2010 IEEE Conference on Computer Vision and Pattern Recognition (CVPR), San Francisco, CA, USA, 13–18 June 2010; pp. 998–1005.

18. Salti, S.; Tombari, F.; Di Stefano, L. SHOT: Unique signatures of histograms for surface and texture description. *Comput. Vis. Image Underst.* **2014**, *125*, 251–264. [CrossRef]
19. Hinterstoisser, S.; Lepetit, V.; Rajkumar, N.; Konolige, K. Going further with point pair features. In Proceedings of the European Conference on Computer Vision, Amsterdam, The Netherlands, 8–16 October 2016; Springer: Cham, Switzerland, 2016; pp. 834–848.
20. Rusu, R.B.; Bradski, G.; Thibaux, R.; Hsu, J. Fast 3d recognition and pose using the viewpoint feature histogram. In Proceedings of the 2010 IEEE/RSJ International Conference on Intelligent Robots and Systems (IROS), Taipei, Taiwan, 18–22 October 2010; pp. 2155–2162.
21. Aldoma, A.; Tombari, F.; Rusu, R.B.; Vincze, M. OUR-CVFH–oriented, unique and repeatable clustered viewpoint feature histogram for object recognition and 6DOF pose estimation. In Proceedings of the Joint DAGM (German Association for Pattern Recognition) and OAGM Symposium, Graz, Austria, 28–31 August 2012; Springer: Cham, Switzerland, 2012; pp. 113–122.
22. Wohlkinger, W.; Vincze, M. Ensemble of shape functions for 3d object classification. In Proceedings of the 2011 IEEE International Conference on Robotics and Biomimetics (ROBIO), Phuket, Thailand, 7–11 December 2011; pp. 2987–2992.
23. Tombari, F.; Salti, S.; Di Stefano, L. A combined texture-shape descriptor for enhanced 3D feature matching. In Proceedings of the 2011 18th IEEE International Conference on Image Processing (ICIP), Brussels, Belgium, 11–14 September 2011; pp. 809–812.
24. Hinterstoisser, S.; Lepetit, V.; Ilic, S.; Holzer, S.; Bradski, G.; Konolige, K.; Navab, N. Model Based Training, Detection and Pose Estimation of Texture-Less 3D Objects in Heavily Cluttered Scenes. In Proceedings of the Asian Conference on Computer Vision, Daejeon, Korea, 5–9 November 2012; Lee, K.M., Matsushita, Y., Rehg, J.M., Hu, Z., Eds.; Springer: Berlin/Heidelberg, Germany, 2013; pp. 548–562.
25. Kehl, W.; Tombari, F.; Navab, N.; Ilic, S.; Lepetit, V. Hashmod: A hashing method for scalable 3D object detection. *arXiv* **2016**, arXiv:1607.06062.
26. Hodaň, T.; Zabulis, X.; Lourakis, M.; Obdržálek, Š.; Matas, J. Detection and fine 3D pose estimation of texture-less objects in RGB-D images. In Proceedings of the 2015 IEEE/RSJ International Conference on Intelligent Robots and Systems (IROS), Hamburg, Germany, 28 September–3 October 2015; pp. 4421–4428.
27. Zhang, H.; Cao, Q. Texture-less object detection and 6D pose estimation in RGB-D images. *Robot. Auton. Syst.* **2017**, *95*, 64–79. [CrossRef]
28. Kehl, W.; Manhardt, F.; Tombari, F.; Ilic, S.; Navab, N. SSD-6D: Making RGB-based 3D detection and 6D pose estimation great again. In Proceedings of the IEEE Conference on Computer Vision and Pattern Recognition (CVPR), Honolulu, HI, USA, 21–26 July 2017; pp. 1521–1529.
29. Rad, M.; Lepetit, V. BB8: A Scalable, Accurate, Robust to Partial Occlusion Method for Predicting the 3D Poses of Challenging Objects without Using Depth. In Proceedings of the International Conference on Computer Vision, Venice, Italy, 22–29 October 2017.
30. Xiang, Y.; Schmidt, T.; Narayanan, V.; Fox, D. PoseCNN: A Convolutional Neural Network for 6D Object Pose Estimation in Cluttered Scenes. *arXiv* **2017**, arXiv:1711.00199.
31. Do, T.T.; Cai, M.; Pham, T.; Reid, I. Deep-6DPose: Recovering 6D Object Pose from a Single RGB Image. *arXiv* **2018**, arXiv:1802.10367.
32. Kehl, W.; Milletari, F.; Tombari, F.; Ilic, S.; Navab, N. Deep learning of local RGB-D patches for 3D object detection and 6D pose estimation. In Proceedings of the European Conference on Computer Vision, Amsterdam, The Netherlands, 8–16 October 2016; Springer: Cham, Switzerland, 2016; pp. 205–220.
33. Hinterstoisser, S.; Lepetit, V.; Wohlhart, P.; Konolige, K. On Pre-Trained Image Features and Synthetic Images for Deep Learning. *arXiv* **2017**, arXiv:1710.10710.
34. Newcombe, R.A.; Izadi, S.; Hilliges, O.; Molyneaux, D.; Kim, D.; Davison, A.J.; Kohi, P.; Shotton, J.; Hodges, S.; Fitzgibbon, A. KinectFusion: Real-Time Dense Surface Mapping and Tracking. In Proceedings of the 2011 10th IEEE International Symposium on Mixed and Augmented Reality, Basel, Switzerland, 26–29 October 2011; pp. 127–136.

35. Brachmann, E.; Krull, A.; Michel, F.; Gumhold, S.; Shotton, J.; Rother, C. Learning 6d object pose estimation using 3d object coordinates. In Proceedings of the European Conference on Computer Vision, Zurich, Switzerland, 6–12 September 2014; Springer: Cham, Switzerland, 2014; pp. 536–551.

36. Brachmann, E.; Michel, F.; Krull, A.; Ying Yang, M.; Gumhold, S. Uncertainty-driven 6d pose estimation of objects and scenes from a single rgb image. In Proceedings of the IEEE Conference on Computer Vision and Pattern Recognition, Las Vegas, NV, USA, 27–30 June 2016; pp. 3364–3372.

37. Tejani, A.; Tang, D.; Kouskouridas, R.; Kim, T.K. Latent-class hough forests for 3D object detection and pose estimation. In Proceedings of the European Conference on Computer Vision, Zurich, Switzerland, 6–12 September 2014; Springer: Cham, Switzerland, 2014; pp. 462–477.

38. Garrido-Jurado, S.; noz Salinas, R.M.; Madrid-Cuevas, F.; Marín-Jiménez, M. Automatic generation and detection of highly reliable fiducial markers under occlusion. *Pattern Recognit.* **2014**, *47*, 2280–2292, doi:10.1016/j.patcog.2014.01.005. [CrossRef]

39. Tabb, A.; Ahmad Yousef, K.M. Solving the Robot-World Hand-Eye(s) Calibration Problem with Iterative Methods. *Mach. Vis. Appl.* **2017**, *28*, 569–590. [CrossRef]

40. Cognolato, M.; Graziani, M.; Giordaniello, F.; Saetta, G.; Bassetto, F.; Brugger, P.; Caputo, B.; Müller, H.; Atzori, M. Semi-Automatic Training of an Object Recognition System in Scene Camera Data Using Gaze Tracking and Accelerometers. In Proceedings of the International Conference on Computer Vision Systems, Copenhagen, Denmark, 6–9 July 2017; Liu, M., Chen, H., Vincze, M., Eds.; Springer International Publishing: Cham, Switzerland, 2017; pp. 175–184.

41. Redmon, J.; Farhadi, A. YOLO9000: Better, Faster, Stronger. In Proceedings of the IEEE Conference Computer Vision and Pattern Recognition (CVPR), Honolulu, HI, USA, 21–26 July 2017; pp. 6517–6525.

42. Ren, S.; He, K.; Girshick, R.; Sun, J. Faster R-CNN: Towards Real-Time Object Detection with Region Proposal Networks. *IEEE Trans. Pattern Anal. Mach. Intell.* **2017**, *39*, 1137–1149, doi:10.1109/TPAMI.2016.2577031. [CrossRef] [PubMed]

43. Liu, W.; Anguelov, D.; Erhan, D.; Szegedy, C.; Reed, S.; Fu, C.Y.; Berg, A.C. SSD: Single Shot MultiBox Detector. In Proceedings of the European Conference on Computer Vision, Amsterdam, The Netherlands, 8–16 October 2016.

44. Lin, T.Y.; Maire, M.; Belongie, S.; Hays, J.; Perona, P.; Ramanan, D.; Dollár, P.; Zitnick, C.L. Microsoft coco: Common objects in context. In Proceedings of the European Conference on Computer Vision, Zurich, Switzerland, 6–12 September 2014; Springer: Cham, Switzerland, 2014; pp. 740–755.

45. Hinterstoisser, S.; Cagniart, C.; Ilic, S.; Sturm, P.; Navab, N.; Fua, P.; Lepetit, V. Gradient response maps for real-time detection of textureless objects. *IEEE Trans. Pattern Anal. Mach. Intell.* **2012**, *34*, 876–888. [CrossRef] [PubMed]

46. Zhao, W.; Chellappa, R.; Phillips, P.J.; Rosenfeld, A. Face Recognition: A Literature Survey. *ACM Comput. Surv. (CSUR)* **2003**, *35*, 399–458. [CrossRef]

47. Goswami, G.; Bharadwaj, S.; Vatsa, M.; Singh, R. On RGB-D Face Recognition Using Kinect. In Proceedings of the IEEE Sixth International Conference on Biometrics: Theory, Applications and Systems, Arlington, VA, USA, 29 September–2 October 2013; pp. 1–6.

48. Hodaň, T.; Haluza, P.; Obdržálek, Š.; Matas, J.; Lourakis, M.; Zabulis, X. T-LESS: An RGB-D Dataset for 6D Pose Estimation of Texture-less Objects. In Proceedings of the IEEE Winter Conference on Applications of Computer Vision (WACV), Santa Rosa, CA, USA, 24–31 March 2017.

49. Tekin, B.; Sinha, S.N.; Fua, P. Real-Time Seamless Single Shot 6D Object Pose Prediction. *arXiv* **2017**, arXiv:1711.08848.

50. Everingham, M.; Van Gool, L.; Williams, C.K.; Winn, J.; Zisserman, A. The pascal visual object classes (voc) challenge. *Int. J. Comput. Vis.* **2010**, *88*, 303–338. [CrossRef]

51. Zhang, H.; Cao, Q. Combined Holistic and Local Patches for Recovering 6D Object Pose. In Proceedings of the IEEE Conference on Computer Vision and Pattern Recognition, Honolulu, HI, USA, 21–26 July 2017; pp. 2219–2227.

52. Tejani, A.; Kouskouridas, R.; Doumanoglou, A.; Tang, D.; Kim, T.K. Latent-Class Hough Forests for 6 DoF Object Pose Estimation. *IEEE Trans. Pattern Anal. Mach. Intell.* **2018**, *40*, 119–132. [CrossRef] [PubMed]

Sensors **2018**, *18*, 2408

53. Doumanoglou, A.; Kouskouridas, R.; Malassiotis, S.; Kim, T.K. Recovering 6D object pose and predicting next-best-view in the crowd. In Proceedings of the IEEE Conference on Computer Vision and Pattern Recognition, Las Vegas, NV, USA, 27–30 June 2016; pp. 3583–3592.
54. Michel, F.; Kirillov, A.; Brachmann, E.; Krull, A.; Gumhold, S.; Savchynskyy, B.; Rother, C. Global hypothesis generation for 6D object pose estimation. In Proceedings of the IEEE Conference on Computer Vision and Pattern Recognition, Honolulu, HI, USA, 21–26 July 2017
55. Tan, D.J.; Navab, N.; Tombari, F. Looking beyond the simple scenarios: Combining learners and optimizers in 3d temporal tracking. *IEEE Trans. Vis. Comput. Graph.* **2017**, *23*, 2399–2409. [CrossRef] [PubMed]

sensors

MDPI

Article

ED-FNN: A New Deep Learning Algorithm to Detect Percentage of the Gait Cycle for Powered Prostheses

Huong Thi Thu Vu *,†, Felipe Gomez †, Pierre Cherelle, Dirk Lefeber and Ann Nowé
and Bram Vanderborght

Robotics & MultiBody Mechanics Research Group (R& MM) and Artificial Intelligence Lab, Vrije Universiteit
Brussel and Flanders Make; Pleinlaan 2, 1050 Brussel, Belgium; felipe.gomez.marulanda@vub.ac.be (F.G.);
pierre.cherelle@vub.ac.be (P.C.); dlefeber@vub.ac.be (D.L.); ann.nowe@vub.ac.be (A.N.);
Bram.Vanderborght@vub.be (B.V.)
* Correspondence: vu.huong@vub.ac.be
† These authors contributed equally to this work.

Received: 11 June 2018 ; Accepted: 20 July 2018 ; Published: 23 July 2018

Abstract: Throughout the last decade, a whole new generation of powered transtibial prostheses and exoskeletons has been developed. However, these technologies are limited by a gait phase detection which controls the wearable device as a function of the activities of the wearer. Consequently, gait phase detection is considered to be of great importance, as achieving high detection accuracy will produce a more precise, stable, and safe rehabilitation device. In this paper, we propose a novel gait percent detection algorithm that can predict a full gait cycle discretised within a 1% interval. We called this algorithm an exponentially delayed fully connected neural network (ED-FNN). A dataset was obtained from seven healthy subjects that performed daily walking activities on the flat ground and a 15-degree slope. The signals were taken from only one inertial measurement unit (IMU) attached to the lower shank. The dataset was divided into training and validation datasets for every subject, and the mean square error (MSE) error between the model prediction and the real percentage of the gait was computed. An average MSE of 0.00522 was obtained for every subject in both training and validation sets, and an average MSE of 0.006 for the training set and 0.0116 for the validation set was obtained when combining all subjects' signals together. Although our experiments were conducted in an offline setting, due to the forecasting capabilities of the ED-FNN, our system provides an opportunity to eliminate detection delays for real-time applications.

Keywords: gait phase prediction; gait event detection; lower limb prosthesis; exoskeleton; gait recognition

1. Introduction

Gait phase detection is a non-trivial problem for the new generation of powered prostheses and exoskeletons that are under development [1]. Gait phase detection algorithms are used to create and improve control strategies that permit prosthetic devices such as those presented in References [2–5] to work with more precision, safety, and stability. The objective of gait event detection algorithms is to detect non-delayed events in order to build control strategies for improving gait movement. For example, authors in [2] focused on a treatment to fix the amputee's foot in the lifted position by an orthosis. The technology known as functional electrical stimulation (FES) facilitates the artificial generation of action potentials in subcutaneous efferent nerves during the swing phase of the paretic foot by applying tiny electrical pulses via skin electrodes or implanted electrodes. By modulating the frequency or dimensions of these pulses, one can control the contraction of paretic muscles and induce movements in the affected limbs based on gait phase transitions. The foot pitch and roll angles are assessed in real-time by means of an inertial measurement unit (IMU). They detected four phases based on measuring the angular velocity and accelerometer to control the pitch and roll of the foot. Another example is the Ankle Mimicking Prosthetic Foot (AMP-Foot) [4]. This device greatly relies on the accuracy of a gait event detection algorithm to precisely control the torque of the motor in the device. This device stores energy in the springs during the first event of gait from initial-contact (IC) to foot-flat (FF), then releases the energy stored in the push-off (PO) spring and transmits it to the ankle joint by controlling the direct-current (DC) of the motor at the moment of heel-off (HO). This joint effort provides a peak torque and power output to the amputee, producing a toe-off (TO) event. After the TO, the amputee enters a swing phase where the torque of the motor is returned to zero magnitude, allowing the foot to go back to its initial position, resulting in an IC event. Knowing when all these events take place allows the device to take action at the right moment. Due to the different gait terminologies used in different articles, in this paper we follow the wording of Figure 1. In this figure, a *gait cycle percent* is defined as a sample from the continuous space of the gait cycle, an *event* is viewed as a discrete representation of the percentage space, often labelled as IC, FF, HO and TO; a *gait period* is considered as an interval between events, and a *phase* is considered as a union of several periods that represent different stages of the gait cycle. Lastly, a full gait cycle is composed of a *stance phase* and a *swing phase*. A more detailed description of different gait events is provided in Section 3. In the gait percent detection literature, a large set of techniques for improving the performance of event and phase detection can be found. These include threshold-based methods [3,6–8], time-frequency analysis [9,10], peak heuristic algorithms [9,11,12], machine learning (ML) models [13–21], and combinations of these [22]. ML algorithms are among the most popular techniques to detect phases in off-line data (i.e., stored data) and for real-time data (i.e., data gathered in real time). For instance, authors in References [17,19] detected four event-phases using hidden Markov models (HMMs). Evans and Arvind [23] increased the number of event-phases to five, and applied a hybrid method that combined fully connected neural networks (FNNs) and HMMs. The model accuracy of these algorithms is dependent on the type of sensors used to gather the gait event signals. Currently, wearable sensors are widely used for gait phase recognition systems: wearable sensors such as foot switches [14,24,25], foot pressure insoles [6,16,26,27], electromyography (EMG) [28,29], IMUs [3,8,9,15,30–34], and joint angular sensors [20,21] are used specifically for gait detection. A review in [35] showed that foot switches and foot pressure insoles yield the highest accuracy for gait phase detection algorithms. However, these sensors are very sensitive to the placement of the insole, which can influence the accuracy and reliability of the model. Additionally, they have a short lifespan, as they are often exposed to shock forces of the gait. Consequently, foot switches and foot pressure sensors are not considered suitable for daily activity applications. In contrast, Joshi et al. used EMG sensors to accurately extract up to eight gait phases [29]. EMGs are sensors that measure specific muscle activities occurring during a task. Regardless of the amount of information that can be extracted from the EMG signals, a heavy pre-processing step (e.g., a complex combination of filters) is required before it can be directly used in a learning algorithm. Furthermore, these sensors are susceptible to artifacts generated by moisture that builds between the skin and the sensors, and to the way in which they are placed on the skin of the subject. Recently, IMUs including gyroscopes, accelerometers, and magnetometers have become more popular, as they are not affected by most of the limitations of the aforementioned sensors. IMUs are low-cost, low-energy, durable, and can be easily mounted on different parts of the human body. Moreover, a human walking gait is a periodic cycle where IMUs can measure the angular velocities and accelerations of the walking gait. As a result, these signals are composed of rich information that can be used to accurately predict the gait events. Similar to EMG signals, IMU signals are very sensitive to movement artifacts. This means that in some cases these signals may require a strong pre-processing step before they can be directly used for learning. We hypothesise that deep learning algorithms are best suited for gait phase detection using IMU signals, as they perform well on signals that have a medium-low signal-to-noise ratio.

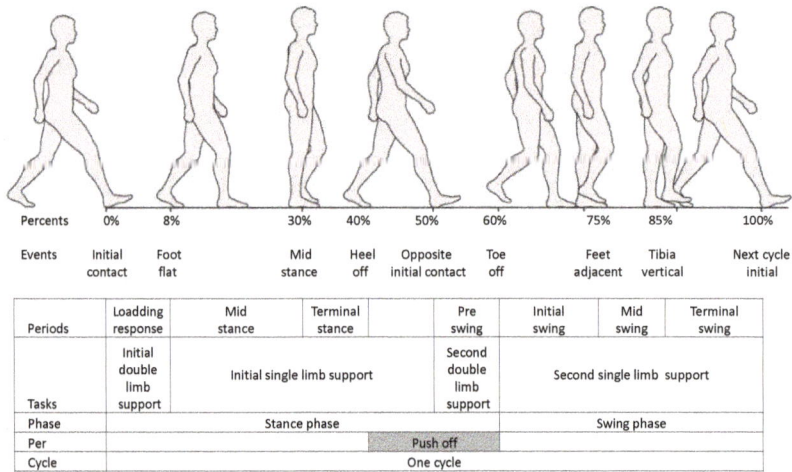

Percents	0%	8%		30%	40%	50%	60%		75%	85%	100%
Events	Initial contact	Foot flat		Mid stance	Heel off	Opposite initial contact	Toe off		Feet adjacent	Tibia vertical	Next cycle initial

Periods	Loadding response	Mid stance	Terminal stance		Pre swing	Initial swing	Mid swing	Terminal swing
Tasks	Initial double limb support	Initial single limb support			Second double limb support	Second single limb support		
Phase	Stance phase					Swing phase		
Per				Push off				
Cycle	One cycle							

Figure 1. A gait cycle is described as a dynamic and continuous occurrence of eight phases from the heel-contact at 0% to the next heel-contact at 100% percent of the gait cycle. Phase 0 is initial double-limb support, which appears during the first 10% of the cycle. Phase 1 is mid-stance, which appears from 10% to approximately 30% of the gait cycle. The following 10% of the gait cycle is terminal-stance. The propulsion phase or toe-off occurs after foot flat from 40% of the gait. This stage pushes the body forwards and prepares for swing phase from approximately 60% of the gait cycle. Single-limb support occurs from foot flat until 50% of the gait-related opposite initial contact limb, typically at 50% of the gait cycle. The second double-limb support occurs from the opposite limb at 50% until the toe leaves the ground at 60% of the gait cycle. Then, the second single-limb support completes the cycle. The following phases are early swing at approximately 60% to 75% of the gait cycle, mid swing at approximately 75% to 85% of the gait cycle, and late swing at approximately 85% to 100% of the gait cycle. Adapted from [36].

This paper introduces a novel gait percent detection model based on deep learning (DL) algorithms that can predict a full gait cycle discretised within a 1% interval. Currently, most studies can accurately detect four to eight phases. However, for real-world applications, it may be not sufficient for controlling active prosthetic devices. As a result, a more densely sampled gait phase is needed in order to obtain more important gait information and give more possibilities of control. The purpose of this study is to open an opportunity for future active devices such as below-the-knee prosthetics to take full control of the gait by accurately predicting the gait percentage in densely sampled phases.

2. Related Work

Due to the capacity of IMU sensors to measure the velocities and accelerations of motion, they are generally used in the fields of gait phase detection, gait event detection, and gait detection. In this section, we analyse and evaluate the performances and time delays of several recent gait phase detection systems in the fields of transtibial prostheses and exoskeletons.

Evans and Arvind [23] presented a method for the detection of five gait phases based on a feed-forward neural network (FNN) embedded in the hidden Markov model (HMM) model. However, their sensor system had to use seven IMUs mounted on the foot, shank, both sides of the thigh, and one on the pelvis. An implementation of complex threshold rules was further applied in exoskeletons in References [8,37]. The study in [8] could detect seven gait phases, and Boutaayamou et al. could detect four events in their study, with a temporal accuracy of around 10 ms [37]. However, systems in [8,37] were required to use four sensors which were attached to the leg segments.

Several papers have proposed applications for robotic prostheses that use signals from one IMU attached to the shank or the foot [12,17,38,39] to detect four phases or events. Mannini et al. [17] and Muller et al. [39] proposed two different models to detect four gait events in real-time using one IMU. The former proposed an HMM and the latter proposed a *finite state automaton* to model the transitions between phases. Nevertheless, both cases showed a time delay limitation in on-line detection. For example, authors in [17] presented an average error latency of 62 ± 47 ms for FS and 86 ± 61 for HO. Moreover, authors in [39] reported delays when subjects were wearing or not wearing shoes. For example, they reported approximately 0.1 ± 0.05 s for the TO and 0.01 ± 0.07 s for the IC. Similar to our work, the recent study of Quintero et al. [40] worked on estimating the continuous progression of the gait cycle by extracting the relationship between the thigh angle and velocity extracted from one IMU. They transformed the angle–velocity relationship to polar coordinates in order to predict the gait percentage. However, a comparison between their results and ours would be unfeasible, as they only visually reported the accuracy of their technique. A study in [12] recently announced an effective algorithm that detects four gait events (e.g, IC, TO, mid-swing (MSw) and mid-stance (MSt)) based on a set of heuristic rules using one gyroscope attached to the shank of subjects performing activities of daily living such as normal walking, fast walking, ramp ascending, and ramp descending. However, this algorithm is limited to an off-line setting and to a non-detection of the push-off event, which is considered to be an important phase before toe-off. Although they state that their algorithm also works in an on-line setting, they do not show any evidence that supports their results. In summary, in the course of our literature review, we encountered gait event detection systems that limit their experiments to one IMU and to gait phases that were partitioned between two [9,41,42] and three phases [3]. In this limited framework, they illustrated that a high accuracy can be achieved. For instance, authors in [41,42] illustrated an accuracy of 100% in IC and TO event detection. Zhou et al. [9] showed an accuracy of above 98% for IC event detection and 95% for TO event detection on three different terrains. We also observed that in cases where the number of phases is increased (i.e., to four phases [12,15,17]), the gait phase detection performance also decreases. For example, the experiment in [15] yielded an average detection accuracy under 95% in every phase, while Mannini et al. [17] showed a long delay of detection from 45 ms to 100 ms. The experiment in [12] showed a mean difference error between the reference and the proposed system of approximately $+4$ ms for IC and -6.5 ms for TO.

Recently, most studies have focused on improving the detection accuracy with the purpose of applying real-time safety walking for amputees while at the same time increasing the number of detectable gait phases for better control of the robotic device. However, we previously saw that increasing the granularity of the phases leads to a significant decrease in the prediction accuracy. To cope with this trade-off, we built a new deep learning architecture called the exponentially delayed fully connected neural network (ED-FNN). This network is made to overcome most of the current limitations in gait phase detection algorithms, such as predicting gait events with one IMU attached to the lower shank with high phase granularity and no detection delay.

3. Materials and Methods

This section discusses the algorithms and materials that were used for our experiments section. First, the concept of *gait classification* is introduced. Second, we describe how the walking gait is typically discretised. Finally, a description of the ED-FNN architecture is provided.

3.1. The Division of the Human Walking Gait

The human walking gait is defined as a periodic cycle involving two legs from the initial contact of one foot on the ground to the following occurrence of the heel of the same event with the same foot. Typically, one gait cycle is divided into two main phases, including a stance phase which is approximately 60% of a gait cycle and a swing phase which is approximately 40% of the remaining gait cycle [43]. The heel-contact and the toe-off events mark the beginning of the stance phase and the swing phase, respectively.

The granularity of one step is equally divided into two, three, four, five, six, seven, and eight gait periods depending on the specific type of application. This is done to define the label of the different periods of the gait cycle [35]. An example of the different periods is shown in Figure 1, which illustrates eight periods that were summarised into one gait cycle based on 100 percent of the gait [35,43].

In our study, we classified a full gait cycle discretised within a 1% interval based on real-time measurements of the gait cycle. Doing so will allow prosthetic devices to have a wider spectrum of control in the cycle. For example, the prosthesis AMP-Foot 3 plus [4] stores energy in the springs at the mid-stance period and starts to inject positive energy at the terminal stance period by using a DC motor. The detection of the terminal stance period is important for injecting energy at the right time. At the initial swing and mid-swing periods, the trajectory for the next work situation (i.e., the next walking phase) needs to be set. It is of great importance to command the control of the prosthesis just before gait events happen in order to avoid action delay on the device. Therefore, an algorithm for detecting 100% of the gait is required for the next concept of gait phase detection. For applications that do not require a granular gait cycle, the phases can be mapped to the fundamentals of human gait phases (Table 1) in order to control the prosthesis when needed.

Table 1. One gait cycle can be subdivided into eight typical phases based on the gait fundamentals shown in Section 3.1. In this table *Dorsi Assist* means assisting the foot to bend up, *Plantar Assist* means assisting the foot to bend down, and *No Assist* means that there is no assistant for the movement. FF: foot-flat.

Label	Phase	Percentage	Function	Controlling
0	Initial Contact	0 to 8	Loading, weight transfer	Dorsi Assist
1	Mid Mid-stance (FF)	8 to 30	Support of entire body weight:	No Assist
2	Terminal Mid-stance (FF)	30 to 40	Center of mass moving forward	No Assist
3	Push Off	40 to 50	Push Off	Plantar Assist
4	Pre-swing, double-limb support, push off	50 to 60	Unloading and preparing for swing	Plantar Assist
5	Initial swing	60 to 75	Foot Clearance	Dorsi Assist
6	Midswing	75 to 85	Limb advances in front of body	Dorsi Assist
7	Terminal Swing	85 to 100	Preparation for weight transfer	Dorsi Assist

According to the fundamentals of human gait phases, the stance phase begins with IC from 0% to 10% of the gait cycle. Initial double-limb support appears during the first 10% of the gait cycle. The foot flat occurs from 10% until the heel leaves the ground at 40% of the gait cycle. Mid-stance appears at approximately 30% of the gait. Single-limb support occurs from foot flat until 50% of the opposite initial contact which is approximately at 50% of the gait cycle. The second double-limb support occurs from the opposite limb at 50% until the toe leaves the ground at 60% of the gait cycle. Then the second single-limb support starts until the cycle is complete. The following periods are early swing at approximately 60–75% of the gait cycle, mid-swing at approximately (75–85% of the gait cycle, and late swing at approximately 85–100% of the gait cycle. The fundamentals of human gait phases are shown in Figure 1.

3.2. Percent Segmentation Method for the Gait Cycle

We propose a model that can predict the gait percentage that was equally divided into 100 one-percent fragments. To achieve this, we reused the method from [44] to segment input signals and label output targets. We began by extracting the lengths of the walking steps from one heel-contact to the next. Then, we sampled each heel-strike window with an interval of 10 ms. This resulted in several signals that were stored as a matrix X of dimension $\mathbb{R}^{p \times (s*d)}$, where p is the percentage value, s is the number of sensors, and d is the number of dimensions in an IMU sensor.

3.3. Gait Prediction Model

We developed an exponentially delayed fully connected neural network (ED-FNN) that accurately detects and forecasts gait percentage that was densely discretised. In Section 2 we showed that Evans and Arvind [23] implemented a combination of FNN and HMM that had been previously used to detect phases that had been partitioned into five events. However, coarse discretisation of the gait is not sufficient to fully control prosthetic devices for real-world applications. In this section we will describe the ED-FNN and show that this algorithm manages to simulate recurrent neural networks for regression problems.

3.3.1. Fully Connected Neural Networks (FNNs)

An FNN is a collection of artificial neurons called computational units. These units are grouped as a set of layers that are arranged in a hierarchical structure. FNNs are divided into *input layer*, *hidden layers*, and *output layer*. The function of the input layer is to directly process the data given by the user and forward it to the first hidden layer to learn complex representation of the data. This forward process is repeated in the following hidden layers, allowing them to learn more specific characteristics of the input data. The function of the output layer is to process the output of the last hidden layer and generate a prediction that agrees with the ground truth of the given input data. This network architecture is called a *fully connected network* because every unit in a layer is connected to every other unit in the following layer. These connections are represented as weights and biases that express the importance of a respective input to the output. The activation of each unit in a layer of the network is computed by the following equation:

$$\mathbf{a}^l = \sigma\left(\theta^l a^{l-1} + b^l\right),\tag{1}$$

where $\theta \in \mathbb{R}^{n \times k}$ is a matrix denoting the weights between layers l and $l-1$, $a^{l-1} \in \mathbb{R}^k$ denotes the activation units of the previous layer ($l-1$), and σ is a predefined activation function. In the literature, several activation functions (e.g., sigmoid [45], rectified linear unit (ReLU) [46], and softmax [47]) can be found. For our model, we chose a ReLU activation function as shown in Equation (2) due to its properties of avoiding saturation in the error gradients:

$$\sigma(z) = max(0, z).\tag{2}$$

Modifying the weights and biases in every activational layer ($\mathbf{a}^l$) will lead the overall model to obtain a desirable output. To learn these weights and biases, the objective of the network is quantified by means of a cost function. Several cost functions for neural networks (NN) are found in the literature, such as the mean square error (MSE) [48] and cross-entropy [49]. For the purpose of gait percentage detection, we use the mean square error (MSE) of Equation (3) to minimise the cost between the ground truth of the data and the prediction of our model:

$$\underset{(\theta,b)}{MSE}(\mathbf{x}) = \frac{1}{2N} \sum_{x_i \in \mathbf{x}}^{N} ||h_{(\theta,b)}(x_i) - y(x_i)||^2,\tag{3}$$

where $h_{(\theta,b)}(x_i)$ is the prediction of the FNN and θ and b are the weights and the biases of the network. This equation indicates that if the MSE is close to zero, then the weights and the biases reflect a good representation of the given data. In an NN, *gradient-based methods* are used to back-propagate the error from the output layer to every weight of the hidden units. The error gradient indicates the direction in which the weights and biases of the units need to be updated. (Stochastic) gradient descent [50], conjugate gradient [51], and Adam [52] are the most popular gradient-based methods that are used in NNs.

3.3.2. Exponentially Delayed Fully Connected Neural Network (ED-FNN)

Because FNNs do not hold any notion of time as they only consider the current example x, a machine learning model needs to rely not only on the signal taken at time t but also on the history of signals $X_d = [x_{t-k}, \cdots, x_t]$ to estimate present and future gait percentages $\mathbf{y} = [y_t, y_{t+1}, \cdots, y_{t+n}]$. In this case, n indicates the number of gait events to estimate in the future, and k specifies the number of IMU samples to take from the past. Recurrent neural networks (RNNs) are known to simulate a historical behaviour by introducing *memory* that encodes information about what has been observed in the past. Figure 2 shows the architecture of an RNN.

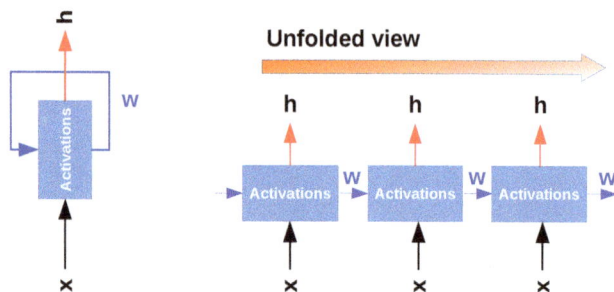

Figure 2. This figure illustrates the information flow in a recurrent neural network (RNN). The left image shows an RNN as an infinite loop network where the model outputs are fed back as inputs. The right figure is an unfolded representation of an RNN [53].

In a simple fully connected neural network (FNN), information flows back and forth from the lower to the higher layers of the network, allowing it to learn higher-order representations of the input data. A similar process is observed in RNNs, with the distinction that the network not only depends on the inputs X, but also on the activations of the hidden units at previous time steps. As a result, these networks will learn to map the sequence of inputs $\mathbf{x} = [x_{t-k}, \ldots x_t]$ into output of sequences $\mathbf{o} = [o_{t+1}, \ldots, o_{t+n}]$. RNN algorithms such as long short-term memory (LSTM) networks have shown great success in many problems that contain temporal information (e.g., IMU signals) [54]. Based on the characteristics of RNNs, it is clear that these networks have the capacity to accurately predict dense gait events. However, we found that they generally require substantially more data than standard FNNs, and they are computationally expensive, which poses limitations when working with microprocessors. Consequently, in our research we created an NN architecture that simulates the "memory" of RNNs and removes the aforementioned limitations.

To simulate a "memory" of an RNN, the input tensor X was delayed according to the following equation:

$$\mathbf{D} = \left[W_{(t-d):t}\right]_{t=d}^{T-1-f},\tag{4}$$

where $W[\cdot]_{start:end}$ is a window sequence from $start$ to end, $[x_t]_A^B$ is a vector of elements x_t that ranges from A to B, T indicates the number of samples in the signal, and f is the number of percentage values that will be predicted in the future. This equation constructs a tensor D of dimensions $\mathbb{R}^{p \times s_d \times s}$, where p indicates percent index, s_d indicates the delayed samples, and s indicates the number of IMU sensors. This tensor is illustrated in Figure 3. Furthermore, the tensor D was reshaped into a matrix $X_d \in \mathbb{R}^{p \times k}$ in order to use this tensor directly in the NN. In this case, $k = s_d * s * d$ refers to the product of the delayed samples, IMU sensors, and the dimensions of the IMU sensors. Finally, we generated the output matrix $Y_d \in \mathbb{R}^{p \times f}$, where p indicates the percentage index and f is the number of samples in the future. Using the matrix, we oblige the network to always predict f percentage values in advance.

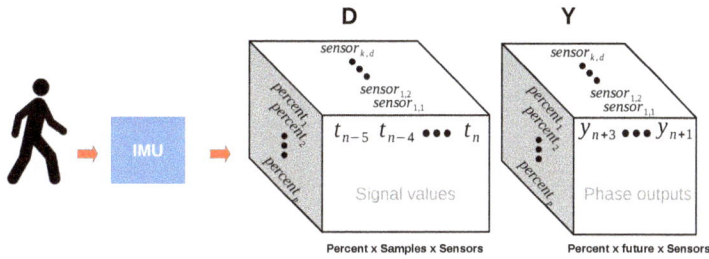

Figure 3. This figure illustrates how the matrix D was created. Every sample in the inertial measurement unit (IMU) is delayed by n times (in this case five times). The output matrix is shifted n times into the future (in this case three times).

Using the matrix X_d directly in an FNN with a *short* delay d will yield bad predictions on the overall trend, but it will be good at predicting fast changes in the percentage. This is because a one-to-one mapping between neighbouring samples X_d and percent samples y_d does not exist. In contrast, using a X_d with long delays will be better at predicting the overall trend of the percentage, but less good at predicting fast changes. A good trade-off between large and short delays will yield an optimal input space to predict the gait percentage. One of the reasons why LSTM works well for time series is due to its ability to choose which of the delay samples are important for the overall prediction. As a result, in order to simulate this behaviour, we introduced the concept of *exponential windows*, which allows us to make trade-offs between short and long delays. An exponential window is defined as:

$$\psi(X_d, \delta) = X_d[t - exp(k)]_{k=0}^{\delta}. \tag{5}$$

This equation uses the delayed matrix X_d and re-samples it using a smaller delay δ. Consequently, we not only obtain a window that includes the knowledge of samples that are close by, but also obtain knowledge of samples that are far away. Note that the number of samples which are close in time are more densely sampled than those that are far away from t. This means that we are including information that encompasses both fast changes in the percentage and samples that contribute to the prediction of the cycle trend. Furthermore, applying this *exponential window* allows us to decrease the number of input units that are needed to predict the gait percentages, which results in a substantial decrease of computational power. Additionally, due to the forecasting properties of the network, we also removed the delay limitations that arise when using algorithms on micro-controllers with low memory and CPU power.

The network that was used in our experiments is made of one fully connected (fc) layer as input, two hidden layers of six units, and an output layer of one unit. Furthermore, we found that training one IMU sensor per input layer and concatenating them later decreased the variance between several runs. Figure 4 illustrates the architecture of the NN. The concatenation of the weights and biases for each layer was performed based on the following equation:

$$\begin{aligned}
a^{lc} &= [a_0^{l-1}, \cdots, a_s^{l-1}], \\
\theta^{lc} &= [\theta_0^l, \cdots, \theta_s^l], \\
b^{lc} &= [b_0^l, \cdots, b_s^l],
\end{aligned} \tag{6}$$

where s is the number of sensors that were modelled by different FNNs. Note that the activation of the concatenated layer can be computed by modifying Equation (1) as:

$$a^l = \sigma\left(\theta^{lc} a^{lc} + b^{lc}\right).$$ (7)

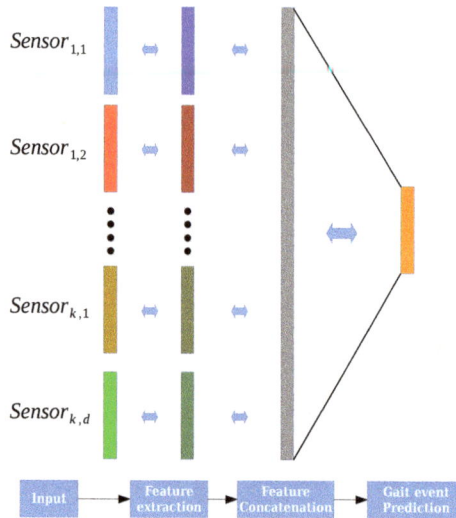

Figure 4. This figure illustrates the exponentially delayed fully connected neural network (ED-FNN) architecture. Initially, the network individually receives each sensor input from the matrix X in Equation (5). Then, the network separately extracts the features of each sensor and concatenates them into a single feature vector. Finally, the output layer uses the feature vector to forecast the gait events of the cycle.

Regarding the framework of the ED-FNN, in order to keep the network architecture constrained by the computation complexity, we chose six hidden units per layer. We found that this number of units satisfied the computational limitations and the accuracy of the model. Moreover, with regard to the sampling of Equation (5), we chose an exponential window of 1.6. We found that this window yielded the best results for every subject in the dataset. Further optimisation of the hyper-parameters could be done in future work by means of cross-validation techniques.

3.3.3. Performance Metric for the ED-FNN

To train the ED-FNN model, we used the MSE to minimise the cost between the prediction and the ground truth. In addition to these metrics, we also calculated the mean absolute error (MAE) and the coefficient of determination (R^2). The absolute error is computed as:

$$MAE = \frac{\sum_i |\hat{y}_i - y_i|}{n},$$ (8)

where $\hat{y}_i$ is the predicted percentage of the model and y is the ground truth. Furthermore, we computed R^2 as:

$$SS_{tot} = \sum_i (\bar{y}_i - y)^2,$$
$$SS_{res} = \sum_i (\hat{y}_i - y)^2,$$ (9)
$$R^2 = 1 - \frac{SS_{res}}{SS_{tot}},$$

where $\bar{y}_i$ is the mean of the ground truth and SS_{tot} and SS_{res} indicate the *total sum of squares* and *the residual sum of squares*, respectively. This metric is generally used to measure the correlation between the prediction and the ground truth. If there is a correlation of 1, it means that the prediction fully represents the ground truth.

4. Experiments

This experimental section is divided into three parts. First, Section 4.1 explains the electronic board that was used to read the IMU signals. Section 4.2 describes the number of subjects and situations that were performed in the experiments. Finally, Section 4.3 explains how the signals were recorded and pre-processed.

4.1. Experimental Electronic Board Prototype, Experiment Protocol, Measurement System

In our experiments, an electronic board with one embedded IMU sensor was used. The board was an Adafruit Feather M0 Bluefruit LE (using ATSAMD21G18 ARM Cortex M0 processor, clocked at 48 MHz and at 3.3 V logic, 256 KB of FLASH ROM and 32 KB of RAM). Moreover, the charging battery unit was designed to measure and monitor the voltage of the battery so we could detect when it needed to be recharged. The board also supports a Bluetooth Low Energy component. This addition makes it convenient for transferring data to the computer or designing mobile applications so that amputees can easily monitor or even control their prosthesis. To measure the gait signals, we used one IMU and two FSRs. The IMU (MPU 6000–Invensense) consists of a gyroscope sensor and an accelerometer. This provides tri-axis signals of angular velocity and tri-axis acceleration of the lower shank. Furthermore, the IMU was connected to the microcontroller via the SPI interface for the purpose of high-speed signal transfer to the commuter (up to 1 MHz). The gyroscope resolution was set at a full range scale of ± 2000 degrees/sec with a sensitivity of ± 16 g LSB/degree/s. Moreover, the resolution of the accelerometer was set at a full range scale of ± 16 g with a sensitivity of 2048 LSB/g (g = 9.8 m/s^2). Regarding the FSRs, we placed two of these sensors under the toe and the heel of the subject to detect the impact of the foot with the ground. FSR signals were used as references for classifying gait events and to build a dataset for training the ED-FNN algorithm. After pre-prossessing the IMU signals with the FSRs, we removed these sensors for the training and prediction of the model. All signals were recorded synchronously at intervals of 10 ms, then transmitted directly to the computer. This electronic board was first used for collecting the data and creating a dataset. Additionally, we embedded the gait percent detection algorithm combined with a device control program for the prosthesis.

4.2. Subjects

The data were extracted from seven healthy subjects with IMUs fixed with a belt placed on the subjects' lower shank, and two FSRs were placed under the toe and heel. Participants were five males and two females. Their age ranged from 25 to 33 years, their height ranged from 160 to 185 cm, and their weight ranged from 48 to 80 kg. We recorded different scenarios in two environments. In the first environment, subjects walked on a treadmill with a 0 degree inclination. In the second environment, subjects walked outside on a 0 and a 15-degree inclination. On the treadmill, each subject was required to walk four different trials with different speeds. The speeds were divided into 2.2 m/s, 2.6 m/s, 3.2 m/s, and 3.8 m/s respectively. For outside walking, subjects performed normal speed (approx. 3.2 m/s) and fast speed (approx. 4.0 m/s). All trials were recorded for an interval of five minutes. The number of steps of each subject was categorised as an example in the training data. On average, each participant walked 275.0 steps in the overall experiment. All IMU signals obtained from each participant were mixed together to build a larger dataset for training the network model. This resulted in a dataset of 2313 walking cycles, as shown in Table 2. Merging the dataset allowed the model to increase the chances of extracting the important features that are relevant in different walking gaits.

Table 2. The number of samples and cycles in the dataset.

Subjects	The Number of Samples	The Number of Cycles
Subject 1	19,805	162
Subject 2	47,089	449
Subject 3	46,367	434
Subject 4	21,531	189
Subject 5	19,149	170
Subject 6	15,858	181
Subject 7	25,166	258
Subject 8	15,858	181
Data on the treadmill	78,473	451
Dataset (all samples and cycles)	269,491	2313

4.3. Off-Line Data Analysis

As mentioned earlier, FSRs were used to extract gait cycles and phases by measuring the heel-strike and the toe impact with the ground. The location of FSRs under the sole can be seen in the right image of Figure 5. The IMU's position on the subject's lower shank can also be seen in the right image.

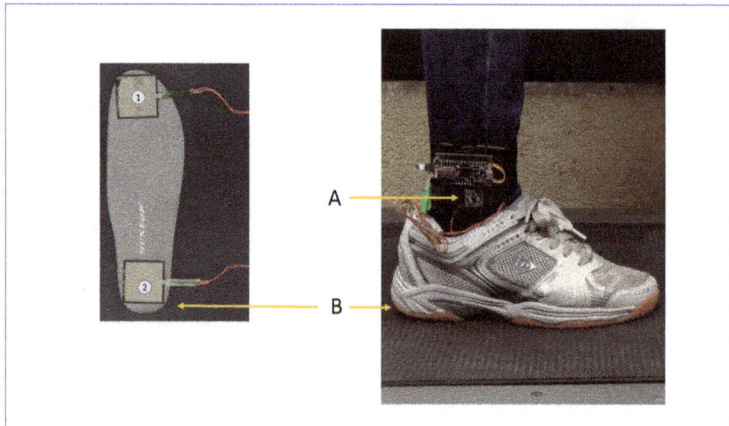

Figure 5. Sensor positions of the IMU and the FSR on the foot. Arrow (A) illustrates the position of the IMU, and arrow (B) the position of FSRs under the sole.

In Section 2 we described that the methodologies that divide the gait cycle into foutr or more phases (see Table 1) yield acceptable results to classify the gait percentage. However, for real-world applications, this phase granularity may not be sufficient for controlling active prosthetics. For this reason, we want to open an opportunity for future active prosthetics to be able to fully control their devices by accurately predicting the gait percentage in a densely discretised gait. An example can be observed in Figure 6, which illustrates a gait cycle that was discretised within a 1% interval in the real-time estimation of the gait cycle. Lastly, in comparison with [3,34,37,55], our model does not require complex pre-processing steps, as it can deal with data with a high signal-to-noise ratio.

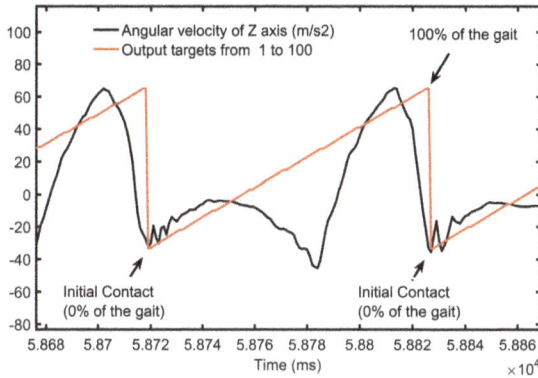

Figure 6. The figure shows one gait cycle discretised with a 1% interval. The division was based on measuring cycle latency, from an initial-contact (IC) at 0% to the next at 100%.

5. Results

The development of a gait percent detection model based on ED-FNN was described above. The model was trained with the angular velocity and acceleration signals in the sagittal plane of the foot. These signals were taken from healthy subjects walking at different speeds on flat ground and on a 15-degree inclined terrain. The performance of our method was evaluated on individual and group bases. Over several runs, we computed the mean loss and variance to determine the overall performance of the model. Additionally, we generated a validation set to validate the generalisation accuracy of our model.

For each subject, we computed the mean absolute error (MAE), mean square error (MSE), and the coefficient of determination (R^2). Table 3 gives an overview of the average error of all subjects and the joined dataset. These values are shown as training and validation errors. Furthermore, due to the number of subjects in the experiments, the visualisation of the results was divided into two parts. The first part illustrates the learning process of the MSE prediction in four different plots: two for one single subject and two for all-subjects' signals combined. The second part individually illustrates the overall MSE performance for every subject by means of a violin plot. The following table provides a summary of the performance in Figures 7–9.

Table 3. The average error across every subject in the dataset and the error obtained when learning in the joined dataset. To convert these errors into percentages (except R^2), the equations for MSE and MAE are given in Equations (10) and (11), respectively. Individual errors for each subject are shown in Figure 9.

	t-MSE	v-MSE	t-MAE	v-MAE	t-R^2	v-R^2
			Error			
Average	0.005 ± 0.0003	0.005 ± 0.0004	0.021 ± 0.001	0.023 ± 0.002	0.93 ± 0.003	0.881 ± 0.015
Joined	0.006 ± 0.0003	0.011 ± 0.0017	0.021 ± 0.001	0.041 ± 0.002	0.91 ± 0.004	0.828 ± 0.022

$$MSE_\% = \sqrt{MSE} \times 100 \qquad (10)$$

$$MAE_\% = MAE \times 100 \qquad (11)$$

5.1. Results: Part 1

In this section we illustrate the results for one subject and all the subjects' signals combined. Figure 7 shows the prediction and learning process of one subject. The $Gyroscope_{y,z}$ signals can be observed in the bottom plot in Figure 7a. Additionally, the prediction of the phase signal can be observed in the top plot. Notice that the prediction almost perfectly follows the trend of the ground truth. Moreover, we can see that this accuracy is reflected in Figure 7b, which shows the MSE of the learning. In this plot we can observe that the MSE reached an average loss of 0.003 in the training set and a value of 0.0028 for the validation set. The reason why the model performed better in the validation set is because it was slightly under-fitting the data. This could be solved by increasing the size of the NN. Figure 8a illustrates the gait prediction for all subjects' signals joined together. In this plot we can observe similar results to those in Figure 7a, with a difference in the accuracy of the MSE. Here we can see that the validation set performed less well than in one subject alone. This was expected, as the signals of each subject slightly vary. Furthermore, we can observe that the prediction still followed the trend of the ground truth despite the decrease of accuracy. Based on these results, we can conclude that the network managed to generalise well.

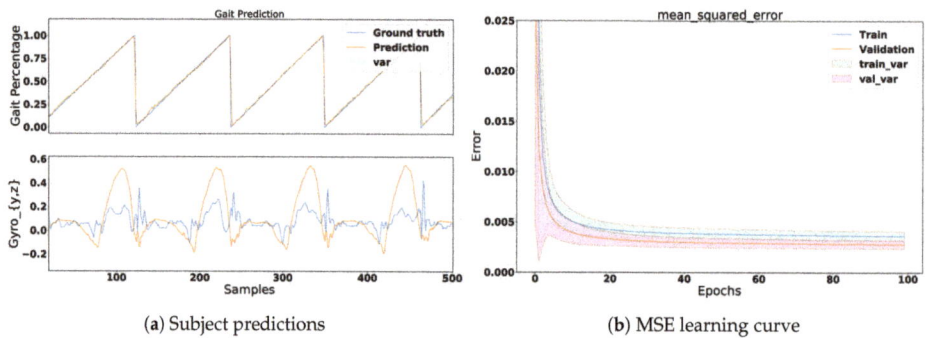

(a) Subject predictions (b) MSE learning curve

Figure 7. This figure shows the prediction and the results of the learning process for one subject. (a) The ground truth and mean prediction of the gait phase discretisation divided into 100 portions normalised between 0 and 1 (0 equals to 0 percent and 1 equals 100 percent of the gait cycle). The bottom figure shows the *y* and *z* signals of the gyroscope sensor; (b) The mean and variance of the mean square error (MSE) learning curve. The average of MSE reached a loss of 0.003 in the training set and 0.0662 in the validation set.

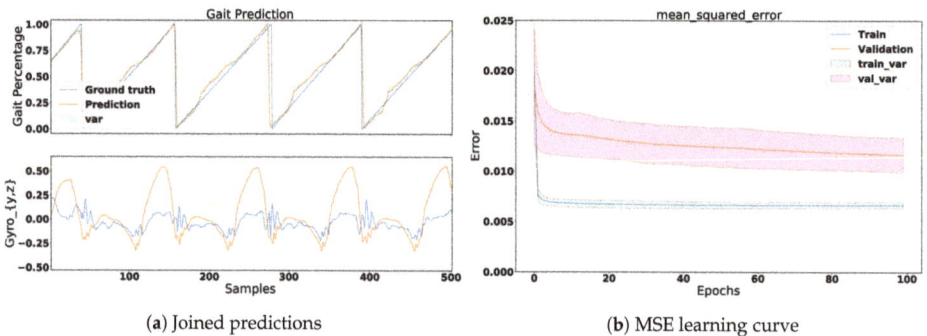

(a) Joined predictions (b) MSE learning curve

Figure 8. This figure shows the prediction and learning process results of the joined signal for several subjects. Similar to Figure 7, (a) shows the comparison between the prediction and the ground truth and (b) illustrates the learning curve of the MSE. The average MSE reached a loss of 0.006 in the training set and an average of 0.0115 in the test set.

5.2. Results: Part 2

This section describes the MSE violin plot for every individual subject in the experiments. In Figure 9, the colours of the plot show the distributions of the training and the test MSEs. Each violin in the plot belongs to one subject in the experiments, and the last violin plot belongs to all the subjects' signals combined. These results show that the ED-FNN accuracy was consistent over every subject in the dataset. Furthermore, thanks to the concatenation of the layers in our network, we can observe that the variance between different runs was small. As a result, it increased the robustness and reliability of our model.

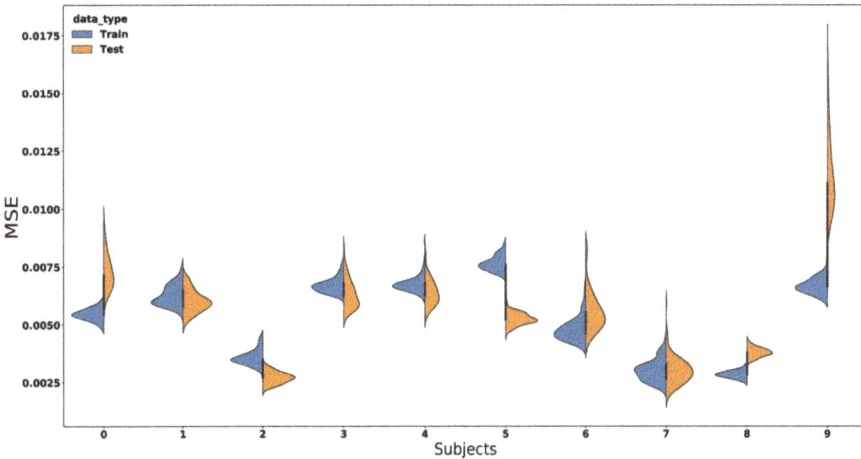

Figure 9. This figure illustrates the MSE for every subject in the experiments. Each number in the *x*-axis from 1 to 7 represents a subject in the experiments. Violin plots 7, 8, and 9 were of two 15 degree incline walks and to all subjects' signals combined, respectively. Every violin plot consists of two distributions (i.e., Train—blue and Test—orange) and the mean of the MSE. The distributions illustrate the MSE variance over 100 runs. This figure shows that the ED-FNN managed to accurately predict the gait cycle over several subjects.

5.3. Reference System

We observed that our algorithm was able to accurately predict 100 percent of the gait cycle. To our best knowledge, this setting has never been done in previous studies. Furthermore, we showed that the MSE managed on average to achieve a 0.003 in both validation and training sets. We expect the real-time performance to be very similar to the off-line performance. It is difficult to compare our algorithm with other studies, as our setting is not standard. However, Table 4 lays out the performance of other existing gait phase prediction systems using one IMU.

Table 4. This table compares existing gait event predictions with our method using one IMU. Every method was applied on lower limbs. FF: foot-flat; HO: heel-off; IC: initial contact; TO: toe-off.

Author	Detectable Events or Phases	Performance	Metric	Detection
Ledoux et al. [42] (2018)	IC and TO	$-1.7\% \pm 0.6$ stride (IC), $-1.8\% \pm 0.6$ stride (TO)	Detection delays	On-line
Zakria et al. [12] (2017)	IC and TO	3.92 ms $\pm$ 1.56 (IC), -1.81 ms $\pm$ 4.03 (TO)	Time difference	Off-line

Table 4. *Cont.*

Author	Detectable Events or Phases	Performance	Metric	Detection
Maqbool et al. [41] (2016)	IC and TO	15.44 ms ± 25.2 (IC), −28.44 ms ± 16.2 (TO)	Time difference	On-line
Zhou et al. [9] (2016)	IC and TO	95% (TO: upstairs), 99% (IC: upstairs), 99% (TO: downstairs) 98% (IC: downstairs)	Detection precision	On-line
Mannini et al. [17] (2014)	IC, FF, HO, TO	62 ms ± 47 (IC), −3 ms ± 53 (FF), 86 ms ± 61 (HO), 36 ms ± 18 (IC),	Time difference	On-line
Muller et al. [39] (2015)	Detected four phases	100 ms ± 50 (TO), 50 ms ± 79 (IC)	Time difference	On-line
Quintero et al. [40] (2017)	100 gait percent	Reported visually	Theory	Off-line
Our method	100 gait percent	2.1% ± 0.1	MAE—No delay	Off-line

6. Conclusions and Future Work

Over the recent decades, gait phase detection algorithms have become a challenging topic for researchers due to their extensive application in assistance devices. One example is improving the gait phase detection accuracy in prosthetics so that amputees can safely use these devices. To date, many gait phase detection methods have been developed. However, these methods use only one sensor, and often detect gait cycles on a low granular domain of the phase space. In order to take full control of current prosthetics, we proposed a robust walking gait percent detection method that can detect 100 percent of the gait cycle for walking on flat ground and on a 15-degree inclination. Because other similar methods have shown that the accuracy of gait phase detection algorithms is suitable for ambulatory applications, we showed that our method yielded state-of-the-art performance for these applications. In summary, our study obtained the following outcomes:

- A compact system using one IMU mounted on the lower shank.
- A model that is capable of learning highly discretised percentages of the gait cycles.
- An average mean square error of approximately 0.003 in both training and validation sets for single subjects.
- A model that generalises toward several subjects with an average MSE of 0.006 in the training set and 0.01 in the validation set.
- A model that is consistent over several subjects. (i.e., low variance between several runs).
- A model with powerful forecast capabilities that introduces a no-delay prediction method within 10 ms.

It is important to note that our tests were performed on ARM chips, which are known to underperform when conducting heavy mathematical computations. Because our experiments were purely offline, computational cost was not a concern. However, our model was built with the purpose of working on scenarios where computational power is an issue. The ED-FNN requires less samples than a normal FNN or RNN. Consequently, it takes less computational power for prediction. Additionally, we also included a forecasting option that allows the network to predict future percentages in the case of a delay in the system. Furthermore, we would like to point out that we are not learning on the ARM chips directly. Instead, we learn the models in a normal computer and then make the predictions in the ARM chip. Lastly, in the near future we plan to leave aside the Beagle Bone Black and use other more powerful alternatives, such as ODROID-C2 boards or neuromorphic computation.

For future work we will focus on improving the prediction accuracy and evaluation with different walking conditions such as stair walking and real-time implementation with the newest prosthesis version of the AMP-Foot [4]. Additionally, the hyper-parameters of the ED-FNN model will be improved by means of cross-validation methods.

Author Contributions: Conceptualization, F.G. and H.V.; Methodology, F.G. and H.V.; Software, F.G.; Validation, F.G. and H.V.; Formal Analysis, F.G. and H.V.; Investigation, H.V.; Resources, B.V., P.C. and D.L.; Data Curation, H.V. and F.G.; Writing—Original Draft Preparation, F.G. and H.V.; Writing—Review & Editing, F.G. and H.V.; Visualization, F.G. and H.V.; Supervision, B.V., A.N., and P.C.; Project Administration, B.V., A.N., P.C. and D.L.; Funding Acquisition, B.V. and H.V.

Funding: This research was funded by the European Commission ERC Starting grant SPEAR (No. 337596), and by the Vietnamese Government for university and college lecturers on doctoral training during 2010–2020 Decision of the Prime Minister No. 911/QD-TTg dated 17 June 2010.

Acknowledgments: The authors gratefully thank Timothy Verstraeten for his helpful comments and mathematical insights. Additionally, authors would like to express their gratitude to Astrid Van Hecke and Michal Rudz-ki for reviewing the grammar and the spelling. Finally, we would like to acknowledge the detailed assessment from the anonymous reviewers which improved the clarity and quality of the paper.

Conflicts of Interest: The authors declare no conflict of interest.

Ethical Statements: All subjects gave their informed consent for inclusion before they participated in the study. The study was conducted in accordance with the Declaration of Helsinki, and the protocol was approved by the Ethics Committee of B.U.N 143201526629.

References

1. Ferris, D.P.; Sawicki, G.S.; Daley, M.A. A physiologist's perspective on robotic exoskeletons for human locomotion. *Int. J. Humanoid Robot.* **2007**, *4*, 507–528. [CrossRef] [PubMed]
2. Qi, Y.; Soh, C.B.; Gunawan, E.; Low, K.S.; Thomas, R. Assessment of foot trajectory for human gait phase detection using wireless ultrasonic sensor network. *IEEE Trans. Neural Syst. Rehabil. Eng.* **2016**, *24*, 88–97. [CrossRef] [PubMed]
3. Kotiadis, D.; Hermens, H.J.; Veltink, P.H. Inertial gait phase detection for control of a drop foot stimulator. *Med. Eng. Phys.* **2010**, *32*, 287–297. [CrossRef] [PubMed]
4. Cherelle, P.; Grosu, V.; Flynn, L.; Junius, K.; Moltedo, M.; Vanderborght, B.; Lefeber, D. The Ankle Mimicking Prosthetic Foot 3—Locking mechanisms, actuator design, control and experiments with an amputee. *Robot. Auton. Syst.* **2017**, *91*, 327–336. [CrossRef]
5. Flynn, L.L.; Geeroms, J.; van der Hoeven, T.; Vanderborght, B.; Lefeber, D. VUB-CYBERLEGs CYBATHLON 2016 Beta-Prosthesis: Case study in control of an active two degree of freedom transfemoral prosthesis. *J. Neuroeng. Rehabil.* **2018**, *15*, 3. [CrossRef] [PubMed]
6. Catalfamo, P.; Moser, D.; Ghoussayni, S.; Ewins, D. Detection of gait events using an F-Scan in-shoe pressure measurement system. *Gait Posture* **2008**, *28*, 420–426. [CrossRef] [PubMed]
7. Lau, H.; Tong, K. The reliability of using accelerometer and gyroscope for gait event identification on persons with dropped foot. *Gait Posture* **2008**, *27*, 248–257. [CrossRef] [PubMed]
8. Meng, X.; Yu, H.; Tham, M.P. Gait phase detection in able-bodied subjects and dementia patients. In Proceedings of the 2013 35th Annual International Conference of the IEEE Engineering in Medicine and Biology Society (EMBC), Osaka, Japan, 3–7 July 2013; pp. 4907–4910.
9. Zhou, H.; Ji, N.; Samuel, O.W.; Cao, Y.; Zhao, Z.; Chen, S.; Li, G. Towards Real-Time Detection of Gait Events on Different Terrains Using Time-Frequency Analysis and Peak Heuristics Algorithm. *Sensors* **2016**, *16*, 1634. [CrossRef] [PubMed]
10. Khandelwal, S.; Wickström, N. Gait event detection in real-world environment for long-term applications: Incorporating domain knowledge into time-frequency analysis. *IEEE Trans. Neural Syst. Rehabil. Eng.* **2016**, *24*, 1363–1372. [CrossRef] [PubMed]
11. Goršič, M.; Kamnik, R.; Ambrožič, L.; Vitiello, N.; Lefeber, D.; Pasquini, G.; Munih, M. Online phase detection using wearable sensors for walking with a robotic prosthesis. *Sensors* **2014**, *14*, 2776–2794. [CrossRef] [PubMed]

12. Zakria, M.; Maqbool, H.F.; Hussain, T.; Awad, M.I.; Mehryar, P.; Iqbal, N.; Dehghani-Sanij, A.A. Heuristic based gait event detection for human lower limb movement. In Proceedings of the 2017 IEEE EMBS International Conference on Biomedical & Health Informatics (BHI), Jeju Island, Korea, 11–15 July 2017; pp. 337–340.

13. Mannini, A.; Sabatini, A.M. Machine learning methods for classifying human physical activity from on-body accelerometers. *Sensors* **2010**, *10*, 1154–1175. [CrossRef] [PubMed]

14. Bae, J.; Tomizuka, M. Gait phase analysis based on a Hidden Markov Model. *Mechatronics* **2011**, *21*, 961–970. [CrossRef]

15. Mannini, A.; Sabatini, A.M. A hidden Markov model-based technique for gait segmentation using a foot-mounted gyroscope. In Proceedings of the 2011 Annual International Conference of the IEEE Engineering in Medicine and Biology Society (EMBC), Boston, MA, USA, 30 August–3 September 2011; pp. 4369–4373.

16. Crea, S.; De Rossi, S.M.; Donati, M.; Reberšek, P.; Novak, D.; Vitiello, N.; Lenzi, T.; Podobnik, J.; Munih, M.; Carrozza, M.C. Development of gait segmentation methods for wearable foot pressure sensors. In Proceedings of the 2012 Annual International Conference of the IEEE Engineering in Medicine and Biology Society (EMBC), San Diego, CA, USA, 28 August–1 September 2012; pp. 5018–5021.

17. Mannini, A.; Genovese, V.; Sabatini, A.M. Online decoding of hidden Markov models for gait event detection using foot-mounted gyroscopes. *IEEE J. Biomed. Health Inf.* **2014**, *18*, 1122–1130. [CrossRef] [PubMed]

18. Taborri, J.; Scalona, E.; Rossi, S.; Palermo, E.; Patanè, F.; Cappa, P. Real-time gait detection based on Hidden Markov Model: is it possible to avoid training procedure? In Proceedings of the 2015 IEEE International Symposium on Medical Measurements and Applications (MeMeA), Torino, Italy, 7–9 May 2015; pp. 141–145.

19. Taborri, J.; Scalona, E.; Palermo, E.; Rossi, S.; Cappa, P. Validation of inter-subject training for hidden Markov models applied to gait phase detection in children with cerebral palsy. *Sensors* **2015**, *15*, 24514–24529. [CrossRef] [PubMed]

20. Liu, D.X.; Wu, X.; Du, W.; Wang, C.; Xu, T. Gait phase recognition for lower-limb exoskeleton with only joint angular sensors. *Sensors* **2016**, *16*, 1579. [CrossRef] [PubMed]

21. Tanghe, K.; Harutyunyan, A.; Aertbeliën, E.; De Groote, F.; De Schutter, J.; Vrancx, P.; Nowé, A. Predicting seat-off and detecting start-of-assistance events for assisting sit-to-stand with an exoskeleton. *IEEE Robot. Autom. Lett.* **2016**, *1*, 792–799. [CrossRef]

22. Gouwanda, D.; Gopalai, A.A. A robust real-time gait event detection using wireless gyroscope and its application on normal and altered gaits. *Med. Eng. Phys.* **2015**, *37*, 219–225. [CrossRef] [PubMed]

23. Evans, R.L.; Arvind, D. Detection of gait phases using orient specks for mobile clinical gait analysis. In Proceedings of the 2014 11th International Conference on Wearable and Implantable Body Sensor Networks (BSN), Zurich, Switzerland, 16–19 June 2014; pp. 149–154.

24. Agostini, V.; Balestra, G.; Knaflitz, M. Segmentation and classification of gait cycles. *IEEE Trans. Neural Syst. Rehabil. Eng.* **2014**, *22*, 946–952. [CrossRef] [PubMed]

25. Skelly, M.M.; Chizeck, H.J. Real-time gait event detection for paraplegic FES walking. *IEEE Trans. Neural Syst. Rehabil. Eng.* **2001**, *9*, 59–68. [CrossRef] [PubMed]

26. De Rossi, S.M.; Crea, S.; Donati, M.; Reberšek, P.; Novak, D.; Vitiello, N.; Lenzi, T.; Podobnik, J.; Munih, M.; Carrozza, M.C. Gait segmentation using bipedal foot pressure patterns. In Proceedings of the 2012 4th IEEE RAS & EMBS International Conference on Biomedical Robotics and Biomechatronics (BioRob), Rome, Italy, 24–27 June 2012; pp. 361–366.

27. Cherelle, P.; Grosu, V.; Matthys, A.; Vanderborght, B.; Lefeber, D. Design and validation of the ankle mimicking prosthetic (AMP-) foot 2.0. *IEEE Trans. Neural Syst. Rehabil. Eng.* **2014**, *22*, 138–148. [CrossRef] [PubMed]

28. Moulianitis, V.C.; Syrimpeis, V.N.; Aspragathos, N.A.; Panagiotopoulos, E.C. A closed-loop drop-foot correction system with gait event detection from the contralateral lower limb using fuzzy logic. In Proceedings of the 2011 10th International Workshop on Biomedical Engineering, Kos, Greece, 5–7 October 2011; pp. 1–4.

29. Joshi, C.D.; Lahiri, U.; Thakor, N.V. Classification of gait phases from lower limb EMG: Application to exoskeleton orthosis. In Proceedings of the 2013 IEEE Point-of-Care Healthcare Technologies (PHT), Bangalore, India, 16–18 January 2013; pp. 228–231.

30. Mannini, A.; Trojaniello, D.; Cereatti, A.; Sabatini, A.M. A machine learning framework for gait classification using inertial sensors: Application to elderly, post-stroke and huntington's disease patients. *Sensors* **2016**, *16*, 134. [CrossRef] [PubMed]

31. Patterson, M.; Caulfield, B. A novel approach for assessing gait using foot mounted accelerometers. In Proceedings of the 2011 5th International Conference on Pervasive Computing Technologies for Healthcare (PervasiveHealth), Dublin, Ireland, 23–26 May 2011; pp. 218–221.

32. Zheng, E.; Wang, Q. Noncontact capacitive sensing-based locomotion transition recognition for amputees with robotic transtibial prostheses. *IEEE Trans. Neural Syst. Rehabil. Eng.* **2017**, *25*, 161–170. [CrossRef] [PubMed]

33. Zhao, Y.; Zhou, S. Wearable device-based gait recognition using angle embedded gait dynamic images and a convolutional neural network. *Sensors* **2017**, *17*, 478. [CrossRef] [PubMed]

34. Maqbool, H.F.; Husman, M.A.B.; Awad, M.I.; Abouhossein, A.; Iqbal, N.; Dehghani-Sanij, A.A. A real-time gait event detection for lower limb prosthesis control and evaluation. *IEEE Trans. Neural Syst. Rehabil. Eng.* **2017**, *25*, 1500–1509. [CrossRef] [PubMed]

35. Taborri, J.; Palermo, E.; Rossi, S.; Cappa, P. Gait partitioning methods: A systematic review. *Sensors* **2016**, *16*, 66. [CrossRef] [PubMed]

36. Neumann, D.A. *Kinesiology of the Musculoskeletal System: Foundations for Physical Rehabilitation*; Mosby: St. Louis, MO, USA, 2002.

37. Boutaayamou, M.; Schwartz, C.; Stamatakis, J.; Denoël, V.; Maquet, D.; Forthomme, B.; Croisier, J.L.; Macq, B.; Verly, J.G.; Garraux, G.; et al. Development and validation of an accelerometer-based method for quantifying gait events. *Med. Eng. Phys.* **2015**, *37*, 226–232. [CrossRef] [PubMed]

38. Rueterbories, J.; Spaich, E.G.; Andersen, O.K. Gait event detection for use in FES rehabilitation by radial and tangential foot accelerations. *Med. Eng. Phys.* **2014**, *36*, 502–508. [CrossRef] [PubMed]

39. Muller, P.; Steel, T.; Schauer, T. Experimental evaluation of a novel inertial sensor based realtime gait phase detection algorithm. In Proceedings of the Technically Assisted Rehabilitation Conference, Berlin, Germany, 12 March 2015.

40. Quintero, D.; Lambert, D.J.; Villarreal, D.J.; Gregg, R.D. Real-time continuous gait phase and speed estimation from a single sensor. In Proceedings of the 2017 IEEE Conference on Control Technology and Applications (CCTA), Mauna Lani, HI, USA, 27–30 August 2017; pp. 847–852.

41. Maqbool, H.F.; Husman, M.A.B.; Awad, M.I.; Abouhossein, A.; Mehryar, P.; Iqbal, N.; Dehghani-Sanij, A.A. Real-time gait event detection for lower limb amputees using a single wearable sensor. In Proceedings of the 2016 IEEE 38th Annual International Conference of the Engineering in Medicine and Biology Society (EMBC), Orlando, FL, USA, 16–20 August 2016; pp. 5067–5070.

42. Ledoux, E. Inertial Sensing for Gait Event Detection and Transfemoral Prosthesis Control Strategy. *IEEE Trans. Biomed. Eng.* **2018**. [CrossRef] [PubMed]

43. Shorter, K.A.; Polk, J.D.; Rosengren, K.S.; Hsiao-Wecksler, E.T. A new approach to detecting asymmetries in gait. *Clin. Biomech.* **2008**, *23*, 459–467. [CrossRef] [PubMed]

44. Fisher, R.A. The use of multiple measurements in taxonomic problems. *Ann. Hum. Genet.* **1936**, *7*, 179–188. [CrossRef]

45. Ito, Y. Representation of functions by superpositions of a step or sigmoid function and their applications to neural network theory. *Neural Netw.* **1991**, *4*, 385–394. [CrossRef]

46. Duan, K.; Keerthi, S.S.; Chu, W.; Shevade, S.K.; Poo, A.N. Multi-category classification by soft-max combination of binary classifiers. In Proceedings of the International Workshop on Multiple Classifier Systems, Guildford, UK, 11–13 June 2003; Springer: Berlin, Germany, 2003; pp. 125–134.

47. Nair, V.; Hinton, G.E. Rectified linear units improve restricted boltzmann machines. In Proceedings of the 27th International Conference on Machine Learning (ICML-10), Haifa, Israel, 21–24 June 2010; pp. 807–814.

48. Allen, D.M. Mean square error of prediction as a criterion for selecting variables. *Technometrics* **1971**, *13*, 469–475. [CrossRef]

49. Shore, J.; Johnson, R. Axiomatic derivation of the principle of maximum entropy and the principle of minimum cross-entropy. *IEEE Trans. Inf. Theory* **1980**, *26*, 26–37. [CrossRef]

50. Bottou, L. Large-scale machine learning with stochastic gradient descent. In Proceedings of the 19th International Symposium on Computational Statistics (COMPSTAT'2010), Paris, France, 22–27 August 2010; Springer: Berlin, Germany, 2010; pp. 177–186.

51. Møller, M.F. A scaled conjugate gradient algorithm for fast supervised learning. *Neural Netw.* **1993**, *6*, 525–533. [CrossRef]

52. Kingma, D.P.; Ba, J. Adam: A method for stochastic optimization. *arXiv* **2014**, arXiv:1412.6980.

53. LeCun, Y.; Bengio, Y.; Hinton, G. Deep learning. *Nature* **2015**, *521*, 436. [CrossRef] [PubMed]

54. Gers, F.A.; Schmidhuber, J.; Cummins, F. Learning to Forget: Continual Prediction with LSTM. In Proceedings of the 9th International Conference on Artificial Neural Networks (ICANN '99), San Sebastián, Spain, 20–22 June 2007; pp. 850–855.

55. Wang, Q.; Yuan, K.; Zhu, J.; Wang, L. Walk the walk: A lightweight active transtibial prosthesis. *IEEE Robot. Autom. Mag.* **2015**, *22*, 80–89. [CrossRef]

sensors

MDPI

Article

A Vision-Driven Collaborative Robotic Grasping System Tele-Operated by Surface Electromyography

Andrés Úbeda [1,2], **Brayan S. Zapata-Impata** [1,2], **Santiago T. Puente** [1,2], **Pablo Gil** [1,2], **Francisco Candelas** [1,2] **and Fernando Torres** [1,2,*]

[1] Department of Physics, System Engineering and Signal Theory, University of Alicante, 03690 Alicante, Spain; andres.ubeda@ua.es (A.Ú.); brayan.impata@ua.es (B.S.Z.-I.); santiago.puente@ua.es (S.T.P.); pablo.gil@ua.es (P.G.); Francisco.Candelas@ua.es (F.C.)

[2] Computer Science Research Institute, University of Alicante, 03690 Alicante, Spain

* Correspondence: fernando.torres@ua.es; Tel.: +34-965-90-9491

Received: 27 June 2018; Accepted: 16 July 2018; Published: 20 July 2018

Abstract: This paper presents a system that combines computer vision and surface electromyography techniques to perform grasping tasks with a robotic hand. In order to achieve a reliable grasping action, the vision-driven system is used to compute pre-grasping poses of the robotic system based on the analysis of tridimensional object features. Then, the human operator can correct the pre-grasping pose of the robot using surface electromyographic signals from the forearm during wrist flexion and extension. Weak wrist flexions and extensions allow a fine adjustment of the robotic system to grasp the object and finally, when the operator considers that the grasping position is optimal, a strong flexion is performed to initiate the grasping of the object. The system has been tested with several subjects to check its performance showing a grasping accuracy of around 95% of the attempted grasps which increases in more than a 13% the grasping accuracy of previous experiments in which electromyographic control was not implemented.

Keywords: surface electromyography; computer vision; grasping; assistive robotics

---

1. Introduction

Nowadays, robots can perform a variety of tasks to help human operators in their work [1]. The use of robots to collaborate with people with disabilities in industrial environments is a growing sector. For instance, several studies analyse the execution of manufacturing tasks by disabled people [2,3]. In this line, robotic assistive technologies have been successfully introduced following two different approaches. They are used to assist humans who have motor disabilities to perform daily activities. Typical examples are prosthetics devices and exoskeletons for motor substitution, or smart homes where household tasks are performed and controlled by automatic systems. These technologies also provide novel rehabilitation therapies to recover motor function and reduce further complications. Essentially, assistive technologies seek to improve the well-being of humans with disabilities [4].

The inclusion of assistive robotics in industrial applications contributes to the improvement of occupational health of human operators. Tele-operation systems increase the degree of assistance in dangerous manipulation tasks. Their goal is to make a system capable of mimicking and scaling the movements of a human operator in the control of a manipulator avoiding the risks of handling dangerous products or carrying out dangerous actions. Before including assistive technologies in industrial tasks, several teleoperation aspects must be considered. One of them is the feedback to the user, therefore the use of haptic interfaces [5] is critical to obtain a more natural feeling of the robot operation. Another important aspect is the additional assistance given to the user in the performance of the assigned task; focused, for instance, on the possibility of providing an amputee with the capability of performing bimanual tasks [6]. The need of interacting with the environment requires of vision systems to recognise the working place and provide a proper manipulation of the products [7].

A good option to achieve a proper tele-operated robotic manipulation is to implement solutions based on techniques that provide reliable control signals from the human operator. Surface electromyography (sEMG) allows a system to record the electrical activity of muscle contractions in a non-invasive way [8]. The use of this information to control external devices is called myocontrol. Myocontrol techniques have been usually developed to obtain a reliable actuation of assistive devices in the field of prosthetics. This actuation ranges from simple binary control commands to complex multidimensional control [9,10].

Complex techniques have been applied to multi-finger prosthetic devices and robotic hands. However, myocontrol is generally limited to a few hand grips and still unreliable in realistic environments [11]. To avoid these limitations, several approaches have been recently proposed. One option is to provide a proper sensory feedback to the subject to close the control loop [12,13]. However, this option is still limited to the low accuracy in the classification of complex biomechanical tasks. Another alternative is the introduction of multimodal control of the robotic actuation which may provide a good solution to the unreliability of multidimensional control. In this case, another control method, such as gaze-tracking or electrooculography, is combined with myocontrol to increase reliability and speed [14,15]. Its main disadvantage is the increased workload on the user as both interaction methods must be controlled simultaneously.

To solve the problems arisen from the previously described solutions, we propose the use of a shared control of the end effector of the robot arm. To achieve this, complex positioning and grasping tasks are performed by an alternative system and sEMG processing provides high-level commands. In this case, myocontrol will be combined with a vision-based grasping system.

Grasping is one of the most significant tasks which is performed by humans in everyday manipulation processes. In recent works, robots have been provided with the ability to grasp objects [16,17]. It is often possible to see robots autonomously grasping objects in many industrial applications in which the environment is not dynamic and where both geometry and pose of objects are known. Therefore, the proper pose of the robotic hand or gripper to grasp the object is computed only once. This process is repeated whenever it is needed. More recently, robots are beginning to be self-sufficient and they are reaching a great level of autonomy to work without human intervention in unstructured scenarios or with dynamics in which the kind of objects or their poses are unknown, for example in industrial applications as in [18] and in storage and logistic applications [19].

Many grasp methods have been made possible by the advances in visual perception techniques of the environment, both 2D [20] and 3D [21]. In general, both techniques combine computer vision algorithms and traditional machine learning, the first for the extraction of object features of the scene and the second for the recognition of the objects by comparison and classification of extracted features with features from a dataset of known objects. Thereby, visual perception has allowed robots to have the ability of grasping in a similar way to humans, though under certain conditions, making use of object recognition algorithms [22–24] and pose estimation algorithms [25,26]. Recently, a significant number of new approaches have been proposed to localize robotic grasp configurations directly from sensor data without estimating object pose using training databases of real objects [27] or synthetic objects (CAD models) as in [28].

However, currently it is still not possible to compare the ability of robots and humans to grasp objects in a generic way, for each and every situation. The main drawback of applying visual perception techniques to accomplish a completely autonomous grasping is the great variability of the kind of objects (geometric shape, pose and visual appearance such as color or texture) that can be present in an environment. This demands a large datasets of training data to implement a robust algorithm to avoid ambiguity in both recognition and location processes of the objects in the scene. The proposed system may solve both the more relevant issues of grasping and the complexity of multidimensional myoelectric control, by combining the visual-driven system with simple electromyographic analysis, based on ON/OFF sEMG commands.

2. System Architecture

2.1. Vision-Guided Robotic Grasping System

The system architecture is composed of a PA-10 industrial robot arm (Mitsubishi, Tokyo, Japan). This robot has seven degrees of freedom (DoF). The robot arm is controlled as a slave in a client-server software architecture managed from a Robot Operating System (ROS) framework. The PA-10 is connected to a server module installed on a computer acting as the PA-10 controller, and both elements are communicated via the Attached Resource Computer NETwork (ARCNET) protocol. The robot is always waiting for commands generated from the orders given by the computer vision algorithm running in the slave module. This module is also responsible for the planning and simulation of trajectories computed from the information obtained from the vision algorithm and from the data supplied by the sEMG system. In addition, the robot arm has an Allegro hand (Wonik Robotics, Seoul, Korea) attached to its end effector with a payload of 5 kg. It is a low cost and highly adaptive multi-finger robotic hand composed of 4 fingers and 16 independent torque-controlled joints, 4 for each finger. The Allegro hand is connected to the slave module via the Controller Area Network (CAN) protocol. The implementation of the system, with its different components, can be seen in Figure 1.

Additionally, the architecture of the system includes a RealSense Camera SR300 (Intel, Santa Clara, CA, USA). It is a depth-sensing camera that uses coded-light methodology for close-range depth perception. With this sensor, the system can acquire 30 colour frames per second with 1080 p resolution. SR300 is able to capture depth in a scenario from a distance between 0.2 m and 1.5 m. It is ideal to obtain shapes of real-world objects using point clouds.

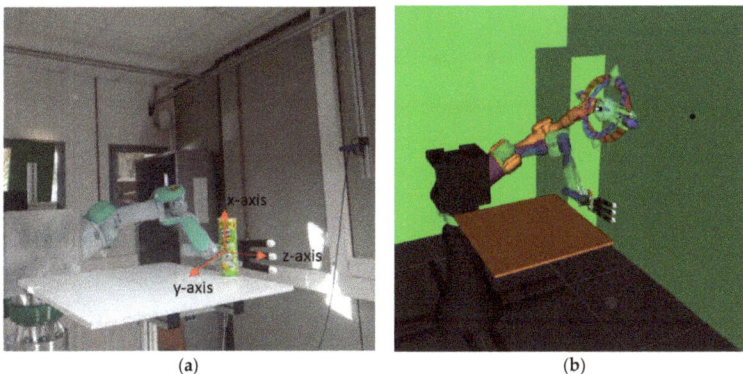

(a) (b)

Figure 1. Pre-grasping pose of the robotic system computed by the vision algorithm. (**a**) Real robotic system in which the grasps are executed. (**b**) Simulation system where the movement is planned and the robotic hand pose is evaluated.

2.2. Electromyography -Based Movement Control System for Robotic Grasping

After positioning the robot hand in front of the object, subjects perform a fine control of the grasping action by reorienting the end effector left or right and then provide the control output for the final approach to the object and subsequent robot hand closing. To obtain these control outputs surface electromyography has been recorded from the forearm during the performance of wrist flexion and extension.

To record surface electromyography (sEMG) signals a Mini DTS 4-channel EMG wireless system (Noraxon, Scottsdale, Arizona, USA) has been used (Figure 2). Two sEMG bipolar channels have been located over the *flexor digitorum superficialis* (FDS) and the *extensor carpi radialis longus* (ECR) of the forearm. Signals have been acquired with a sample frequency of 1500 Hz, then low-pass filtered below 500 Hz, full-wave rectified and, finally, smoothed with a mean filter of 50 ms (Figure 3).

Three different states have been classified from the filtered sEMG signal corresponding to a weak wrist flexion, a weak wrist extension and a strong wrist flexion. To classify these states, two thresholds have been defined to identify weak contractions (flexion on the FDS and extension on the ECR). Additionally, a higher threshold has been defined for strong contractions of the FDS (Figure 3). A ROS message is sent with the decoded output commands to the robotic system. This classification is performed every 0.5 s.

Weak flexion and extension is used to adjust the end effector in the z-axis (direction of the hand) with an initial step of 5 cm. These corrections can be performed through several control commands. When the robot end effector changes direction, the initial step is reduced to a 50%, which allows a fine adjustment of the position of the robot end effector avoiding a loop between end locations. Finally, when the operator thinks that the robot hand is properly positioned a strong flexion is used to perform the final approach to the object and the subsequent grip action.

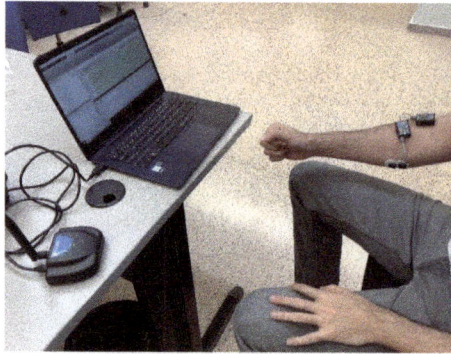

Figure 2. Surface electromyography (sEMG) system acquiring data from a subject.

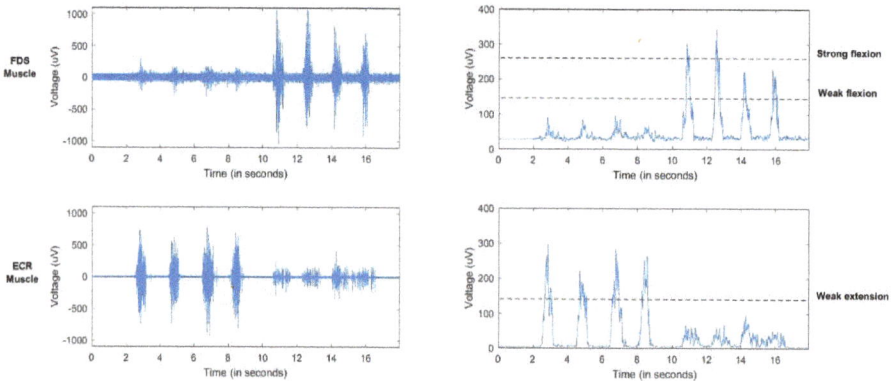

Figure 3. EMG raw signal for several flexion/extension wrist movements (**left**). Processed EMG signal and estimative thresholds (**right**).

3. Proposed Method for Grasping

The proposed method consists of two phases. First, the vision algorithm detects the presence of unknown objects on the scene, segments the scenes to obtain clusters of each object (each cluster is a point cloud) and then, it computes grasping points on the surface of each of the objects (Figure 4).

The method is flexible to obtain grasping points of objects even changing the scenario providing that objects are located on a table or flat surface. Once the vision algorithm provides the robot with the optimal grasping points of the object, the robot plans the trajectory in order to position the robot hand to grasp the object. Occasionally, the grasping of the object is not optimal. For this reason, the method adds a second phase which is used to plan fine hand robot-object interactions. In this step, EMG-based teleoperation of the robot hand-arm is performed to accomplish a successful and stable grasp without slipping and avoiding damage to the object.

3.1. Grasping Points and Pose Estimation

The algorithm calculates pairs of contact points for unknown objects given a single point cloud captured from a RGBD sensor with eye-to-hand configuration. Firstly, the point cloud is segmented in order to detect the objects present in the scene. Then, for each detected object, the algorithm evaluates pairs of contact points that fulfil a set of geometric conditions. Basically, it approximates the main axis of the object using the major vector obtained by running a Principal Component Analysis (PCA) extraction. Then, it calculates the centroid in the point cloud. With this information, it is possible to find a cutting plane perpendicular to the main axis of the object through its centroid. The candidate contact areas are at the opposite edges of the surface of the object that are close to the cutting plane. A standard grasping configuration consists of one point from each of these two areas. Figure 4 shows all these steps graphically.

These candidate areas, in which the robot hand can be positioned, contain multiple potential points so the vision algorithm evaluates a great variety of grasping configurations for the robot hand, using a custom metric that ranks their feasibility. Thereby, the best-ranked pair of contact points is selected, since it is likely to be the most stable grasp, given the view conditions and the used robotic hand. The algorithm takes into account four aspects: the distance of the contact points to the cutting plane, the geometric curvature at the contact points, the antipodal configurations and the perpendicularity to the contact points.

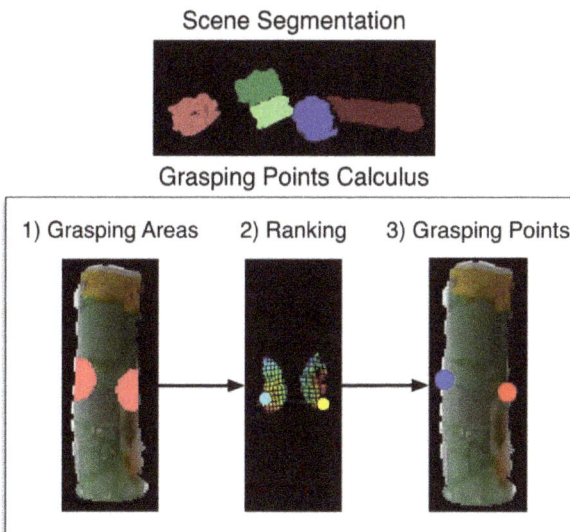

Figure 4. Steps of the method for calculating a pair of contact points. Scene Segmentation: clouds of the detected objects. Grasping Points Calculus, executed for each detected object: (1) grasping areas with potential contact points, (2) curvature values and a pair of evaluated contact points, (3) best ranked pair of contact points.

The first one, distance of the contact points to the cutting plane, is important because it is assumed that the grasping of the object is more stable as the robotic hand grasps closer to the centroid of the object, which is an approximation of its centre of mass. This way, the inertial movements caused throughout the manipulation process of the object are more controllable. The second aspect, the curvature, is considered to avoid the grasps of unstable parts on the object surface. The goal is to place the fingertips on planar surfaces instead of highly curved areas that are prone to be more unstable. Grasping objects on non-planar areas can cause a slip and fall of a grasped object when it is being manipulated, for example, if the robot arm executes a lifting movement. Regarding to the third aspect, contact points should be located on places where the robotic fingers can apply opposite and collinear forces (antipodal configuration). Finally, it is desirable to have contact points that are connected by a line perpendicular to the main axis of the object. That is, the contact points are equally distanced from the cutting plane.

The aforementioned aspects are used to define a quality metric to evaluate the candidate contact point and to propose the best grasp points to carry out a successful grasp of the object on the scene. Accordingly, this quality metric ranks with greater values the grasping configurations that place the robotic hand with its palm point towards the object, its fingertips perpendicular to the axis of the object, parallel to the cutting plane and close to the centroid of the object. Notice that this operation is performed for every detected object. Consequently, the final pose of the robot hand is calculated using the best ranked grasping configuration and the approximated main axis of the object.

Our vision algorithm only computes pairs of contact points. This is assumed to avoid the method being dependent on the type of robotic hand mounted at the end of the robotic arm. Two points are the minimum required for a simple robotic gripper but also, any multi-finger robotic hand can adapt its grasping configuration to two points on the object surface. In the experiments, we use an Allegro hand with four fingers, one of which acts as the thumb. In practice, it is assumed that the grasps will be done with three fingers. This number has been limited to three because the Allegro hand size is often bigger than the object size which will be grasped.

In order to perform three-finger grasps, the algorithm takes into account the following criterion: one of the contact points corresponds to the place the thumb must reach during a grasp, while the other contact point remains between the first two fingers (index and middle). This means that the first and second finger wrap around the second contact point. In this way, the grasp adapts its configuration to only two contact points even though the hand uses three fingers. In addition, the robotic hand is oriented perpendicular to the axis of the object, meaning that it adapts to the pose of the object.

When the human operator has selected the desired object that will be grasped, the robotic system guided by the vision algorithm performs the following steps to reach it:

1. First, the robotic hand is moved to a point 10 cm away from the object. This is a pre-grasping position which is used to facilitate the planning of the following steps. The pre-grasping position is computed, from location (position and orientation) of contact points on the object surface, by the vision algorithm previously described.
2. Second, the robotic hand is moved forward facing the object with its palm and the fingers opened. In this step the hand reaches the point in which, after closing, it would place the fingertips on the calculated contact points.

The correctness of this position depends on the calibration of the camera position with regards to the world's origin as well as lighting conditions and reflectance properties of the objects in the scene. Owing to this, the proposed method performs the correction of the robot hand using the sEMG signals. But also, sEMG can be used to accomplish a proper grasp of objects in a complex manipulation.

3.2. Collaborative System with Both Visual and Electromyography Data

The proposed solution has been implemented using the ROS in order to develop nodes in charge of different responsibilities but keeping a communication framework among them. One node has been created, called *pointcloud_listener*, where point clouds are read and processed to perform the calculus of the grasp contacts. This node publishes a custom ROS message called *GraspConfiguration* where the point clouds of the objects and the calculated grasp contacts are stored.

Another node, called *allegro_control_grasp*, subscribes to this topic and reads the published contact points to generate a grasp pose for the robotic gripper. Then, it proceeds to plan a trajectory following the steps listed in the previous section. MoveIt! [29] has been used to perform this trajectory planning. Once it reaches the grasping position, the EMG control starts. To do so, it subscribes to a topic called /*emgsensor*/*move* where the correcting movements are published.

These corrections are published by a third node called *emg_reader*, which processes the sEMG signals in order to provide messages of type *geometry_msgs*/*Quaternion*. This type of ROS message allows us to describe the direction of movement for the arm that the operator wants to perform in order to correct the position of the robotic gripper. Thus, using one of the axis of the Quaternion, we can specify in which axis we want to move the gripper. The w term is set to 1 when we detect the grasping pattern in the EMG signal so the *allegro_control_grasp* node closes the gripper and continues to lift and carry the object.

It is important to note that this message is constantly published by the *emg_reader* node but the *allegro_control_grasp* only reads them after performing a correction. This means that messages published during the physical movement of the robot are ignored and, as soon as it stops, the control returns to wait for a new message in the topic. Figure 5 shows a scheme of the nodes and their interactions through ROS.

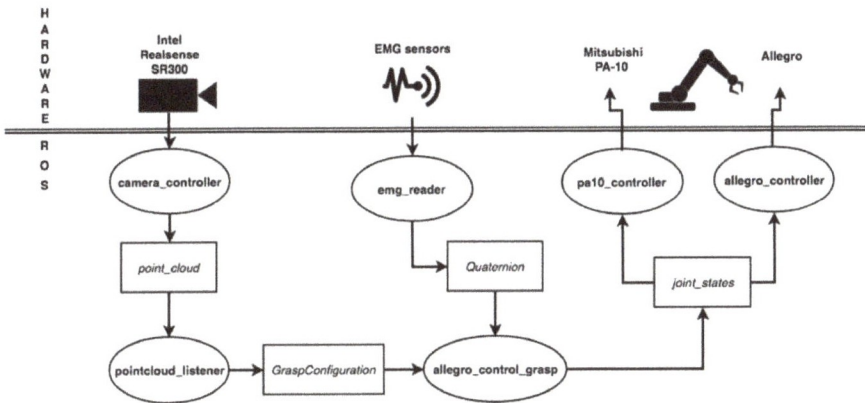

Figure 5. Scheme of the proposed method implemented in Robot Operating System (ROS) showing communication modules among different steps.

4. Experiments and Discussion

4.1. Test Design

Six subjects (age 24.5 ± 6.2 years old, four male and two female) without previous experience on myoelectric control participated in the experimental tests. First, subjects were asked to perform several wrist flexion and extensions at different force levels and thresholds were visually chosen from the processed sEMG signals of the FDS and ECR. After selecting the proper thresholds, subjects were asked to freely perform wrist contractions and the classification output was shown to them until they felt comfortable with the myoelectric setup.

The experimental tests were divided into three sets of grasping activities, each one for a different positioning of the object. The object, a cylindrical plastic can (23 cm height, 8 cm diameter), was placed vertically (position 1), horizontally (position 2) and in a diagonal orientation (position 3). Each grasping activity was performed five times for each position and subject. Subject 5 did not perform the last set (position 3) of grasping tasks due to fatigue and technical problems.

During the grasping activity, the visual-driven robot arm positioned the robotic hand facing the side of the object and then, subjects were asked to readjust the z-axis (weak wrist extension or flexion) and then grasp the object voluntarily with a strong wrist flexion. The accuracy of classifying sEMG signals was measured by counting correct sEMG commands (classification success), no detections (if muscle contraction was present but the control command was not generated) and errors in the classification output. No detections were manually counted from the visualization of correct contractions that did not reach the selected thresholds. Errors were counted as wrong generated commands. Grasping accuracy was measured by counting correct graspings of the object, i.e., if the object did not flip or fall from the robotic hand.

4.2. Results and Evaluation

Tables 1–3 show the results obtained on sEMG performance (classification success, no detection, classification error) and grasping performance in terms of accuracy (ACC), i.e., percentage of correct grasps. sEMG accuracy was obtained by dividing successful classifications by performed contractions.

Table 1. sEMG performance and grasping accuracy for object position 1.

Subject	Success	Error	No Detection	sEMG ACC	Grasping ACC
A01	10	0	0	100%	100%
A02	10	0	1	91%	100%
A03	10	1	2	77%	100%
A04	8	1	0	89%	100%
A05	10	0	0	100%	80%
A06	6	2	1	67%	80%
Average	9.00	0.67	0.67	87.23%	93.33%
Standard deviation	1.67	0.82	0.82	13.20%	10.33%

Table 2. sEMG performance and grasping accuracy for object position 2.

Subject	Success	Error	No Detection	sEMG ACC	Grasping ACC
A01	8	1	0	89%	100%
A02	10	1	1	83%	100%
A03	10	0	1	91%	100%
A04	8	1	0	89%	100%
A05	10	1	3	71%	100%
A06	10	0	2	83%	100%
Average	9.33	0.67	1.17	84.46%	100.00%
Standard deviation	1.03	0.52	1.17	7.12%	0.00%

Table 3. sEMG performance and grasping accuracy for object position 3.

Subject	Success	Error	No Detection	sEMG ACC	Grasping ACC
A01	10	0	1	91%	80%
A02	10	0	0	100%	100%
A03	10	1	0	91%	100%
A04	10	0	1	91%	100%
A06	8	1	0	89%	80%
Average	9.60	0.40	0.40	92.32%	92.00%
Standard deviation	0.89	0.55	0.55	4.38%	10.95%

From the results, it can be concluded that both sEMG and grasping accuracy is high. sEMG errors or no detections do not always affect grasping accuracy as the robot hand is quite well positioned with the visual-driven system alone. It is interesting to notice that for object position 2 the grasping is always successful. This is possibly due to the fact that the object is placed horizontally to the ground and, as it is cylindrical, it sometimes rolls until touching the thumb of the hand when the hand is repositioned. Nevertheless, grasping for the remaining object positions is also very accurate (93.33% ± 10.33% for position 1 and 92.00% ± 10.95% for position 3). Regarding sEMG classifications, errors are fewer than no detections. A possible solution to reduce these errors is a longer training of the subjects (in these tests, subjects were naïve to myoelectric control systems). Another option could be the use of a more conservative threshold selection. This will prevent the appearance of errors but would probably increase the no detections increasing the time taken to perform the grasping.

The results of a previous experiment, in which only the visual-driven system was used, are compared, in Table 4, to the results of the proposed sEMG-based system. Visual-driven tests are automatic, so there is no direct implication of a human operator in the positioning of the robot and the following grasping. The error for experiments without EMG represents two kind of errors. One of them is due to the slipping of the object during the grasping tasks. Other errors occurred because the hand position is not properly fit with vision techniques. Both cases are mostly solved when sEMG control is added to the grasping system. This way, sEMG can be used to correct the hand pose and its grasps, showing an increase in grasping accuracy close to a 9% using the same cylindrical object. Besides, the accuracy increases up to a 15% if it is compared with other grasping experiments using other cylindrical objects Consequently, the average increase in accuracy is around 13.8% considering the 81 trials without sEMG.

Table 4. Comparison of the grasping accuracy for the proposed (visual data + sEMG) compared to the previous method (only visual data).

Subject	Trials	Success	Error	Grasping ACC
with sEMG	85	81	4	95.29%
without sEMG (same object)	15	13	2	86.66%
without sEMG (other cylindrical objects)	66	53	13	80.30%

5. Conclusions

In this paper, we propose a method based on combining both computer vision and sEMG techniques to allow a human operator to carry out grasping tasks of objects. The proposed method has been demonstrated and validated by several human operators with different ages and sex. To do this, our method uses a vision algorithm to estimate grasping points on the surface of the detected object and moves the robotic hand-arm system from any pose to a pre-grasping pose according to the object. Then, sEMG signals from arm muscles of human operators are measured, processed and transformed into movements of the robotic hand-arm system. Thereby, the human operator can readjust the robotic hand to properly grasp the object. The results show an increase of around a 9% in grasping accuracy compared to the use of the visual-driven system alone with the same object and around a 15% with similar cylindrical objects.

The proposed method evaluates a simple ON/OFF myocontrol classification algorithm based on a threshold selection with a very high reliability and that could be easily translated into an industrial environment with the introduction of low-cost sEMG devices such as the MYO Thalmic bracelet or Arduino-based acquisition systems. Additionally, specific expertise is not needed to instrument the sEMG system, as the location of electrodes on flexor and extensor muscles is straight-forward. This is a first approach towards bridging the gap between human operators with and without disabilities in industrial works in which grasping and manipulation tasks are required. In the future, we hope to integrate more signals to control additional degrees of freedom during the movement to generate better grasps and more complex manipulation tasks.

Author Contributions: All authors conceived the proposed method. A.Ú., B.S.Z.-I. performed the experiments; all authors analyzed the data and results; A.Ú., B.S.Z.-I., S.T.P. and P.G. wrote the paper; all authors reviewed and helped with clarifying the paper.

Funding: This work was funded by the Spanish Government's Ministry of Economy, Industry and Competitiveness through the DPI2015-68087-R, by the European Commission's and FEDER funds through the COMMANDIA (SOE2/P1/F0638) action supported by Interreg-V Sudoe and by University of Alicante through project GRE16-20, Control Platform for a Robotic Hand based on Electromyographic Signals.

Acknowledgments: Authors would like to thank John A. Castro Vargas for his help during the performance of the experiments and all the students that participated in the experiments.

Conflicts of Interest: The authors declare no conflict of interest.

References

1. Paperno, N.; Rupp, M.; Maboudou-Tchao, E.M.; Smither, J.A.A.; Behal, A. A Predictive Model for Use of an Assistive Robotic Manipulator: Human Factors versus Performance in Pick-and-Place/Retrieval Tasks. *IEEE Trans. Hum.-Mach. Syst.* **2016**, *46*, 846–858. [CrossRef]
2. Treurnicht, N.F.; Blanckenberg, M.M.; van Niekerk, H.G. Using poka-yoke methods to improve employment potential of intellectually disabled workers. *S. Afr. J. Ind. Eng.* **2011**, *22*, 213–224. [CrossRef]
3. Kochan, A. Remploy: Disabled and thriving. *Assem. Autom.* **1996**, *16*, 40–41. [CrossRef]
4. Maja, J.M.; Brian, S. Socially Assistive Robotics. In *Springer Handbook of Robotics*, 2nd ed.; Bruno, S., Oussama, K., Eds.; Springer: Berlin/Heidelberg, Germany, 2016; pp. 1973–1994. ISBN 978-3-319-32550-7.
5. Li, H.B.; Zhang, L.; Kawashima, K. Operator dynamics for stability condition in haptic and teleoperation system: A survey. *Int. J. Med. Robot. Comp. Assist. Surg.* **2018**, *14*, e1881. [CrossRef] [PubMed]
6. Ureche, L.P.; Billard, A. Constraints extraction from asymmetrical bimanual tasks and their use in coordinated behavior. *Robot. Auton. Syst.* **2018**, *103*, 222–235. [CrossRef]
7. Kasaei, S.H.; Oliveira, M.; Lim, G.H.; Lopes, L.S.; Tome, A.M. Towards lifelong assistive robotics: A tight coupling between object perception and manipulation. *Neurocomputing* **2018**, *291*, 151–166. [CrossRef]
8. Chowdhury, R.H.; Reaz, M.B.I.; Ali, M.A.B.M.; Bakar, A.A.A.; Chellappan, K.; Chang, T.G. Surface Electromyography Signal Processing and Classification Techniques. *Sensors* **2013**, *13*, 12431–12466. [CrossRef] [PubMed]
9. Geethanjali, P. Myoelectric Control of Prosthetic Hands: State-of-the-art Review. *Med. Devices* **2016**, *9*, 247–255. [CrossRef] [PubMed]
10. Rojas-Martínez, M.; Alonso, J.; Jordanic, M.; Romero, S.; Mañanas, M. Identificación de Tareas Isométricas y Dinámicas del Miembro Superior Basada en EMG de Alta Densidad. In *Revista Iberoamericana de Automática e Informátia Industrial*; Elsevier: New York, NY, USA, 2017; Volume 4, pp. 406–411.
11. Connan, M.; Ruiz Ramírez, E.; Vodermayer, B.; Castellini, C. Assessment of a Wearable Force—And Electromyography Device and Comparison of the Related Signals for Myocontrol. *Front. Neurorobot.* **2016**, *10*, 17. [CrossRef] [PubMed]
12. Dosen, S.; Markovic, M.; Somer, K.; Graimann, B.; Farina, D. EMG Biofeedback for Online Predictive Control of Grasping Force in a Myoelectric Prosthesis. *J Neuroeng. Rehabil.* **2015**, *12*, 55. [CrossRef] [PubMed]
13. Schweisfurth, M.A.; Markovic, M.; Dosen, S.; Teich, F.; Graimann, B.; Farina, D. Electrotactile EMG feedback improves the control of prosthesis grasping force. *J. Neural Eng.* **2016**, *13*, 5. [CrossRef] [PubMed]
14. Chin, C.A.; Barreto, A. The Integration of Electromyogram and Eye Gaze Tracking Inputs for Hands-Free Cursor Control. *Biomed. Sci. Instrum.* **2007**, *43*, 152–157. [PubMed]
15. Nam, Y.; Koo, B.; Cichocki, A.; Choi, S. GOM-Face: GKP, EOG, and EMG-based Multimodal Interface with Application to Humanoid Robot Control. *IEEE Trans. Biomed. Eng.* **2014**, *61*, 453–462. [CrossRef]
16. Bagnell, J.A.; Cavalcanti, F.; Cui, L.; Galluzzo, T.; Hebert, M.; Kazemi, M.; Klingensmith, M.; Libby, J.; Liu, T.Y.; Pollard, N.; et al. An integrated system for autonomous robotics manipulation. In Proceedings of the IEEE Intelligent Robots and Systems, San Francisco, CA, USA, 25–30 September 2011; pp. 2955–2962.
17. Rusu, R.B. Semantic 3D Object Maps for Everyday Manipulation. *KI-Künstl. Intell.* **2010**, *24*, 345–348. [CrossRef]

Sensors **2018**, *18*, 2366

18. Wahrmann, D.; Hildebrandt, A.C.; Schuetz, C.; Wittmann, R.; Rixen, D. An Autonomous and Flexible Robotic Framework for Logistics Applications. *J. Intell. Robot. Syst.* **2017**, 1–13. [CrossRef]

19. Schwarz, M.; Milan, A.; Selvam-Periyasamy, A.; Behnke, S. RGB-D object detection and semantic segmentation for autonomous manipulation clutter. *Int. J. Robot. Res.* **2017**. [CrossRef]

20. Bo, L.; Ren, X.; Fox, D. Unsupervised feature learning for RGB-D based object recognition. In *Experimental Robotics*; Desai, J., Dudek, G., Khatib, O., Kumar, V., Eds.; Springer: Heildelberg, Germany, 2013; Volume 88, pp. 387–402. ISBN 978-3-319-00064-0.

21. Ulrich, M.; Wiedemann, C.; Steger, C. Combining scale-space and similarity-based aspect graphs for fast 3d object recognition. *IEEE Trans. Pattern Anal. Mach. Intell.* **2012**, *34*, 1902–1914. [CrossRef] [PubMed]

22. Mateo, C.M.; Gil, P.; Torres, F. Visual perception for the 3D recognition of geometric pieces in Robotic manipulation. *Int. J. Adv. Manuf. Technol.* **2016**, *83*, 1999–2013. [CrossRef]

23. Wohlhart, P.; Lepetit, V. Learning descriptors for object recognition and 3D pose estimation. In Proceedings of the IEEE Conference on Computer Vision and Pattern Recognition, Boston, MA, USA, 7–12 June 2015; pp. 3109–3118.

24. Zapata-Impata, B.S.; Mateo, C.M.; Gil, P.; Pomares, J. Using geometry to detect grasping points on 3D unknown point cloud. In Proceedings of the 14th International Conference on Informatics in Control, Automation and Robotics, Madrid, Spain, 26–28 July 2017; pp. 154–161.

25. Ten Pas, A.; Gualtieri, M.; Saenko, K.; Platt, R. Grasp Pose Detection in Point Clouds. *Int. J. Robot. Res.* **2017**, *36*, 1455–1473. [CrossRef]

26. Kehl, W.; Manhardt, F.; Tombari, F.; Illic, S.; Navab, N. SSD-6D: Making RGB-based 3D detection and 6D pose estimation great again. In Proceedings of the International Conference on Computer Vision, Venice, Italy, 22–29 October 2017.

27. Levine, S.; Pastor, P.; Krizhevsky, A.; Ibarz, J.; Quillen, D. Learning hand-eye coordination for Robotic grasping with deep learning and large scale data collection. *Int. J. Robot. Res.* **2017**, 421–436. [CrossRef]

28. Mahler, J.; Liang, J.; Niyaz, J.; Laskey, J.; Doan, R.; Liu, X.; Ojea, J.A.; Goldberg, K. Dex-net 2.0: Deep learning to plan robust grasps with synthetic point clouds and analitic grasp metrics. *arXiv*, **2017**, arXiv:1703.09312v3.

29. Chitta, S.; Sucan, I.; Cousins, S. MoveIt! [ROS Topics]. *IEEE Robot. Autom. Mag.* **2012**, *19*, 18–19. [CrossRef]

sensors

MDPI

Article

Virtual Sensor of Surface Electromyography in a New Extensive Fault-Tolerant Classification System

Karina de O. A. de Moura * and Alexandre Balbinot

Electrical Engineering, Instrumentation Laboratory, Federal University of Rio Grande do Sul (UFRGS), Avenue Osvaldo Aranha 103, Porto Alegre, RS 90035-190, Brazil; alexandre.balbinot@ufrgs.br
* Correspondence: koalves@gmail.com; Tel.: +55-2151-3308-4440

Received: 14 March 2018; Accepted: 26 April 2018; Published: 1 May 2018

Abstract: A few prosthetic control systems in the scientific literature obtain pattern recognition algorithms adapted to changes that occur in the myoelectric signal over time and, frequently, such systems are not natural and intuitive. These are some of the several challenges for myoelectric prostheses for everyday use. The concept of the virtual sensor, which has as its fundamental objective to estimate unavailable measures based on other available measures, is being used in other fields of research. The virtual sensor technique applied to surface electromyography can help to minimize these problems, typically related to the degradation of the myoelectric signal that usually leads to a decrease in the classification accuracy of the movements characterized by computational intelligent systems. This paper presents a virtual sensor in a new extensive fault-tolerant classification system to maintain the classification accuracy after the occurrence of the following contaminants: ECG interference, electrode displacement, movement artifacts, power line interference, and saturation. The Time-Varying Autoregressive Moving Average (TVARMA) and Time-Varying Kalman filter (TVK) models are compared to define the most robust model for the virtual sensor. Results of movement classification were presented comparing the usual classification techniques with the method of the degraded signal replacement and classifier retraining. The experimental results were evaluated for these five noise types in 16 surface electromyography (sEMG) channel degradation case studies. The proposed system without using classifier retraining techniques recovered of mean classification accuracy was of 4% to 38% for electrode displacement, movement artifacts, and saturation noise. The best mean classification considering all signal contaminants and channel combinations evaluated was the classification using the retraining method, replacing the degraded channel by the virtual sensor TVARMA model. This method recovered the classification accuracy after the degradations, reaching an average of 5.7% below the classification of the clean signal, that is the signal without the contaminants or the original signal. Moreover, the proposed intelligent technique minimizes the impact of the motion classification caused by signal contamination related to degrading events over time. There are improvements in the virtual sensor model and in the algorithm optimization that need further development to provide an increase the clinical application of myoelectric prostheses but already presents robust results to enable research with virtual sensors on biological signs with stochastic behavior.

Keywords: biomedical signal modelling; virtual sensor; cross-correlation; self-recovery; fault-tolerant sensor; signal disturbance

1. Introduction

Advances in the intuitive and natural prosthetic control with multiple degrees of freedom could significantly improve the quality of life of amputees [1–3]. However, a robotic hand based on non-invasive techniques is still a challenge in real application [4], especially when dealing with long-term usage. The myoelectric signal changes over time explain the difficulty of applying an intuitive and natural system. The myoelectric signals of a human being may change for reasons such as electrode conductivity changes that may be due to transpiration or environmental humidity, muscle fatigue, atrophy or hypertrophy, electrode displacement on the skin, changes in movement execution and the force intensity applied by the user [5–7].

The development of systems adaptive to these changes in the sEMG signal is necessary to obtain more intuitive and natural control myoelectric prostheses, but most of the recent developments are not adapted to such changes [4–11]. The sEMG signal is a stochastic time-series based on the physiological muscle properties and the subject muscle contraction form [12]. The same stochastic behaviour that provides the sEMG signal hinders its estimation.

Several solutions have been developed to reduce the interference in the acquired biomedical signals. However, a residual interference of these interferences still presents [13–20]. The signal contamination by motion artifacts causes data irregularities. Nonetheless, the effects of motion artifacts can be reduced by proper design of the electronic circuitry and set-up, but not eliminated [21]. The ECG interference is difficult to remove with conventional filters because the contamination overlaps with the sEMG signal in both the time domain and frequency domain [22]. For these reasons, it is essential to design classification systems that are robust enough to operate on signals containing such artifacts or can detect these artifacts so that signals containing them are discarded.

Virtual sensors are an emergent and intelligent tool which have been successfully used in other fields [23–26]. Usually, they are used to replace physical sensors [25,27,28]. They can also be used as part of fault detection methodologies, where their output is compared to the corresponding sensor [24,29–32]. The concept of virtual sensors is also present in studies in the context of wearable sensors [27,33] and physiological signals [34]. In Ref. [33], a multi-layer task model based on Hidden Markov Model (HMM) was presented and applied in the context of gait analysis. This virtual sensor approach research confirms the application effectiveness while maintaining high efficiency and accuracy. The virtual sensing service presented in [35] was used to estimate human body temperatures of various parts of the body by integrating human physiological models with measurable sensor data. In [34], a virtual sensor estimated the respiratory rate performance for time intervals, starting from a single-lead electrocardiogram signal.

The operation principle of these sensors is based on the mathematical model estimation of the collected data. The novel approach of the extensive fault-tolerant motion classification system consists of the use of the virtual sensor concept to reduce the impact over time of the sEMG signal degradation, combined with a fault-tolerant signal quality analysis detector. This study evaluated the five most common contaminants in sEMG signals [7,11,36]: motion artifacts, amplifier saturation, electrode displacements, power line interference and ECG interference. The study in [11] of signal contamination insertion and detection presents a one-class Support Vector Machine (SVM) successfully employed to detect a variety of contamination in sEMG signals with different SNR levels. There are also studies [11,36] that specifically identify which contaminant is present in the signal. Moreover, a signal quality analysis system in conjunction with re-training of the classifier in the removal of the contaminated channel is presented in [7].

The purpose of a virtual sensor model is to produce a signal output model independent of the physical acquisition of the signal of interest. The surface electromyography signal modeling is designed combining concepts of multichannel and their cross-correlation to replace degraded signal channels. This approach is referred to as multichannel cross-correlation. The objective of this new processing system is to maintain the classification accuracy after some signal degradation without the need for any retraining or calibration.

In this research, two types of sEMG signal modeling are evaluated: Time-Varying Autoregressive Moving Average (TVARMA) and the Time-Varying Kalman filter (TVK). The TVARMA models have already been used to improve non-stationary signal models [37]. Furthermore, the Kalman filter model has been used extensively in other fields [37–42], where it is considered an extremely efficient and flexible signal processing tool, and it is also employed in other virtual sensor enforcement [31].

Moreover, the system was evaluated in the characterization of seventeen hand-arm segment movements and the resting positions through the sEMG signals of twelve surface electrodes positioned on the upper limb. Regarding the number of subjects, ten non-amputee subjects and ten amputee subjects from the Non-Invasive Adaptive Hand Prosthetics (NINAPro) database were used to evaluate the mean classification accuracy of the system presented in this work.

2. Methods

2.1. System Overview

The processing and classification of the sEMG signals can be divided into the following seven experimental procedures to simplify the designed method: database loading, contaminated signal simulation, pre-processing, fault-tolerant detection, virtual sensor, feature extraction, and classification. The experimental system starts with the loading of NINAPro sEMG database. Before the common step where the data is preprocessed (filtering, rectification, and normalization), the stage of simulated contamination signals is composed of sixteen cases of contaminated electrode studies for each one of five types of contamination simulated from real examples of signal contamination.

The researchers in [43–45] reported differences in the classification accuracy with the processing with pre-recorded data and real-time performance. Signal processing with pre-recorded data typically presents a higher movement accuracy. Consequently, the fault-tolerant detection, virtual sensor production, and feature extraction were performed online by scanning each of twelve-channel sliding windows to providing a suitable solution to the replicate real-life situation. However, the classification step was performed offline to perform the statistical analysis of the results. The features extracted of each case study are saved separately for further classification.

The sensor fault-tolerant detector (SFTD) performs quality analysis for sEMG signals using a two-class SVM, where the training occurs with selected signals. The training occurs in advance with sEMG signals with real acquired noise, several samples of clean sEMG signals, and their contamination based on motion artifacts, amplifier saturation, electrode displacements, power line interference and ECG interference. Only a few samples from the database of both intact and amputated subjects were used for the SFTD training. The objective was to train the detection sensor for the different amplitude variations of the sEMG signal. The same trained SFTD is then used for all subjects analyzed. When the SFTD detects that there is contamination in more than 70% of the signal for the last 3 s analyzed, the virtual sensor performs the signal modeling of interest for TVARMA and TVK models. One limitation of the response of the SFTD is that virtual sensor activation does not occur for noise bursts lasting less than 2.1 s.

The features were selected based on results obtained in other studies [46,47]. The features Mean Absolute Value (MAV), Root Mean Square (RMS), Wave Length (WL), Maximum Fractal Length (MFL) and Power (PWR) are extracted. Each procedure is detailed in the next subsections. The experimental procedures can be seen in Figure 1a. All data saved from each subject have the same number of samples. The features extracted were used in the multi-class non-linear SVM classification in different analyses of setting classification cases, according to Figure 1b. The first analysis case corresponds to the usual classification, where 50% of the signal data without degradation were used for training and the other 50% for the test. The signal without degradation or the original signal was referred in this study as the clean signal. For cases 2, 4 and 6, the classifier training still remains using 50% of the clean signal. However, the other 50% of the corresponding data of the test are replaced with the signals with degradation inserted (case 2), the virtual sensor signal with the TVARMA model (case 4) and the TVK model (case 6). Case 3 analyzes the retraining of the classifier and test without the degraded channel detected by the SFTD. The last cases analyze the re-training of the classifier and test with the clean signal dataset with the replacement of the degraded windows by the virtual sensor with the TVARMA model (case 5) and the TVK model (case 7). In these two last cases, the training and the test were performed with the replacement of the signal by the models at all points detected by the SFTD.

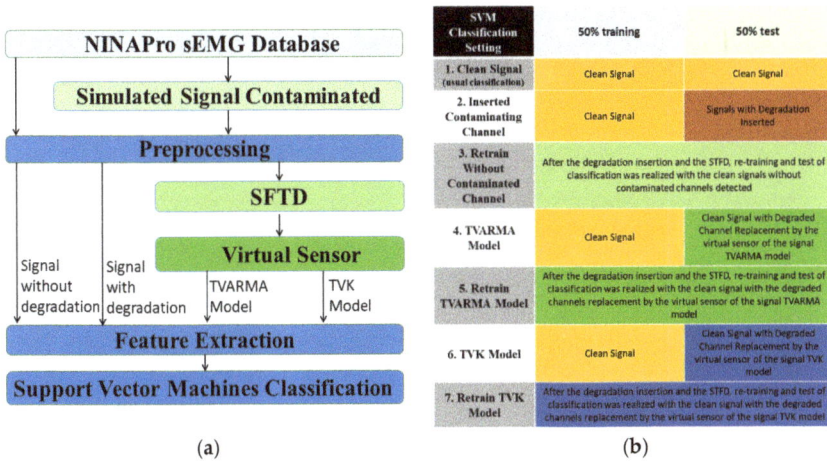

SVM Classification Setting	50% training	50% test
1. Clean Signal (usual classification)	Clean Signal	Clean Signal
2. Inserted Contaminating Channel	Clean Signal	Signals with Degradation Inserted
3. Retrain Without Contaminated Channel	After the degradation insertion and the STFD, re-training and test of classification was realized with the clean signals without contaminated channels detected	
4. TVARMA Model	Clean Signal	Clean Signal with Degraded Channel Replacement by the virtual sensor of the signal TVARMA model
5. Retrain TVARMA Model	After the degradation insertion and the STFD, re-training and test of classification was realized with the clean signal with the degraded channels replacement by the virtual sensor of the signal TVARMA model	
6. TVK Model	Clean Signal	Clean Signal with Degraded Channel Replacement by the virtual sensor of the signal TVK model
7. Retrain TVK Model	After the degradation insertion and the STFD, re-training and test of classification was realized with the clean signal with the degraded channels replacement by the virtual sensor of the signal TVK model	

(a) (b)

Figure 1. The experimental procedures (**a**) Flowchart of the method performed; (**b**) Description of SVM classification setting.

2.2. NINAPro sEMG Database

The NINAPro project provides to the scientific community a database of sEMG signals. This database utilizes 12 active wireless electrodes of the DelsysTM TrignoWireless System® [48]. The twelve electrodes are placed on the forearm, with eight uniformly spaced electrodes just beneath the elbow at a fixed distance from the radiohumeral joint, two on the flexor digitorum and the extensor digitorum, and two electrodes on the main activity spots of the biceps and the triceps. Figure 2 demonstrated the positions of the electrodes on the arm. NINAPro data is acquired using a NI-DAQ PCMCIA 6024E platform (National Instruments, Austin, TX, USA) at a rate of 2 kHz, 12 bits and with a lower than 750 nV RMS [49].

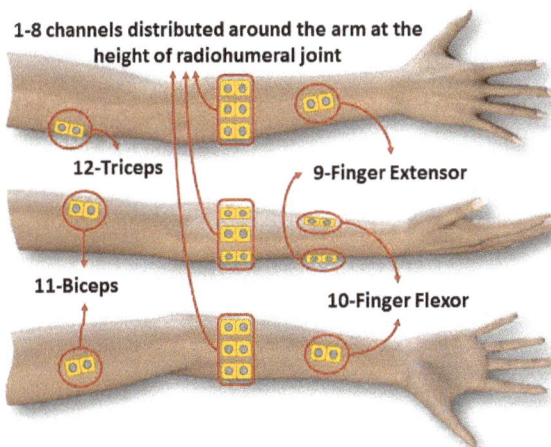

Figure 2. The positions of the electrodes.

The timestamp was based on the virtual model of the orientation screen after the realignment treatment of each movement's limits and the start and end time adjustment [50]. The timestamp generated from the stimulus videos was used to segment the signal. This study also does not monitor the subject-applied force and does not use any feedback procedure.

During a signal acquisition session, a subject was required to perform six repetitions of 17 distinct movements (i.e., a total of 102 movements) which were interspersed with periods of rest in which the subject's hand was in the resting position. The movement is performed for 5 s, interspersed with pauses of 3 s to allow the volunteer to rest. The rest position and the 17 distinct movements are presented in Figure 3.

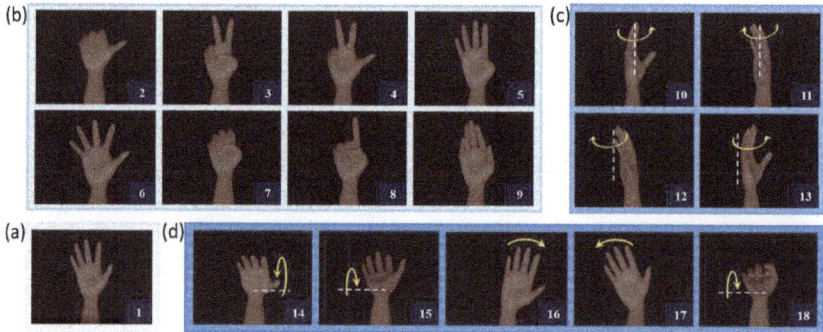

Figure 3. The hand-arm segment movements: (**a**) rest position; (**b**) hand movements; (**c**) Rotational movements; and (**d**) Wrist movements. The sequence of movements from 1 to 18: rest position; thumb up; flexion of ring and little finger, thumb flexed over middle and little finger; flexion of ring and little finger; thumb opposing base of little finger; abduction of the fingers; fingers flexed together; pointing index; fingers closed together; wrist supination and pronation (rotation axis through the middle finger); wrist supination and pronation (rotation axis through the little finger); wrist flexion and extension; wrist radial and ulnar deviation and wrist extension with closed hand.

The first ten non-amputee subjects (aged 29 to 45 years) and ten amputee subjects (aged 32 to 67 years) of the NINAPro database were evaluated in this study. One amputee volunteer of NinaPRO database had an extremely noisy signal and was consequently excluded from the analysis. In these subjects, the SFTD detected similarities among the types of degradation evaluated in almost all the channel windows. The number of volunteers kept the same the number of amputee subjects and non-amputee subjects. The clinical characteristics of the analyzed amputated subjects provided by NINAPro are listed in Table 1.

Table 1. Clinical characteristics of the amputated subjects.

Subject	Age	Handedness	Amputated Hand (s)	Amputation Cause	Remaining Forearm (%)	Year Since Amputation
1	32	Right	Right	Accident	50	13
2	35	Right	Left	Accident	70	6
3	50	Right	Right	Accident	30	5
4	34	Right	Right and Left	Accident	40	1
5	67	Left	Left	Accident	90	1
6	32	Right	Left	Accident	40	13
7	33	Right	Right	Accident	50	5
8	44	Right	Right	Accident	90	14
9	59	Right	Right	Accident	50	2
10	45	Right	Right	Cancer	90	5

2.3. Preprocessing

The pre-processing step aims to perform the digitized signal segmentation, rectification, and normalization. In real-time applications of prostheses, the classification is frequently based on the features extracted from the segmentation by sliding windows [51]. In this type of segmentation, the analysis window slides along in increments, adding new collected data and discarding the oldest data [52]. Majority voting strategies are commonly used to minimize the classifier output error when dealing with sliding windows [51,52]. Other studies suggest that the controller delay has to be as approximately 100 ms [29] and the top limit must be of roughly 300 ms [8,48,53].

The study in [48] analyzed the mean classification accuracy according to the window length variation in the NINAPro database. The study used sliding windows of 100, 200, and 400 ms with an increment of 10 ms for all analysis windows and they obtained the best accuracy with a window length of 400 ms. However, they did not analyze whether the effects of changing window length and increment variation were significant regarding the subject and the movement change.

Some researchers have already demonstrated that classification accuracy increases when the pattern recognition is performed on larger data windows [47,50,52,53]. However, this ends up increasing the time which is required to collect and process a more extensive dataset [52]. Thereby, a more significant amount of data results in features with lower statistical variance, which increases classification accuracy. The optimal mean classification accuracy for the different sliding windows is dependent on both the classifier and the feature extracted. Therefore, comparing results between studies with different characteristics are difficult. Thereby, this study chooses to use the sliding window size of 300 ms with the increment of 75 ms most common in the literature.

The Butterworth filter used on the signals was a digital band-pass filter of order 20 with a frequency range of 20 to 500 Hz. The sampling frequency was 2 kS/s. The normalization was performed separately for each channel considering all channel data. This method is not suitable for online processing. The online normalization must be standardized by a calibration procedure capturing the muscle signals in rest time and a moment of maximum voluntary contraction (MVC). However, the NINAPro database does not contain this information for calibration [54–56].

2.4. Sensor Fault-Tolerant Detector (STFD)

The SFTD performs a quality analysis of sEMG signals based on the presence or absence of contaminants in the sEMG signal. The SFTD uses a two-class SVM that classifies a signal with or without contamination. The disturbances were simulated by the Matlab® R2016b software using real signal contamination data samples as standard. In Figure 4 a comparison between each contaminant insertion in a clean sEMG signal sample with examples of a sEMG signal sample with acquisition noise can be observed. Each artificial contamination was approximated to a real acquisition of contaminated signal, except for ECG interference. The sEMG signals with actual acquisition of ECG interference are not possible in the forearm.

The correct detection of the occurrence of disturbances in the sEMG signal can allow the application of techniques to reduce the impact on movement classification accuracy. The contamination detection by the SFTD in sEMG signals was based on the results described in [11]. Their research already used the SVM trained only with clean signals and tested with artificially contaminated signals with different SNR levels. Their results show that a one-class SVM could be successfully employed to detect a variety of contaminants in sEMG signals. Differently, this proposed study sought to deepen the SFTD knowledge with the signals training with contamination and observed in real acquisitions of sEMG signals without varying the SNR level.

The data used for the SFTD training are composed of samples from some channels of the NINAPro database and real samples of contamination. The training occurred in advance of all signal processing. The signal quality test occurs for each sliding window of the 12 electrode channels. The developed method does not compromise online processing and can be applied to any subject. The operation analysis of this fault-tolerant detection requires a broad system for study the identification accuracy of the 5 types of contaminants. Therefore, it is essential to specify the simulation of the contaminated signal.

2.5. Signal Contamination Simulation

Some studies identify which contaminants are present in the signal [36]. However, this information is not relevant if the focus is to find a unique solution for all contamination cases. The classifier retraining disregarding the contaminated channels obtained significant results to maintain classification accuracy. Nevertheless, other studies observed that there is a processing cost to update the classifier only with the channels without contamination [7,57]. Also, there is not yet a thorough study in the processing cost to retrain the classifier after the end of the contaminant occurrence.

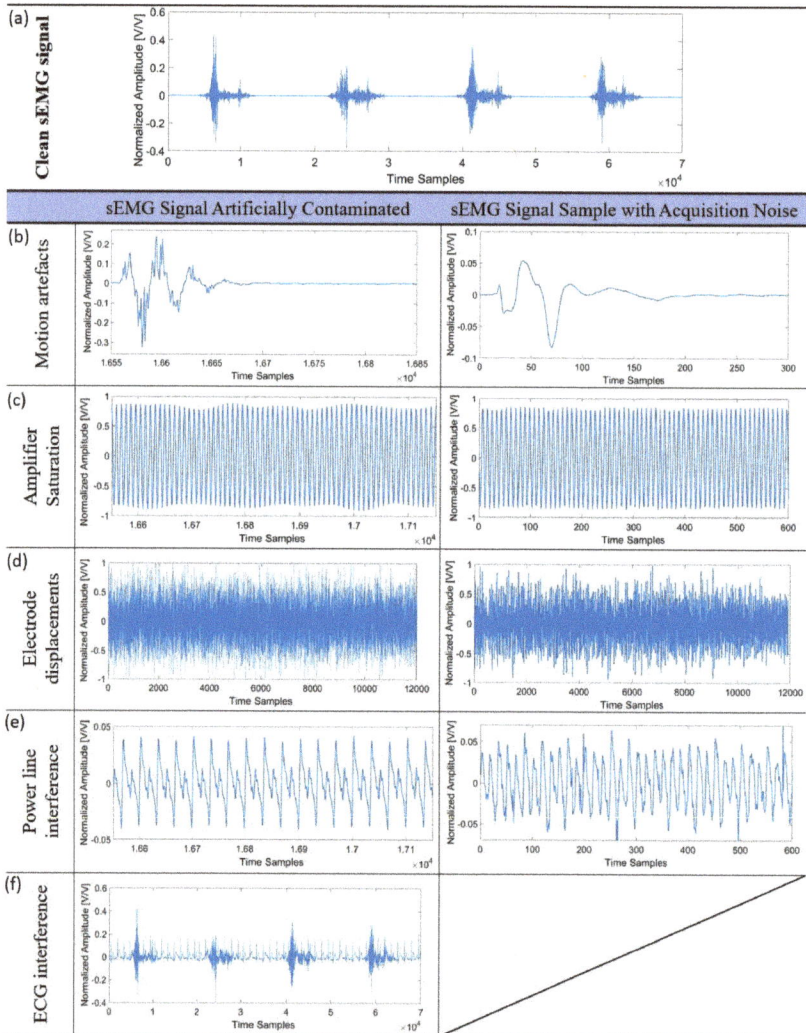

Figure 4. The comparison of each contaminant insertion in the clean sEMG signal of the subject 1 sample in four repetitions of movement 7. The clean sEMG signal sample in (**a**) is artificially contaminated by Motion artefacts in the first column in (**b**), by Amplifier Saturation in the first column in (**c**), by Electrode displacements in the first column in (**d**), by Power line interference in the first column in (**e**) and by ECG interference in the first column in (**f**). The second column in (**b**–**e**) are the sEMG signal samples with the acquisition of real noise.

The performance analysis of this new approach which includes the virtual sensor in conjunction with STDF was carried out through different study cases for each contaminant type. Sixteen cases of contamination were examined for each contaminant type, where, in each case, artifacts were inserted into one channel or a combination of channels. The channels which were contaminated in each case are described in Table 2. The research in [58] related the impact of channels combination in the mean classification accuracy of the NINAPro database, demonstrating that include the signals of the biceps and triceps are a positive influence on the movement recognition. The channel combinations analyzed was select for the influence verifies of each channel and the influence of acquisition region of the upper limb. It essential note that this study not evaluated the variation effect of SNR and not changing the gain of the artifact signal before insertion.

Table 2. Channels contaminated in sixteen cases for each contaminant type.

Case	Channels Contaminated	Case (Cont.)	Channels Contaminated (Cont.)
1	The 1st electrode of eight uniformly spaced electrodes	9	Flexor digitorum electrode
2	The 2nd electrode of eight uniformly spaced electrodes	10	Extensor digitorum electrode
3	The 3rd electrode of eight uniformly spaced electrodes	11	Bicep electrode
4	The 4th electrode of eight uniformly spaced electrodes	12	Tricep electrode
5	The 5th electrode of eight uniformly spaced electrodes	13	Flexor digitorum and extensor digitorum electrodes
6	The 6th electrode of eight uniformly spaced electrodes	14	Bicep and tricep electrodes
7	The 7th electrode of eight uniformly spaced electrodes	15	Flexor digitorum, extensor digitorum, bicep and tricep electrodes
8	The 8th electrode of eight uniformly spaced electrodes	16	All eight uniformly spaced electrodes

The simulated contamination covers the entire channel period for each channel degradation case study before window segmentation. The motion artifact signals simulated to contaminate sEMG signal were estimated models based on the real signals acquired according to the tests described in other studies [7,36]. The electrode displacement artificially contamination was generated by Added White Gaussian Noise (AWGN) of 15 dB [36]. The amplifier saturation was implemented by addition of six sine waves with a random frequency between 200 and 240 Hz as analyzed in the sEMG signal sample with acquisition amplifier saturation noise and based on another study [11].

A sine wave, which has a frequency of 60 Hz, its harmonics, and an amplitude of 0.4 V, was added to the signals to simulate power line interference. The ECG artificial interference occurred with ECG database available from PhysioBank ATM of Physionet (http://www.physionet.org) with the same sampling frequency of sEMG signal. The ECG interference does not occur in the electrode position in this work that is why the acquisition of real noise is not demonstrated. However, the interference detection was analyzed for the application of myoelectric protheses and the ECG interference is present in severe cases of left upper limb amputation, depending on where the electrodes are positioned and whether Targeted Muscular Reinnervation (TMR) occurred. The ECG signal was normalized and added to sEMG signal with maximum amplitude was established at 0.2 for the detection tests.

2.6. Virtual Sensor

The initial idea of this study was conceived through studies of cross-correlation coefficient analysis applications involving investigation of the crosstalk among different sEMG channels [59] and analysis of the degree of synchronization between the surface electromyography recordings for the two muscles [60]. We approached the idea that there is some cross-correlation coefficient between the muscular fibers or muscles of the same hand-arm segment in the performance of the movement and this idea was called multichannel cross-correlation (MCC).

The cross-correlation (r_{xy}) between two sEMG signal channels (x, y) was normalised (c_{xy}) for the number of channels (M), which is the same for all channels. The novel approach called multichannel cross-correlation for utilization on the virtual sensor provides the base of the signal modelling of TVARMA and TVK models. The MCC is a matrix which can be obtained through the percentage contribution (p_{xy}) of each cross-correlation coefficient between the M acquired channels with Equation (1):

$$p_{xy} = \begin{cases} \dfrac{c_{xy}}{\sum_{i=1}^{M} c_{xy,i}}, & x \neq y \\ 0, & x = y \end{cases} \tag{1}$$

For 12 sEMG signal channels, x and y vary with the interval (1,12) providing a multichannel cross-correlation matrix of order 12×12 as can be seen in (2):

$$p_{xy} = \begin{bmatrix} 0 & p_{2,1} & p_{3,1} & \cdots & p_{12,1} \\ p_{1,2} & 0 & p_{3,2} & \cdots & p_{12,2} \\ p_{1,3} & p_{2,3} & 0 & \cdots & p_{12,3} \\ \vdots & \vdots & \vdots & \ddots & \vdots \\ p_{1,12} & p_{2,12} & p_{3,12} & \cdots & 0 \end{bmatrix} \tag{2}$$

The process of signal quality analysis and the signal modeling by the virtual sensor occurs for each examined window. Virtual sensor activation is performed by the SFTD. The SFTD adjusts the MCC matrix, removing contaminated channels when detected contaminated signals above 70% of the windows in the last 3 s. After, the SFTD transmit the matrix and the signals for the virtual sensor. The correlation coefficient between sensor responses has already been used differently as the input of one virtual sensor [61]. The virtual sensor replaces sEMG contaminated signal channel based on the SFTD analysis to the sEMG signal modeling using the TVARMA and TVK models. This operating logic can be seen in Figure 5.

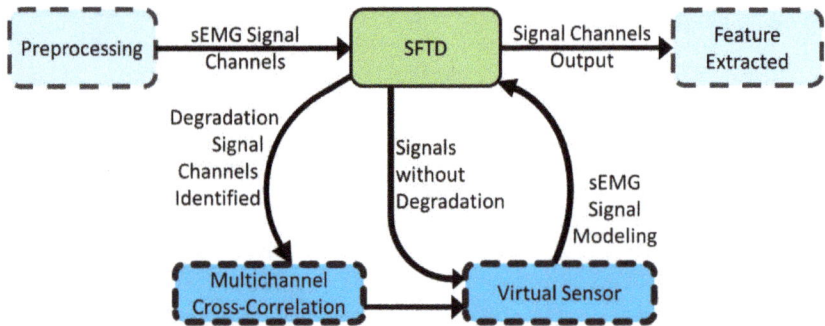

Figure 5. The operating logic of the SFTD with the virtual sensor.

2.6.1. TVARMA Model

The proposed sEMG signal TVARMA model for the virtual sensor at time n can be represented in (3) and was based on the sEMG model proposed in [37,62]. Each sEMG signal channel segment of the virtual sensor $y_{channel}(n)$ is established as a linear combination of the previous output samples, plus the previous input samples $u_{channel}(n)$ and an error term of the noise error present $e_{channel}(n)$, which is independent of past samples:

$$y_{channel}(n) = \sum_{i=1}^{P} a(i,n) \cdot y_{channel}(n-i) + \sum_{j=1}^{Q} b(j,n) \cdot u_{channel}(n-j) + e_{channel}(n) \tag{3}$$

where the $a(i,n)$ is the time-varying autoregressive (AR) coefficient, $b(j,n)$ is the time-varying moving average (MA) coefficient and the indexes P and Q are the highest orders of the AR and MA models, respectively. The model has the model orders set to 4 for P and 2 for Q.

The input samples $u_{channel}(n - j)$ can be obtained through Equation (4), which uses the respective channel of interest's column of the MCC matrix and the other channel samples considered as clean signals from sEMG $\vec{y}(n - i)$. The channels with known contamination have the correlation coefficient line set to zero:

$$u_{channel}(n - j) = \left(\vec{P}_{x,channel}\right)^T \cdot \vec{y}(n - i) \tag{4}$$

where $\vec{P}_{x,channel}$ is the respective values column of the percentage contribution of the cross-correlation coefficient between the channel of interest and the other channels.

A finite-order base function $f(n, m)$ was imposed for the functions of parameters $a(i, n)$ and $b(j, n)$ to obtain a better model in Equations (5) and (6). These functions were proposed in other studies of TVARMA models [62,63]:

$$a(i, n) = \sum_{m=0}^{V} \alpha(i, m) f(n, m) \tag{5}$$

$$b(j, n) = \sum_{m=0}^{V} \beta(j, m) f(n, m) \tag{6}$$

where $f(n, m)$ with $m = 0, 1, 2, \ldots, V$ and $n = 0, 1, 2, \ldots, N - 1$ is the base function that needs to be selected. The parameter functions $\alpha(i, m)$ and $\beta(j, m)$ represent the expansion of the parameters with V, which is the maximum number of base sequences [62]. Therefore, the proposed sEMG signals TVARMA model is finally defined as shown in (7):

$$y_{channel}(n) = \sum_{i=1}^{P} \sum_{m=0}^{V} \alpha(i, m) \cdot y_{channel}(n - i) + \sum_{j=1}^{Q} \sum_{m=0}^{V} \beta(j, m) \cdot u_{channel}(n - j) + e_{channel}(n) \tag{7}$$

2.6.2. TVK Model

The TVK model of sEMG signals for the virtual sensor is based on Kalman state estimator that is given a state-space model with satisfactory known inputs $u_{channel}(n)$, white process noise $w(n)$, and white measurement noise $v(n)$. The virtual sensor measurement $y_{channel}(n)$ and the state equation $x(n)$ can be represented by Equation (8):

$$\begin{aligned} x(n + 1) &= Ax(n) + Bu_{channel}(n) + Gw(n) \\ y_{channel}(n) &= Cx(n) + v(n) \end{aligned} \tag{8}$$

where A, B, C, and G are state matrices.

The $y_{channel}(n)$ and $u_{channel}(n)$ are estimated in the same way as reported for the TVARMA models. The estimator generates output estimates $\hat{y}[n|n]$ and state estimates $\hat{x}[n|n]$ using all available measurements up to $y_{channel}(n)$. The state estimator can be represented as in (9):

$$\hat{x}[n + 1|n] = A\hat{x}[n|(n - 1)] + Bu_{channel}(n) + M(y_{channel}(n) - C\hat{x}[n|(n - 1)]) \tag{9}$$

where M is the innovation gain, which updates the prediction $\hat{x}[n + 1|n]$ using the new measurement $y_{channel}(n)$, and it is determined by solving an algebraic Riccati equation in (10), where P solves the corresponding equation:

$$M = PC^T \left(CPC^T + R\right)^{-1} \tag{10}$$

The final estimator $\hat{y}[n|n]$ for the virtual sensor has the following output equation:

$$\begin{bmatrix} \hat{y}[n|n] \\ \hat{x}[n|n] \end{bmatrix} = \begin{bmatrix} C(I - MC) \\ I - MC \end{bmatrix} \hat{x}[n|n - 1] + \begin{bmatrix} CM \\ M \end{bmatrix} y_{channel}[n|n] \tag{11}$$

The matrices of the TVK model were set with the identity matrix. The sEMG signals models were applied to all sliding windows detected by the SFTD for each subject in the sixteen different cases reported for each simulated signal contamination.

2.7. Feature Extraction

The efficiency of the sEMG signal classification system depends on the choice of features [64]. However, the features selected may be redundant or even irrelevant [65]. Each feature is not equally necessary for a specific task. The features were selected based on the result of the classification of other studies [46,53,66] with the same database and the evaluation of characteristics for sEMG [47].

The characteristics were extracted from each sliding window and the features selection in the time domain is directly related to the MMC concept, which analyses the percentage contribution of each cross-correlation coefficient of two sEMG signal channels in the time domain. It is important to emphasize that no feature reduction technique was used. The combination of the MAV, RMS, WL, MFL, and PWR features presented better mean classification accuracy results for the analyzed subjects.

2.8. Classification with Support Vector Machines

The non-linear SVM classification with radial basis function (RBF) kernel was implemented in the sEMG signals classification of all simulated cases. The kernel functions parameters were selected by a search algorithm of the best result for each subject. The multi-class definition technique for classifying movement was eighteen binary classifications of one versus all. In more than one positive class case, the algorithm selects the class that is farthest away from the hyperplane that separates each binary classification.

The majority voting technique is used as a post-processing mechanism considering last three window classification. Also, the k-fold complimentary technique was applied for improving the reliable accuracy test with the small number of samples in which the model is trained and tested. The tests are carried out in all possible different input conditions forming 20 k-folds for the six available movement repetitions. For each k-fold, three of the six repetitions (50% of the dataset) were selected for the training model and three (50% of the dataset) for testing.

The classification procedure was applied to seven different settings, according to the Figure 1b. The classification using the signal without contamination performed the usual training and test in the average of all possible k-folds. The classification analyses of the signal contamination were performed by training the classifier with the data part of the signal without degradation and the test with the other corresponding part in each separate case for the signal with degradation inserted. The classifications using the virtual sensor occur similar to the previous, the data part of the signal without degradation was used for training, but the signals with TVARMA model and TVK model replace the degradation inserted detected for STFD for the test.

The seven different settings include the re-training practice of the classifier for comparative performance analysis. The classification re-training without the degraded channel was performed when 80% of the windows were considered contaminated by the SFTD. The classification with re-training using the two signals modeled was made through the contaminated signal replacement by the virtual sensor in the training and testing. For every sliding window that featured 70% of contamination detected by the SFTD in the last 3 s, the virtual sensor replaced the sEMG signal by the modeled signal. This threshold was established based on another study [57], which introduced the idea of contaminant temporality and the return to reconsideration of the contaminated channel. Each subject was analyzed for different contamination types in different channel arrangements, and each analysis resulted in a specific confusion matrix of all k-folds each classification results.

2.9. Experimental Statistical Analysis

Design and analysis of three-factor experiments were entirely randomized and used for the statistical validation of the test methodology. A Design of Experiments full factorial design [67] was realized with the mean accuracy response variable and the controllable factors: the classification setting type, the contaminant signal inserted and the variation cases of channel contaminations. The model used follows in (12):

$$y\_ijk = \mu + \alpha\_i + \beta\_j + \gamma\_k + (\alpha\beta)\_ij + (\alpha\gamma)\_ik + (\beta\gamma)\_jk + (\alpha\beta\gamma)\_ijk + \varepsilon\_ijk \qquad (12)$$

where y_ijk corresponds to the level i response in repetition j, μ corresponds to the general average, α_i corresponds to the effect of each level i and ε_ijk corresponds to the error of level i in repetition j.

Analyses of variance (ANOVA) and multiple comparisons also were used for providing a statistical test, which makes possible to assert whether the various group's average differences are significant or not. Two averages are significantly different when their intervals are disjointed. When their intervals overlap, they are not significantly different.

3. Results

The signal processing was analyzed for ten intact subjects and ten amputees. It is important to note that the proposed method using a virtual sensor is independent of the pattern algorithm because the virtual sensor is applied before classification. The SFTD detection and the movement classification used the algorithm SVM. However, there is no dependency on this algorithm. This new extensive fault-tolerant system could use any other algorithm. The SVM is widely used in sEMG signals [46,68–72] and was selected to compare and evaluate the effectiveness of this new system.

3.1. SFTD Results

For the correct interpretation of the results, an analysis of the detection of the windows contaminated by the SFTD is necessary. The SFTD detection accuracy of 85.31 ± 24.88% and false-positive recognition for the channels that non-received the artificial contamination of 8.18 ± 17.52% demonstrated the sensibility of sensor detection. All acquired signal has noise, even after preprocessing, since there may be noises in the frequency range of the sEMG signals.

The SFTD has a differentiation in detection depending on the analyzed contaminant. The ECG and power line interference detection have obtained lower precisions. These contaminants present an effect of lower significance than the other three disturbances. The detection of the ECG and power line interferences in channel 12, which corresponds to the electrode positioned on the triceps, obtained a precision lower than 20% of the average of the other channels. The acquired channel 12 also features the low representativity of the sEMG signal compared to the noise level of the channel. This low representativity may explain the accuracy detection decrease of the SFTD since channel 12 does not seem to be significantly relevant to the movement classification.

3.2. Movement Classification Results

Figure 6 presents the mean classification accuracy for all 16 noise insertion cases with different channel combinations.

In Figure 6, it is possible to realize that each contamination type impairs the classification differently in comparison to the clean signal classification (legend 1), in which some contaminations are more expressively than others, such as saturation and displacement of electrodes. For contamination by electrode displacement and saturation, the substitution of the contaminated signal by the signals modeled with TVARMA (legend 4) and TVK (legend 6) recovers at least 30% of the mean accuracy classification. For contamination by movement artifacts, recovery is at least 20% for intact subjects and approximately 4% for amputated subjects. For ECG contamination and power line interference, the contaminated signal replaced by the virtual sensor models only further damages the signal.

In the comparison of the classifier retraining with removal of the contaminated channel (legend 3), the classifier retraining with the contaminated channel entirely replaced by the virtual sensor using the TVARMA model (legend 5) and using the TVK model (legend 6) demonstrate the improvement or recovery of the mean accuracy classification results after the signal contamination. The classifier retraining method with removal of the contaminated channel recovers 50% of the mean accuracy in the intact subjects and 30% in the amputees for electrode displacement and saturation, 30% in the intact subjects and 20% in the amputees for motion artifacts, 7% in the intact subjects and 3% in the amputees for ECG and power line interference. The classifier retraining method of replacing the degraded channel by the signal modeled by the virtual sensor in the TVARMA and TVK models achieves a mean classification accuracy of at most 5.7% below the clean signal classification.

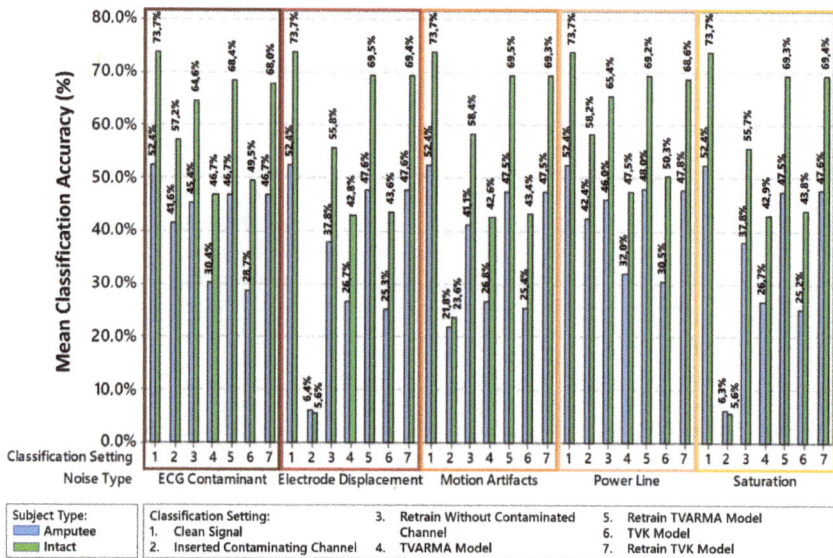

Figure 6. Classification setting comparison for each contaminant insertion type and the clean sEMG signal classification.

3.3. Experimental Statistical Results

The ANOVA obtained results that disregard the possible factors such as subjects and movements because they have already proven that they influence the mean classification accuracy significantly [46,73]. The ANOVA used a 95% confidence interval for all response factors. The significant F-ratio is used to investigate whether the difference among the sample means is significant or just matter of sampling fluctuations. The F-test on main effects and interactions follows directly from the expected mean squares. The calculated value of F-ratio above of the table value of P define if data is significant. The factors classification setting (F = 6992.86 > p-value = 0.0), the contaminant signals inserted (F = 194.02 > p-value = 0.0) and the cases of variation of channel contaminations (F = 1299.08 > p-value = 0.0) indicate that all the factors are individually significant, and their interactions are also significant for mean classification accuracy. That is, each method can achieve better results than another depending on the contaminant and the analyzed combination of noise insertion.

4. Discussion

Analyzing the contaminants detection with the other two similar studies [11,36], SFTD demonstrated very similar performance to the study [11], which also uses SVM and has difficulty detecting the same contaminants with low SNR. It is more difficult to compare with the study [36], that identifies which noise is present in the signal among the contaminants. Since its classification accuracy is based on the correct identification of the noise present among the possible contaminants, it is impossible to compare it with the SFTD. However, the study [36] also uses an SVM algorithm and reinforces the potential of applying this method to identify and mitigate the influence of contamination on acquired sEMG signals. Thereby, after using the SFTD, the presented detection method obtained a suitable solution for identification of contaminants in the sEMG signal. However, the selection of additional features for better detection of ECG and power line interference is a possibility to improve the sensor detection.

The classification setting method analysis for the 16 cases of insertion of contamination in different combinations of channels can be visualized in Figure 7. Regarding the previous analysis, this figure shows that the classifier retraining method without the contaminated channel does not maintain its performance when more than one channel has any contaminant insertion. However, the classifier retraining method with replacement signal by virtual sensor models maintains up to 20% variation for the worst case of degradation (legend 16). The already widely used method of removing the contaminated channel [7,57] has a lower performance in the movements characterization when more than one channel was degraded to the method proposed in this paper with the classifier retraining.

It is important to emphasize that the mean accuracy of the classification settings 5 and 6 are practically overlapped in Figure 7. This overlap demonstrates that it is possible that the chosen model between TVARMA and TVK is not significant. However, the classifier retraining method replacing the signal degraded by the virtual sensor signal increased the mean accuracy in the analysis of all contaminated cases.

The implementation in [57] of a self-adaptive neural network of self-retraining discarding the channels which considered with the looseness or electrodes misplacement was used for the improvement of acquired sEMG signals application. However, there is no artificial contamination, only the contaminations obtained during signal acquisition. The sEMG signal had already intrinsic known contaminants, which were identified using threshold for detection. The presented new system in this work is more complex and evaluates several other interferences. Nonetheless, both demonstrate the possibility of using detection systems to improve the movement classification.

The study in [7] already has a contamination simulation approach like the one implanted in this study. However, this other study tested contact artifacts, loose contacts, and baseline noise with different SNR levels. Although the other study in [7] does not use the same database and they focus on reducing the re-training time of the classifier after their detection sensor module, it is possible to make some comparisons with this work. The number of false-positive occurrences is lower in [7], but this may be a feature of the database used. In a comparison among the noises used, it is possible to affirm that they obtained a less accurate detection for loose contacts.

For the loss of the average accuracy classification after the contamination of one to three channels, this other study had a lower effect of the contaminants in the movement classification for the amputated subjects and non-amputated subjects. The worst case of three contaminated channels presented in the other study has a decrease of classifier accuracy by up to 15%. The comparison the same retraining method discarding contaminated channels of this work with the other study in [7] shows a lower decrease in accuracy for the same noise with more than one contaminated channel for the other study. However, it also showed less influence of the contaminants on the accuracy of the classifier.

Therefore, it is possible to affirm that this new fault-tolerant classification system obtained a suitable solution with better results than the already presented studies. Also, the other studies did not evaluate so many contaminants and cases of channel contamination. This new classification system when using classifier retraining with the virtual sensor signal recovered the mean classification accuracy to the maximum of 5.7% below the clean signal classification, and the worst case with eight contaminated channels obtained a maximum decrease in the mean classification accuracy of 15%.

This statistical analysis can be observed previously, however, the proximity of the results to the retraining methods with the virtual sensor models attained very close results. Thus, a new ANOVA was needed using only the retraining classifier methods with the TVARMA and TVK models for the virtual sensor in the classification setting type factor. This ANOVA results obtained showed that individual mean results are significant (F = 0.62 > *p*-value = 0.43), but their interactions with the other factors are not (F(combination channels, classification setting) = 0.21 < *p*-value = 0.999), F(combination channels, noise type) = 0.51 < *p*-value = 0.999), F(classification setting, noise type) = 0.25 < *p*-value = 0.908)). In other words, one model always has better results than the other when it changes the contaminant type or the variation of cases of channel contaminations and both together.

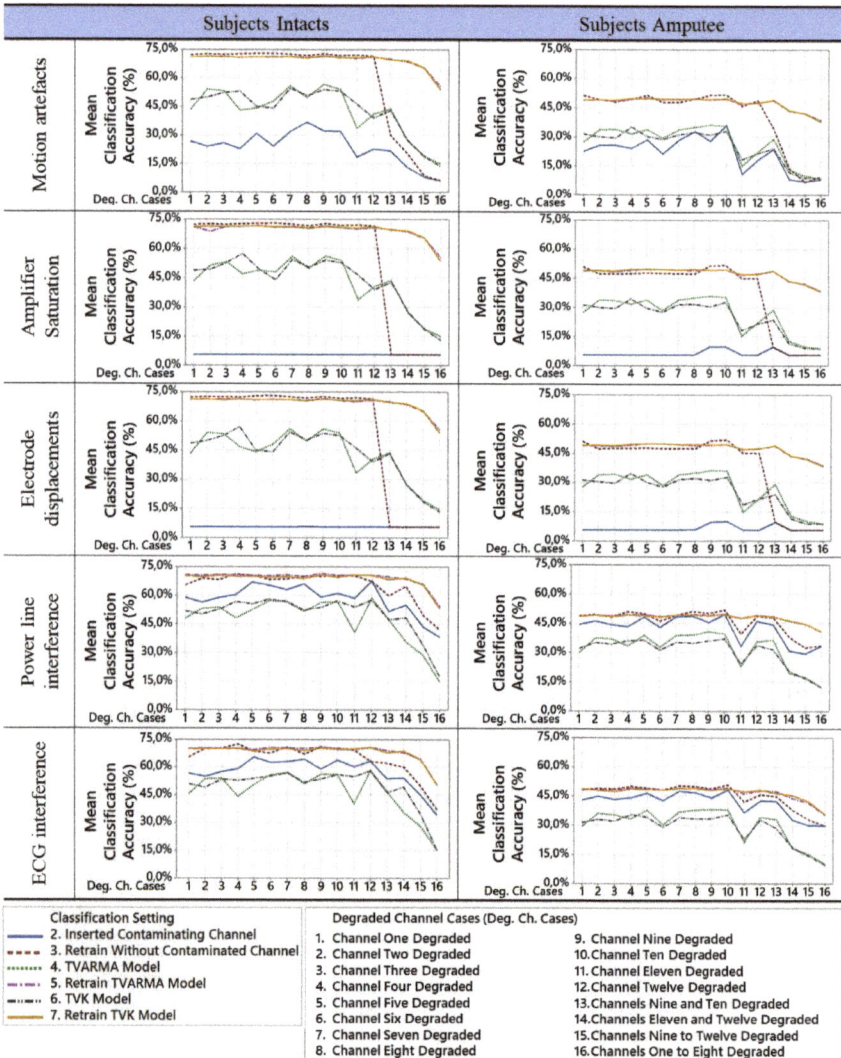

Figure 7. Degradation channel cases comparison for each classification setting.

5. Conclusions

The proposed method can maintain over 60% of the classification accuracy of the clean sEMG signal when contaminated with electrode displacement and saturation and 20% when contaminated with motion artifacts, without a retraining or calibration procedure. For classifier retraining methods, the replacement of the contaminated channel by the virtual sensor model demonstrates the best results for all types of contaminants.

The statistical experimental results determined that the variation of the classification setting, the contaminant signal inserted and the cases channel contaminations significantly affect the mean classification accuracy. The comparison and statistical analysis of the virtual sensor method with the TVARMA and TVK models demonstrate that the TVARMA model obtains more significant mean accuracy for the variation of the contaminant signal inserted, for the variation of cases channel contaminations and both together.

The SFTD accuracy was obtained using a training base of intact subjects and amputees. It is may be possible to improve this accuracy by dividing the intact subjects and amputees for training SFTD. However, it is necessary the controlled acquisition of contaminants in amputees, and calibration may be required, because of the uniqueness of each amputee's muscular integrity.

The quality analysis and virtual sensor model run online at the same time for both TVARMA and TVK models. The processing time of the virtual sensors with the SFTD was not computed. The classification is performed offline later to generate the results of the statistical tests. The algorithm of the proposed system can be optimized for the user online application in the next steps, considering the results analyzed in this work. Classification accuracy cannot be determined in most surveys since different movements and subjects are evaluated. The use of a public database such as NINAPro enables the scientific community to evaluate the contribution of this innovation. It is important to note that proposed technic of adaptation can be a form of complementation with several other approaches such as user adaptation with online feedback [74].

The proposed system is a complementary technique to increase the clinical impact of the myoelectric prosthesis or other applications with the same stochastic behavior signals. For example, EEG for seizure detection, ECG for QRS detection or arrhythmia classification. However, the next step in this research will be the improvement of virtual sensor model adaptation for frequency features and consequently the classification results enhancement. In parallel, an adjustment of SFTD method should be developed for ECG and power line interference.

Author Contributions: Karina de O. A. de Moura conceived the methodology, as well as designed and performed the experiments, processed and analyzed the data, performed the statistical analysis. Alexandre Balbinot contributed to the concept and ideas, implementation of the proposed and checked the system methodology. All authors participated in the result analysis, and the paper revision.

Funding: This research supported by National Counsel of Technological and Scientific Development (CNPq).

Acknowledgments: This research was performed at the Electro-Electronic Instrumentation (IEE) laboratory of the Federal University of Rio Grande do Sul. The authors would like to thank all the collaboration of laboratory colleagues.

Conflicts of Interest: The authors declare no conflict of interest.

References

1. Deijs, M.; Bongers, R.M.; Ringeling-van Leusen, N.D.M.; van der Sluis, C.K. Flexible and static wrist units in upper limb prosthesis users: Functionality scores, user satisfaction and compensatory movements. *J. Neuroeng. Rehabilit.* **2016**, *13*, 26. [CrossRef] [PubMed]
2. Engdahl, S.M.; Christie, B.P.; Kelly, B.; Davis, A.; Chestek, C.A.; Gates, D.H. Surveying the interest of individuals with upper limb loss in novel prosthetic control techniques. *J. Neuroeng. Rehabilit.* **2015**, *12*, 53. [CrossRef] [PubMed]
3. Zhang, D.; Zhao, X.; Han, J.; Zhao, Y. A comparative study on PCA and LDA based EMG pattern recognition for anthropomorphic robotic hand. In Proceedings of the 2014 IEEE International Conference on Robotics and Automation (ICRA), Hong Kong, China, 31 May–7 June 2014; pp. 4850–4855. [CrossRef]
4. Blana, D.; Kyriacou, T.; Lambrecht, J.M.; Chadwick, E.K. Feasibility of using combined EMG and kinematic signals for prosthesis control: A simulation study using a virtual reality environment. *J. Electromyogr. Kinesiol.* **2016**, *29*, 21–27. [CrossRef] [PubMed]

5. Sensinger, J.W.; Lock, B.A.; Kuiken, T.A. Adaptive pattern recognition of myoelectric signals: Exploration of conceptual framework and practical algorithms. *IEEE Trans. Neural Syst. Rehabilt. Eng.* **2009**, *17*, 270–278. [CrossRef] [PubMed]
6. Liu, J. Adaptive myoelectric pattern recognition toward improved multifunctional prosthesis control. *Med. Eng. Phys.* **2015**, *37*, 424–430. [CrossRef] [PubMed]
7. Zhang, X.; Huang, H. A real-time, practical sensor fault-tolerant module for robust EMG pattern recognition *J. Neuroeng. Rehabilt.* **2015**, *12*, 18. [CrossRef] [PubMed]
8. Jiang, N.; Lorrain, T.; Farina, D. A state-based, proportional myoelectric control method: Online validation and comparison with the clinical state-of-the-art. *J. Neuroeng. Rehabilt.* **2014**, *11*, 110. [CrossRef] [PubMed]
9. Atzori, M.; Müller, H. Control Capabilities of Myoelectric Robotic Prostheses by Hand Amputees: A Scientific Research and Market Overview. *Front. Syst. Neurosci.* **2015**, *9*, 162. [CrossRef] [PubMed]
10. Spanias, J.; Perreault, E.; Hargrove, L. Detection of and Compensation for EMG Disturbances for Powered Lower Limb Prosthesis Control. *IEEE Trans. Neural Syst. Rehabilt. Eng.* **2015**, *4320*, 226–234. [CrossRef] [PubMed]
11. Fraser, G.D.; Chan, A.D.C.; Green, J.R.; MacIsaac, D.T. Automated Biosignal Quality Analysis for Electromyography Using a One-Class Support Vector Machine. *IEEE Trans. Instrum. Meas.* **2014**, *63*, 2919–2930. [CrossRef]
12. Zivanovic, M. Time-Varying Multicomponent Signal Modeling for Analysis of Surface EMG Data. *IEEE Signal Process. Lett.* **2014**, *21*, 692–696. [CrossRef]
13. Tomasini, M.; Benatti, S.; Milosevic, B.; Farella, E.; Benini, L. Power Line Interference Removal for High-Quality Continuous Biosignal Monitoring with Low-Power Wearable Devices. *IEEE Sens. J.* **2016**, *16*, 3887–3895. [CrossRef]
14. Mastinu, E.; Ahlberg, J.; Lendaro, E.; Hermansson, L.; Hakansson, B.; Ortiz-Catalan, M. An alternative myoelectric pattern recognition approach for the control of hand prostheses: A case study of use in daily life by a dysmelia subject. *IEEE J. Transl. Eng. Health Med.* **2018**, *6*. [CrossRef] [PubMed]
15. Zhang, X.; Huang, H.; Yang, Q. Real-time implementation of a self-recovery EMG pattern recognition interface for artificial arms. In Proceedings of the 2013 35th Annual International Conference of the IEEE Engineering in Medicine and Biology Society (EMBC), Osaka, Japan, 3–7 July 2013; pp. 5926–5929. [CrossRef]
16. Favieiro, G.; Cene, V.H.; Balbinot, A. Self-adaptive method for sEMG movement classification based on continuous optimal electrode assortment. *Braz. J. Instrum. Control* **2016**, *4*, 21–26. [CrossRef]
17. Chowdhury, R.H.; Reaz, M.B.; Ali, M.A.B.M.; Bakar, A.A.; Chellappan, K.; Chang, T.G. Surface Electromyography Signal Processing and Classification Techniques. *Sensors* **2013**, *13*, 12431–12466. [CrossRef] [PubMed]
18. Soedirdjo, S.D.H.; Ullah, K.; Merletti, R. Power line interference attenuation in multi-channel sEMG signals: Algorithms and analysis. In Proceedings of the 2015 37th Annual International Conference of the IEEE Engineering in Medicine and Biology Society (EMBC), Milan, Italy, 25–29 Augst 2015; pp. 3823–3826. [CrossRef]
19. Fraser, G.D.; Chan, A.D.C.; Green, J.R.; Abser, N.; MacIsaac, D. CleanEMG—Power line interference estimation in sEMG using an adaptive least squares algorithm. In Proceedings of the 2011 Annual International Conference of the IEEE Engineering in Medicine and Biology Society, Boston, MA, USA, 30 August–3 September 2011; pp. 7941–7944. [CrossRef]
20. Zhang, Y.; Su, S.; Xu, P.; Yao, D. Performance evaluation of Noise-Assisted Multivariate Empirical Mode Decomposition and its application to multichannel EMG signals. In Proceedings of the 2017 39th Annual International Conference of the IEEE Engineering in Medicine and Biology Society (EMBC), Seogwipo, Korea, 11–15 July 2017; pp. 3457–3460. [CrossRef]
21. Nazmi, N.; Abdul Rahman, M.; Yamamoto, S.; Ahmad, S.; Zamzuri, H.; Mazlan, S. A Review of Classification Techniques of EMG Signals during Isotonic and Isometric Contractions. *Sensors* **2016**, *16*, 1304. [CrossRef] [PubMed]
22. Chen, M.; Zhang, X.; Chen, X.; Zhu, M.; Li, G.; Zhou, P. FastICA peel-off for ECG interference removal from surface EMG. *Biomed. Eng. Online* **2016**, *15*, 65. [CrossRef] [PubMed]
23. Sánchez-Molina, J.A.; Rodríguez, F.; Guzmán, J.L.; Ramírez-arias, J.A. Water content virtual sensor for tomatoes in coconut coir substrate for irrigation control design. *Agric. Water Manag.* **2015**, *151*, 114–125. [CrossRef]

24. Ponsart, J.C.; Theilliol, D.; Aubrun, C. Virtual sensors design for active fault tolerant control system applied to a winding machine. *Control Eng. Pract.* **2010**, *18*, 1037–1044. [CrossRef]

25. Ploennigs, J.; Ahmed, A.; Hensel, B.; Stack, P.; Menzel, K. Virtual sensors for estimation of energy consumption and thermal comfort in buildings with underfloor heating. *Adv. Eng. Inform.* **2011**, *25*, 688–698. [CrossRef]

26. Fagiano, L.; Novara, C. A combined Moving Horizon and Direct Virtual Sensor approach for constrained nonlinear estimation. *Automatica* **2013**, *49*, 193–199. [CrossRef]

27. Liu, K.; Inoue, Y.; Shibata, K. Physical-Sensor and Virtual-Sensor Based Method for Estimation of Lower Limb Gait Posture Using Accelerometers and Gyroscopes. *J. Biomech. Sci. Eng.* **2010**, *5*, 472–483. [CrossRef]

28. Ibargüengoytia, P.H.; Delgadillo, M.A.; García, U.A.; Reyes, A. Viscosity virtual sensor to control combustion in fossil fuel power plants. *Eng. Appl. Artif. Intell.* **2013**, *26*, 2153–2163. [CrossRef]

29. Nazari, R.; Seron, M.M.; De Doná, J.A. Fault-tolerant control of systems with convex polytopic linear parameter varying model uncertainty using virtual-sensor-based controller reconfiguration. *Annu. Rev. Control* **2013**, *37*, 146–153. [CrossRef]

30. Nazari, R.; Seron, M.M.; Yetendje, A. Invariant-set-based fault tolerant control using virtual sensors. *IET Control Theory Appl.* **2011**, *5*, 1092–1103. [CrossRef]

31. Tabbache, B.; Benbouzid, M.E.H.; Kheloui, A.; Bourgeot, J.-M. Virtual-Sensor-Based Maximum-Likelihood Voting Approach for Fault-Tolerant Control of Electric Vehicle Powertrains. *IEEE Trans. Veh. Technol.* **2013**, *62*, 1075–1083. [CrossRef]

32. Ho, L.M.; Satzger, C.; de Castro, R. Fault-tolerant control of an electrohydraulic brake using virtual pressure sensor. In Proceedings of the 2017 International Conference on Robotics and Automation Sciences (ICRAS), Hong Kong, China, 26–29 August 2017; pp. 76–82. [CrossRef]

33. Raveendranathan, N.; Galzarano, S.; Loseu, V.; Gravina, R.; Giannantonio, R.; Sgroi, M.; Jafari, R.; Fortino, G. From Modeling to Implementation of Virtual Sensors in Body Sensor Networks. *IEEE Sens. J.* **2012**, *12*, 583–593. [CrossRef]

34. Crema, C.; Depari, A.; Flammini, A.; Sisinni, E.; Vezzoli, A.; Bellagente, P. Virtual Respiratory Rate Sensors: An Example of A Smartphone-Based Integrated and Multiparametric mHealth Gateway. *IEEE Trans. Instrum. Meas.* **2017**, *66*, 2456–2463. [CrossRef]

35. Li, Y.; Pandis, I.; Guo, Y. Enabling Virtual Sensing as a Service. *Informatics* **2016**, *3*, 3. [CrossRef]

36. McCool, P.; Fraser, G.D.; Chan, A.D.C.; Petropoulakis, L.; Soraghan, J.J. Identification of contaminant type in surface electromyography (EMG) signals. *IEEE Trans. Neural Syst. Rehabilit. Eng.* **2014**, *22*, 774–783. [CrossRef] [PubMed]

37. Ting, C.M.; Salleh, S.H.; Zainuddin, Z.M.; Bahar, A. Spectral estimation of nonstationary EEG using particle filtering with application to event-related desynchronization (ERD). *IEEE Trans. Biomed. Eng.* **2011**, *58*, 321–331. [CrossRef] [PubMed]

38. Kautz, T.; Eskofier, B. A Robust Kalman Framework with Resampling and Optimal Smoothing. *Sensors* **2015**, *15*, 4975–4995. [CrossRef] [PubMed]

39. Van de Walle, A.; Naets, F.; Desmet, W. Virtual microphone sensing through vibro-acoustic modelling and Kalman filtering. *Mech. Syst. Signal Process.* **2018**, *104*, 120–133. [CrossRef]

40. LeBreux, M.; Désilets, M.; Lacroix, M. Control of the ledge thickness in high-temperature metallurgical reactors using a virtual sensor. *Inverse Probl. Sci. Eng.* **2012**, *20*, 1215–1238. [CrossRef]

41. Halim, D.; Cheng, L.; Su, Z. Virtual sensors for active noise control in acoustic-structural coupled enclosures using structural sensing: Robust virtual sensor design. *J. Acoust. Soc. Am.* **2011**, *129*, 1390–1399. [CrossRef] [PubMed]

42. Cho, S.; Gao, Z.; Moan, T. Model-based fault detection, fault isolation and fault-tolerant control of a blade pitch system in floating wind turbines. *Renew. Energy* **2018**, *120*, 306–321. [CrossRef]

43. Ortiz-Catalan, M.; Rouhani, F.; Branemark, R.; Hakansson, B. Offline accuracy: A potentially misleading metric in myoelectric pattern recognition for prosthetic control. In Proceedings of the 2015 37th Annual International Conference of the IEEE Engineering in Medicine and Biology Society (EMBC), Milan, Italy, 25–29 August 2015; pp. 1140–1143. [CrossRef]

44. Vujaklija, I.; Shalchyan, V.; Kamavuako, E.N.; Jiang, N.; Marateb, H.R.; Farina, D. Online mapping of EMG signals into kinematics by autoencoding. *J. Neuroeng. Rehabil.* **2018**, *15*, 21. [CrossRef] [PubMed]

45. Vujaklija, I.; Roche, A.D.; Hasenoehrl, T.; Sturma, A.; Amsuess, S.; Farina, D.; Aszmann, O.C. Translating Research on Myoelectric Control into Clinics—Are the Performance Assessment Methods Adequate? *Front. Neurorobot.* **2017**, *11*, 7. [CrossRef] [PubMed]

46. Moura, K.O.A.; Favieiro, G.W.; Balbinot, A. Support vectors machine classification of surface electromyography for non-invasive naturally controlled hand prostheses. In Proceedings of the 2016 38th Annual International Conference of the IEEE Engineering in Medicine and Biology Society (EMBC), Orlando, FL, USA, 16–20 August 2016; pp. 788–791. [CrossRef]

47. Phinyomark, A.; Quaine, F.; Charbonnier, S.; Serviere, C.; Tarpin-Bernard, F.; Laurillau, Y. EMG feature evaluation for improving myoelectric pattern recognition robustness. *Expert Syst. Appl.* **2013**, *40*, 4832–4840. [CrossRef]

48. Gijsberts, A.; Atzori, M.; Castellini, C.; Müller, H.; Caputo, B. Movement error rate for evaluation of machine learning methods for sEMG-based hand movement classification. *IEEE Trans. Neural Syst. Rehabilit. Eng.* **2014**, *22*, 735–744. [CrossRef] [PubMed]

49. Atzori, M.; Gijsberts, A.; Castellini, C.; Caputo, B.; Hager, A.-G.M.; Elsig, S.; Giatsidis, G.; Bassetto, F.; Müller, H. Electromyography data for non-invasive naturally-controlled robotic hand prostheses. *Sci. Data* **2014**, *1*, 140053. [CrossRef] [PubMed]

50. Atzori, M.; Gijsberts, A.; Kuzborskij, I.; Elsig, S.; Mittaz Hager, A.-G.; Deriaz, O.; Castellini, C.; Muller, H.; Caputo, B.; Heynen, S.; et al. Characterization of a Benchmark Database for Myoelectric Movement Classification. *IEEE Trans. Neural Syst. Rehabilit. Eng.* **2015**, *23*, 73–83. [CrossRef] [PubMed]

51. Riillo, F.; Quitadamo, L.R.; Cavrini, F.; Gruppioni, E.; Pinto, C.A.; Pastò, N.C.; Sbernini, L.; Albero, L.; Saggio, G. Optimization of EMG-based hand gesture recognition: Supervised vs. unsupervised data preprocessing on healthy subjects and transradial amputees. *Biomed. Signal Process. Control* **2014**, *14*, 117–125. [CrossRef]

52. Farrell, T.R. Determining delay created by multifunctional prosthesis controllers. *J. Rehabilit. Res. Dev.* **2011**, *48*. [CrossRef]

53. Englehart, K.; Hudgins, B. A robust, real-time control scheme for multifunction myoelectric control. *IEEE Trans. Biomed. Eng.* **2003**, *50*, 848–854. [CrossRef] [PubMed]

54. Winkler, G.; Balbinot, A. Proposal of a Neuro Fuzzy System for Myoelectric Signal Analysis from Hand-Arm Segment. In *Computational Intelligence in Electromyography Analysis—A Perspective on Current Applications and Future Challenges*; IntechOpen: Rijeka, Croatia, 2012; pp. 337–362, ISBN 978-953-51-0805-4. [CrossRef]

55. Balbinot, A.; Júnior, A.; Favieiro, G.W. Decoding Arm Movements by Myoelectric Signal and Artificial Neural Networks. *Intell. Control Autom.* **2013**, 87–93. [CrossRef]

56. De la Rosa, R.; Alonso, A.; Carrera, A.; Durán, R.; Fernández, P. Man-machine interface system for neuromuscular training and evaluation based on EMG and MMG signals. *Sensors* **2010**, *10*, 11100–11125. [CrossRef] [PubMed]

57. Cene, V.H.; Favieiro, G.; Balbinot, A. Upper-limb movement classification based on sEMG signal validation with continuous channel selection. In Proceedings of the 2015 37th Annual International Conference of the IEEE Engineering in Medicine and Biology Society (EMBC), Milan, Italy, 25–29 August 2015; pp. 486–489.

58. Cene, V.H.; Balbinot, A. Optimization of Features to Classify Upper—Limb Movements Through sEMG Signal Processing. *Braz. J. Instrum. Control* **2016**, *4*, 14–20. [CrossRef]

59. Mogk, J.P.M.M.; Keir, P.J. Crosstalk in surface electromyography of the proximal forearm during gripping tasks. *J. Electromyogr. Kinesiol.* **2003**, *13*, 63–71. [CrossRef]

60. Farina, D.; Merletti, R.; Enola, R.M. The extraction of neural strategies from the surface EMG. *J. Appl. Physiol.* **2004**, *96*, 1486–1495. [CrossRef] [PubMed]

61. Shakoorjavan, S.; Akbari, S.; Kish, M.H.; Akbari, M. Correlation of sensory analysis with a virtual sensor array data for odour diagnosis of fragrant fabrics. *Measurement* **2016**, *90*, 396–403. [CrossRef]

62. Zou, R.; Chon, K.H. Robust algorithm for estimation of time-varying transfer functions. *IEEE Trans. Biomed. Eng.* **2004**, *51*, 219–228. [CrossRef] [PubMed]

63. Jachan, M.; Matz, G.; Member, S.; Hlawatsch, F.; Member, S. Time-frequency ARMA models and parameter estimators for underspread nonstationary random processes. *IEEE Trans. Signal Process.* **2007**, *55*, 4366–4381. [CrossRef]

64. Hudgins, B.; Parker, P.; Scott, R.N. A New Strategy for Multifunction Myoelectric Control. *IEEE Trans. Biomed. Eng.* **1993**, *40*, 82–94. [CrossRef] [PubMed]

65. Lin, S.-W.; Ying, K.-C.; Chen, S.-C.; Lee, Z.-J. Particle swarm optimization for parameter determination and feature selection of support vector machines. *Expert Syst. Appl.* **2008**, *35*, 1817–1824. [CrossRef]

66. Tosin, M.; Majolo, M.; Chedid, R.; Cene, V.H.; Balbinot, A. SEMG feature selection and classification using SVM-RFE. In Proceedings of the 2017 39th Annual International Conference of the IEEE Engineering in Medicine and Biology Society (EMBC), Seogwipo, Korea, 11–15 July 2017; pp. 390–393. [CrossRef]

67. Montgomery, D.C. *Design and Analysis of Experiments*, 5th ed.; John Wiley & Sons, Inc.: New York, NY, USA, 2001.

68. Dellacasa Bellingegni, A.; Gruppioni, E.; Colazzo, G.; Davalli, A.; Sacchetti, R.; Guglielmelli, E.; Zollo, L. NLR, MLP, SVM, and LDA: A comparative analysis on EMG data from people with trans-radial amputation. *J. Neuroeng. Rehabil.* **2017**, *14*, 82. [CrossRef] [PubMed]

69. Wei, Y.; Geng, Y.; Yu, W.; Samuel, O.W.; Jiang, N.; Zhou, H.; Guo, X.; Lu, X.; Li, G. Real-time Classification of Forearm Movements Based on High Density Surface Electromyography. In Proceedings of the 2017 IEEE International Conference on Real-Time Computing and Robotics (RCAR), Okinawa, Japan, 14–18 July 2017; pp. 246–251. [CrossRef]

70. Benatti, S.; Milosevic, B.; Farella, E.; Gruppioni, E.; Benini, L. A Prosthetic Hand Body Area Controller Based on Efficient Pattern Recognition Control Strategies. *Sensors* **2017**, *17*, 869. [CrossRef] [PubMed]

71. Al-Angari, H.M.; Kanitz, G.; Tarantino, S.; Cipriani, C. Distance and mutual information methods for EMG feature and channel subset selection for classification of hand movements. *Biomed. Signal Process. Control* **2016**, *27*, 24–31. [CrossRef]

72. Nilsson, N.; Håkansson, B.; Ortiz-Catalan, M. Classification complexity in myoelectric pattern recognition. *J. Neuroeng. Rehabil.* **2017**, *14*, 68. [CrossRef] [PubMed]

73. Favieiro, G.W.; Moura, K.O.A.; Balbinot, A. Novel method to characterize upper-limb movements based on paraconsistent logic and myoelectric signals. In Proceedings of the 2016 38th Annual International Conference of the IEEE Engineering in Medicine and Biology Society (EMBC), Orlando, FL, USA, 16–20 August 2016; pp. 395–398. [CrossRef]

74. Hahne, J.M.; Markovic, M.; Farina, D. User Adaptation in Myoelectric Man-Machine Interfaces. *Sci. Rep.* **2017**, *7*, 4437. [CrossRef] [PubMed]

sensors

⬡ MDPI

Article

Effects of tDCS on Real-Time BCI Detection of Pedaling Motor Imagery

Maria de la Soledad Rodriguez-Ugarte *, Eduardo Iáñez, Mario Ortiz-Garcia and José M. Azorín

Brain-Machine Interface Systems Lab, Miguel Hernández University of Elche,
Avda. de la Universidad S/N Ed. Innova, Elche, 03202 Alicante, Spain; eianez@umh.es (E.I.);
mortiz@umh.es (M.O.-G.); jm.azorin@umh.es (J.M.A.)
* Correspondence: maria.rodriguezu@umh.es; Tel.: +34-965-222-459

Received: 26 January 2018; Accepted: 5 April 2018; Published: 8 April 2018

Abstract: The purpose of this work is to strengthen the cortical excitability over the primary motor cortex (M1) and the cerebro-cerebellar pathway by means of a new transcranial direct current stimulation (tDCS) configuration to detect lower limb motor imagery (MI) in real time using two different cognitive neural states: relax and pedaling MI. The anode is located over the primary motor cortex in Cz, and the cathode over the right cerebro-cerebellum. The real-time brain–computer interface (BCI) designed is based on finding, for each electrode selected, the power at the particular frequency where the most difference between the two mental tasks is observed. Electroencephalographic (EEG) electrodes are placed over the brain's premotor area (PM), M1, supplementary motor area (SMA) and primary somatosensory cortex (S1). A single-blind study is carried out, where fourteen healthy subjects are separated into two groups: sham and active tDCS. Each subject is experimented on for five consecutive days. On all days, the results achieved by the active tDCS group were over 60% in real-time detection accuracy, with a five-day average of 62.6%. The sham group eventually reached those levels of accuracy, but it needed three days of training to do so.

Keywords: transcranial direct current stimulation (tDCS); brain–computer interface (BCI); real-time; pedaling motor imagery; cerebro-cerebellar pathway

1. Introduction

Transcranial direct current stimulation (tDCS) is a modern technique of non-invasive brain stimulation which has the purpose of temporally modulating cortical excitability [1,2]. Currently, its effects are not known with certainty, but they are believed to be dependent on several factors such as intensity applied [3], time of stimulation [4] and size of the electrodes used [5]. The majority of the studies focused their research on applying tDCS to the representation of the upper limbs in the brain to evaluate the performance of the subjects or to improve the quality of life of stroke patients who have had that area affected [6–8]. Only relatively few studies attempted to investigate how tDCS could affect the lower limbs [9,10]. This could be due to the challenge of reaching the area of the brain where the legs are represented, which is located deep in the longitudinal fissure corresponding to the primary motor cortex (M1).

From a cognitive perspective, brain activity during a lower limb complex motor task, such as gait or pedaling, involves the supplementary motor area (SMA), M1, the primary somatosensory cortex (S1) and the premotor area (PM) [11–14]. Moreover, lower limb motor imagery (MI) is also associated with these areas [15]. Hence, if a person imagines a complex motor task, the person will activate a similar neural pathway to that activated when the task is actually being performed. In addition, the cerebellum is a key part during movement coordination, motor learning and cognition [16]. The underlying mechanism of the ascending outputs from the cerebellum relies on sending information to M1 through the dentate nucleus. Some of the axons in this area cross the midline of the brain to terminate in the ventral lateral complex of the thalamus, and then the motor thalamus sends inputs to the M1 and PM areas [17].

On the one hand, research findings have found that tDCS over the cerebellum produces cortical excitability changes in a polarity-specific manner [18]. While cathodal tDCS over the cerebellum decreases the inhibitory tone the cerebellum exerts over M1, anodal tDCS has the opposite effect [19,20]. From a physiological perspective, the principal neuron found in the cortex of the cerebellum is called the Purkinje cell. If the anode is located over the cerebellum, these neurons are excited producing inhibition in the dentate nucleus and resulting in disfacilitation of the motor cortex [21]. On the other hand, cortical excitability over M1 increases when the anode is located over M1 and the cathode over the contralateral hemisphere, or over the contralateral supraorbital region [22,23]. Nevertheless, no research has studied the cerebro-cerebellar pathway where simultaneously the anode is located over M1 and the cathode over the contralateral cerebellum. Doing this could increase the cortical excitability over M1 even more.

Brain–computer interfaces (BCIs) are devices that translate brain waves into commands to control an external device, such as exoskeletons. They can do this, for example, by reading electroencephalographic (EEG) signals from the brain, extracting useful features from those signals, and then using statistical methods to discern between relevant outputs. This technique can improve the rehabilitation process of a person that has suffered a cerebrovascular accident (CVA). The most challenging aspect of using BCIs is to detect neural cognitive processes in real time, so that, as soon as data are received, they are processed. However, researchers usually analyze data offline, where data are studied once the experiment has finished [24,25]. This can produce unrealistic results when compared to a more challenging online analysis, which is more relevant for real-time applications such as rehabilitation therapies involving exoskeletons.

Motor imagery has been detected using EEG-based BCIs in the past, but most studies focused on upper limbs or simple foot movements [26–29]. Much fewer studies concentrated on lower limb complex tasks such as gait or pedaling [30]. In most of these studies, BCIs have exploited in some way the fact that there is a suppression of the mu waves (8–12 Hz) and beta waves (13–30 Hz) around M1 when a motor task is being imagined [31,32]. The literature involving real-time processing and feedback of BCI signals associated to these types of movements is scarcer, and the methods of reporting results are disperse [26,30,33–37]. Nevertheless, there are many relevant applications of detecting lower limb movement in real time. Indeed, in the long run, it would be desirable to design an online BCI where patients with CVA are rehabilitated with the aid of a lower limb exoskeleton which they are able to control in real time. Additionally, if the effects of tDCS prove to be positive (by exciting M1 and facilitating detection), this could help in improving or simply accelerating the recovery of those patients even more.

Thus, the aim of this work is to strengthen the cortical excitability over M1 and the cerebro-cerebellar pathway by means of a new tDCS configuration to better detect lower limb motor imagery in real time using an online BCI that distinguishes between two different cognitive neural states: relax and pedaling MI. To do that, a single-blind study is carried out where people are randomly divided into two groups, sham and active tDCS, and experimented for five consecutive days. The sham group received a fake stimulation and the active tDCS group was given 0.4 mA. Our hypothesis is that the active tDCS group would achieve better detection accuracy results than the sham group.

2. Materials and Methods

2.1. Subjects

Fourteen healthy subjects between 23 and 38 years old (26.8 ± 4.9) took part in this experiment (most of them were MSc students). There were twelve male participants and two female participants. All of them were right-footed. None of the subjects had any previous experience with BCIs or MI; they reported no neurological diseases; none of them were medicated; and they were not suffering the consequences of an intoxication during the time the experiments were carried out. Lastly, all participants gave written informed consent according to the Helsinki declaration. The Ethics Committee of the Office for Project Evaluations (Oficina Evaluadora de Proyectos: OEP) of the Miguel Hernández University of Elche (Spain) approved the study.

2.2. Experimental Protocol

This section explains the experimental protocol. Several studies which treat different problems such as phantom limb pain, Parkinson's disease or apraxia of speech after stroke, applied tDCS for five days and reported positive effects [38–40]. In addition, a study from [41] stated that the lasting effects of tDCS when it is applied for 15 min were up to 1.5 h. Therefore, taking into account these aspects, our stimulation protocol was established as five consecutive days (Monday to Friday) for 15 min to investigate if there was any improvement in developing pedaling MI.

The experiment consisted on recording the EEG signals (more details on Section 2.3) while the user was performing two mental tasks: relax and imagine. During the imagine task, subjects had to visualize a pedaling movement inside their heads. To remove the placebo effect, a single-blind study was designed in which subjects were randomly divided into two groups: sham or active tDCS. The participants sat in front of a screen which fed them with instructions. Each subject performed 1 session every day which consisted of tDCS supply and MI experiment. First, tDCS (sham or active) was administrated for 15 min (more details in Section 2.4). Then, each subject performed 10 trials of the MI experiment. Each trial included each task (relax and imagine) 10 times. The screen provided three types of instructions: *Relax*, *Imagine* and +. *Relax* and *Imagine* tasks lasted 5.8 s and the order appeared at random, but in such a way that no same task appeared more than two times consecutively. This was done to avoid the user to start an expected task beforehand. The symbol + was always shown between tasks and lasted 3 s. During *Relax* and *Imagine*, the subjects were told to avoid blinking, swallowing or any other kind of artifacts. They were told to postpone these until the + symbol appeared. Figure 1 shows the flow diagram of each session's experimental protocol, while Figure 2 shows the experimental setup.

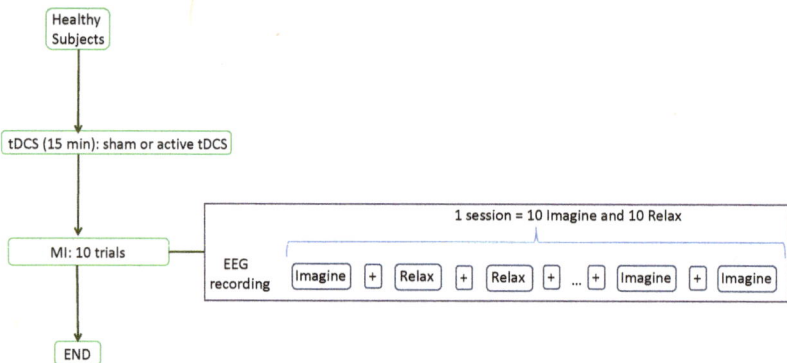

Figure 1. Flow diagram of the experiment for healthy subjects. The subjects were instructed by the screen to perform one of two possible mental tasks: *Relax* or *Imagine*. During *Relax*, subjects had to try not to think about anything, while, during *Imagine*, they had to imagine themselves pedaling. The *Relax* and *Imagine* tasks appeared at random and were always separated by an intermediate period (indicated by the screen with a + symbol). The setup also prevented two tasks of the same type to appear more than two times consecutively.

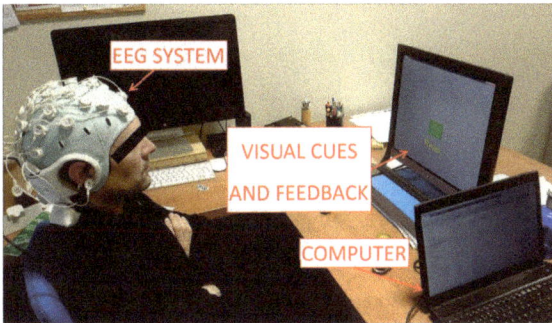

Figure 2. Experimental setup. Subjects sat looking at a screen which fed them with instructions while their EEG signals were recorded. Furthermore, the screen gave feedback about their performance in each task. The participant in the picture gave written informed consent to publish the image.

The first 4 trials were used to train a SVM classifier with which an online BCI was designed. This is explained in Section 2.5. For the remaining trials, the users received real-time positive feedback about their performance using the output from the BCI. That is, if during the relax task, the BCI detected that the subject had executed mental relaxation, then a green bar increased in size (otherwise it stayed the same size); and similarly with the pedaling MI task. The detection accuracy was calculated for each session, but this information was withheld from the subjects until the end of the last day to avoid influencing them.

2.3. EEG Acquisition

The StarStim R32 (Neuroelectrics, Barcelona, Spain) was used to acquire signals from the brain. The device was connected through a USB isolator to the computer. Based on the International 10-10 system, the EEG signals were acquired from 30 channels (P7, P4, CZ, PZ, P3, P8, O1, O2, C2, C4, F4, FP2, FZ, C3, F3, FP1, C1, OZ, PO4, FC6, FC2, AF4, CP6, CP2, CP1, CP5, FC1, FC5, AF3, and PO3) with two reference electrodes (CMS and DRL) at a frequency rate of 500 Hz. The system is shown in Figure 2.

2.4. Supply of tDCS

As mentioned before, the idea was to stimulate the cerebro-cerebellar pathway. To do this, a novel montage which aimed at strengthening the neural activity in M1 was proposed. It involved placing the anode over the primary motor cortex in Cz and the cathode over the right cerebro-cerebellum (two centimeters right and one centimeter down of the inion).

To corroborate that the cerebro-cerebellar pathway was being stimulated with such a choice of electrode placement, an electric field simulation of the brain was performed first. SimNIBS free platform [42] was used for the simulation, and Figure 3 shows the electric field generated by the anode over Cz (M1) and the cathode over the right cerebro-cerebellum. The parameters were set according to the materials utilized in the experiments. Both electrodes had a radius of 1 cm, 3 mm of thickness and 4 mm of space for the conductive gel. The tDCS intensity chosen was 0.4 mA, which produced 0.127 mA/cm^2 of current density. This current density was higher than in most studies (roughly 0.06 mA/cm^2) and it was selected because a previous study reported that a current density of 0.06 mA/cm^2 was not sufficient to reach the representation of the legs in the brain [43]. The current density also lies inside the range of neurological safety that avoids brain damage [44]. In Figure 3 it can be seen that the most affected area is close to the red nucleus and the thalamus. Both areas belong to the pathway of the ascending outputs from the cerebellum to M1 and PM [45], and therefore we expect this configuration to enhance the excitability in the area of interest.

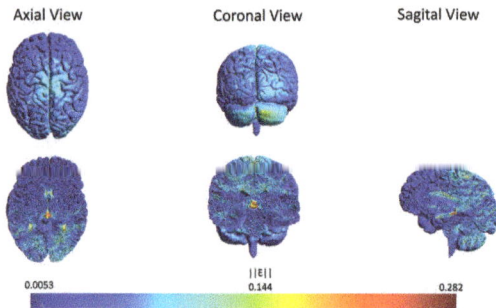

Figure 3. Axial, coronal and sagital view of the tDCS simulation. The scale represents the electric field (V/m) induced by the anode located over Cz and cathode over the right cerebro-cerebellum. The intensity applied was 0.4 mA. The most affected area (red) is close to the red nucleus. The image was generated with SimNIBS.

For the actual experiment, the StartStim R32 supplied anodal tDCS for 15 min at the beginning of each session (one session per day for five consecutive days) through two gel electrodes with a surface area of π cm^2 (1 cm radius). To create a placebo effect, the sham group received a 3 s ramp up until the intensity chosen, followed by 3 s ramp down; then, no stimulation was provided for almost 15 min until again there was a 3 s ramp up followed by a ramp down. Meanwhile, the active tDCS group received a 3 s ramp up until the intensity chosen, followed by constant stimulation throughout 15 min, and finally a 3 s ramp down.

2.5. Brain–Computer Interface (BCI)

As mentioned before, EEG signals were obtained as the subjects performed their relax and pedaling MI tasks. The first two seconds of each task were not considered to avoid influence of the visual cue and assure the total concentration of the subject in the respective task. Signals were processed in 1 s epochs with a 200 ms shift. For each epoch, a 4th order Butterworth high-pass filter with a cut-off frequency of 0.05 Hz was applied to remove the direct current. Then, a Notch filter was used to eliminate the power line interference at 50 Hz. Afterward, a 4th order Butterworth low-pass filter with cut-off frequency of 45 Hz was utilized. Subsequently, based on previous work (e.g., [46,47]), a Laplacian spacial filter was employed as in [48]. This filter eliminates the influence of the other electrodes by means of weighting by their distance. Of these filtered EEG signals, only those coming from nine carefully selected electrodes were considered: Cz, CP1, CP2, C1, C2, C3, C4, FC1 and FC2. These were chosen because the task involved imagination of the lower limbs, so their proximity to the M1, S1, SMA and PM regions of the brain was a deciding factor.

As mentioned above, the first four trials were used to train a support vector machine (SVM) classifier. This classifier is based on hyperplane tasks separation by maximizing the margin between the nearest points of the different tasks [49], with the outcomes obtained using non linear kernels being generally more robust than those of other classifiers [50]. In this work, a radial basis function was used as kernel for the SVM. For every given electrode, the power at each frequency between 6 and 30 Hz (resolution of 1 Hz via Burg's method) was calculated for each epoch. Then, the powers were separated according to the task (relax or pedaling imagery), normalized and averaged across all task-related epochs of the first four trials. Then, for each electrode, the frequency for which the maximum (normalized) power difference between tasks occurred was chosen and designated as the electrode's optimal frequency. Lastly, for each epoch, the feature associated to each electrode was the power at the electrode's optimal frequency, for a total of nine features per epoch. These features were then used to train the SVM classifier.

Therefore, the online BCI designed consisted of filtering the EEG signals of each epoch as described above, finding the nine features (the powers at each electrode's optimal frequency), and classifying the features with the (already trained) SVM. Thus, for each epoch, the BCI predicted whether it corresponded to a relaxed or pedaling state, and it was able to do this in real time (making a prediction every 0.2 s). The remaining six trials were utilized to determine the performance of the user in the day's session by measuring the real-time detection accuracy of the online BCI. The real-time detection accuracy was defined as the percentage of total correct classifications divided by the total number of classifications. As mentioned before, real-time positive feedback was given to the user, so that, if the BCI detected a relaxed state while the screen requested *Relax* (a correct classification), a green bar increased in size (otherwise it did not move), and similarly with the pedaling MI task.

3. Results

3.1. Statistical Analysis

IBM SPSS Statistics 22.0 for Windows (SPSS Inc., Chicago, IL, USA) was used for statistical analysis. First, we wanted to examine the differences in performance between groups (sham and active tDCS). Moreover, we wanted to study, within subjects of each group, the evolution of their performance throughout the five days of the experiment, which we refer to here as plasticity. Therefore, there were two independent variables: group and days; and only one dependent variable: real-time detection accuracy. Thus, a mixed factorial ANOVA was applied, but only after a Mauchly's test of sphericity was completed to verify the equality of variances of the differences within subjects [51]. In addition, pairwise comparisons between groups for each day, and within subjects of each group between days were computed. For every analysis, a *p*-value less than 0.05 was considered statistically significant.

Table 1 shows the results of applying Mauchly's test of sphericity. As it can be seen, variances were significantly different ($p < 0.05$), so data violated the sphericity assumption. Consequently, the correction with the biggest power was applied. In this case, it corresponded to Hyunh–Feldt ($\hat{e} = 0.987$). After applying this correction, a mixed factorial ANOVA was calculated.

Table 1. Mauchly's test of sphericity. Within subjects effect.

	Mauchly's W	df	*p*-Value	Greenhouse-Geisser	Hyunh-Feldt	Lower-Bound
				Epsilon		
days	0.09	9	0.003	0.688	0.987	0.25

3.1.1. Effects of tDCS in MI

Table 2 shows the five-day mean real-time detection accuracies for each subject along with the overall average of the sham and active tDCS groups. In addition, from the mixed factorial ANOVA we obtained that the effects of tDCS in MI were not significant: $F(1, 12) = 0.37$, $p > 0.05$, $r = 0.03$. Moreover, Table 3 shows the comparisons, with Bonferroni adjustment applied for multiple comparisons, between both groups for each day. It can be appreciated that for the first day there was a significant difference ($p < 0.05$) in the real-time detection accuracies (see also Figure 4).

Table 2. Mean real-time detection accuracy.

Subject	Sham	tDCS
1	61.7	66.6
2	66.9	51.8
3	59.6	55.7
4	64.1	55.9
5	51.5	66.9
6	55.2	68.7
7	63.5	72.4
Mean	60.4 ± 5.4	62.6 ± 7.9

Table 3. Pairwise accuracy comparison between tDCS and sham group.

Day	1	2	3	4	5
p-Value	0.04	0.29	1.00	0.74	0.60

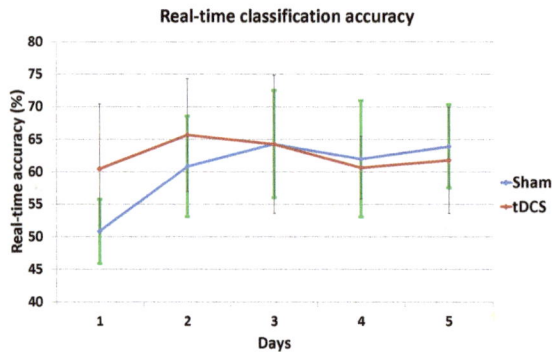

Figure 4. Mean real-time accuracy for all subjects of each group at each day.

3.1.2. MI Plasticity

Figure 4 represents the mean real-time accuracy for each group at each day of the experiment. From the mixed factorial ANOVA, it can be concluded that there was a significant interaction effect between the days and the group of stimulation: $F(3.95, 47.35) = 3.56$, $p < 0.01$, $r = 0.23$. Furthermore, Table 4 shows, for each group, the *p*-values comparing day five and the other days. There was only a significant difference between Day 5 and Day 1 within subjects of the sham group ($p < 0.01$).

Table 4. Comparison between Day 5 and the rest of the days for each group.

Group	Day	Day	*p*-Value
sham	5	1	0.002
		2	1.00
		3	1.00
		4	1.00
tDCS	5	1	1.00
		2	0.78
		3	0.85
		4	1.00

3.2. Optimal Frequencies

The optimal frequencies associated to the BCI model at each electrode on each day are also very useful information. They show where the greatest (normalized) changes in power were occurring, and therefore give a rough idea of the frequency bands that are most important in association with the lower limb motor imagery that is being studied. To present the results of all subjects together, a histogram was made showing, for each day and group, the number of optimal frequencies lying in three important frequency bands: 6–12 Hz (high theta and mu waves), 13–20 Hz (low and mid-range beta waves) and 21–30 Hz (high beta waves). For each group and day, there were a total of 63 optimal frequencies since there were nine electrodes selected for each of the seven subjects in each group. The results are presented in Table 5. The frequency band associated to mu waves seems to be the most preferred.

Table 5. Optimal frequencies histogram for each day and group.

Group	Frequency Range	Day 1	Day 2	Day 3	Day 4	Day 5
sham	(6–12) Hz	27	42	52	36	39
	(13–20) Hz	14	8	10	18	5
	(21–30) Hz	22	13	1	9	19
tDCS	(6–12) Hz	42	48	53	49	47
	(13–20) Hz	11	10	5	11	6
	(21–30) Hz	10	5	5	3	10

3.3. Real-Time Accuracy and ERD of the Best Subjects

As previously mentioned, lower limb motor imagery is thought to be associated to the attenuation of mu and beta waves in M1 [31,32]. This phenomenon is referred to as event-related desynchronization (ERD). To see the changes in ERD, the best subjects of each group were selected based on their five-day real-time detection accuracy (Table 2): Subject 2 of the sham group and Subject 7 of the active tDCS group. Given the results in the previous section and that those electrodes over M1 are thought to be mostly involved, the focus was on the mu waves (8–12 Hz) occurring in the Cz, C1, C2, C3 and C4 electrodes. For an electrode E, and for a fixed frequency f, the ERD was defined as

$$\text{ERD}_E(f) = \left(\frac{P(f) - R(f)}{R(f)}\right) \times 100, \tag{1}$$

where $P(f)$ is the average of the power at the frequency f over all pedaling-epochs, and $R(f)$ is the same but averaged over all relaxing-epochs. Then, the mu band motor cortex ERD for a given day was simply the average of all $\text{ERD}_E(f)$ over $f = 8, 9, 10, 11, 12$ and $E = \text{Cz,C1,C2,C3,C4}$. These results, along with the real-time accuracies of the two best subjects, are shown in Figure 5.

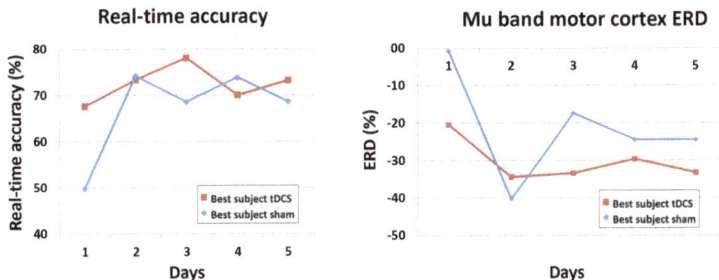

Figure 5. Real-time accuracy and ERD of the best subjects in each group.

4. Discussion

It can be seen in Table 2, as well as concluded from the mixed factorial ANOVA test, that in general there is no significant difference between the active tDCS and sham groups. Indeed, the active tDCS group achieved 62.6% of real-time detection accuracy and the sham group 60.4%. Nevertheless, the mixed factorial ANOVA results also indicated that there was a significant interaction effect between the days and the groups. This is because the sham and tDCS groups differ significantly in the first day, with the tDCS group having 9.6% better real-time accuracy, and also because within the sham group there is a significant variation when comparing the first and last day of the experiment.

Thus, the results show that the positive effects in performance due to tDCS are only relevant in the first few days, possibly only the first day, as the sham group then adapts and achieves the same performance. This is consistent with a study performed by Fernandez et al. [10] where it was observed that adaptation reached a lower limit the first day due to the simplicity of the task, leading to non-significant differences in the days that followed. However, it contradicts a study by Soekadar et al. [20] on upper limbs which suggested that only after three days did changes due to tDCS started to be differentiated from sham. This could be due to several reasons, including the different location of the tDCS electrodes and their surface area, as well as the different nature of the experiment and the way the data were processed. Longer experiment durations (over five days) could also help to discern the root cause of these slight differences. Meanwhile, the results agree in part with those of Wei et al. [8], which only had a one-day experiment showing slight improvement due to tDCS stimulation. Our results also show such slight improvement in favor of the tDCS group on the first day. Naturally, our real-time accuracy results differed from the offline accuracies shown in the study by Wei et al. [8], since they were able to improve on their classifier offline.

From another point of view, the subjects without stimulation showed evidence of brain plasticity, with an overall improvement of 13% in real-time accuracy from the first to the last day. Hence, their brains seem to have adapted very quickly to the task, meaning that if the intention is to eventually develop a therapy that elapses over several days, perhaps it is not necessary to apply tDCS at all. Having said that, tDCS did show evidence of speeding up improvement in the sense that it seems to have had an instantaneous effect in activating the desired neural pathway. Therefore, the results suggest that the active tDCS immediately induces the maximum performance that a subject could reach and it maintains it each day. Meanwhile, the sham group seems to require two to three days of training to reach the same level as the active tDCS group (see Figure 4). Nevertheless, it should be pointed out that these results involved only healthy subjects. When dealing with patients, we expect to see greater differences between the groups due to the greater potential of improvements in the case of rehabilitating patients.

These conclusions are also corroborated with results in Section 3.2. Indeed, looking at the histogram in Section 5 shows that the optimal frequencies lied the most often in the band containing the mu waves: 6–12 Hz. For the tDCS group on every single day at least 66% of the optimal frequencies lied in this band, while for the sham group from the second day onwards at least 57% of the optimal frequencies were in that band. However, on the first day, only 42% of the optimal frequencies of the sham group where in the preferred frequency band. This seems to indicate that the behavior was more disperse among the frequency bands, and could be a contributing factor explaining why the performance of the sham group in the first day was at chance level, while the tDCS group was already performing better on that day. Namely, it is possible that tDCS favored changes in the mu waves, whereas it took at least one day of training for the sham group to focus the motor imagery on that frequency band.

The results from Section 3.3 also confirm the conclusions, albeit at the level of the best subjects in each group. The best subject in the tDCS group started out with real-time accuracies of nearly 70% and remained above 70% from the second day onwards. On the other hand, the best subject in the sham group started at chance level (50%) and from the second day onwards significantly improved and remained at around 70% accuracy. The mu band motor cortex ERD (8–12 Hz in Cz, C1, C2, C3 and

C4) was also interesting. First, as expected, there was presence of ERD each day and for both subjects (negative values because the suppression means that the power while pedaling is lower than when relaxing). On the first day, the sham subject only showed a very subtle ERD, while the tDCS subject had a much more pronounced ERD of around −20%. Then, from the second day onwards, there was an even larger enhancement of the ERD levels of both groups, which remained at an average of about −30%. This shows some level of adaptation of the subjects to the task after the first day.

To address the seemingly low real-time accuracies of around 60% (chance level is 50%), we looked thoroughly at the existing literature to make the appropriate comparisons. It should be noted that the real-time accuracies reported in this study are taking into consideration every single prediction during the experiment (every single epoch is classified). Unfortunately, the current literature on real-time BCIs is somewhat disperse in the way the results are reported [26,30,33–37], but whenever it is possible to compare, our results coincide rather well with those of the literature. Our results are consistent with those of Zich et al. [33] which have a real-time accuracy of 55–65%, [26] with around 65% accuracy (of first 30 sessions), and [30] with around 65% accuracy. Meanwhile, Guger et al. [34] reported the results of the best time point of the best session of each of their three subjects (98%, 93% and 87%), but a careful analysis of their data shows the average real-time classification accuracy is about 80%, 65% and 65% for their three subjects respectively. Prasad et al. [35] averaged the maximum classification accuracies of each task and report them to be 60–75%. All of these studies involve upper limb or simple foot movements with the exception of the study by Liu et al. [30], which involved gait. Therefore, our results actually do not stray far from those found in the literature. Additionally, as pointed out by Prasad et al. [35], these results are reasonable given the fact that all the subjects are novices to BCIs and MI, so their performance is lower than that of experienced users. To improve on the accuracy levels, it might be necessary to change the nature of the motor imagery, since [26] reported significantly better results when doing so. Lastly, in other studies, it was simply not possible to make a fair comparison of the online results [8,36,37].

If the intention is to justify the use of active tDCS over the course of several days or more, then stronger evidence of its effects is needed. In this sense, it could be sensible to change the stimulation montage to one that could possibly lead to more marked differences among the groups. Some possible modifications of the experimental setup would be the number of stimulation anodes and cathodes and their placement, as well as increasing the intensity used whilst keeping safety in mind. From a physiological perspective, we first proposed that the excitation of Purkinje cells in the cerebellum might have the side effect of disfacilitating the motor cortex. However, there is also evidence that their activation can lead to improved motor learning [52]. In fact, anodal stimulation over the cerebellum can speed up learning [19,53,54]. Therefore, we propose an alternative for future use, where two anodes with differing intensities are utilized: one over M1 with relatively high intensity, and another over the cerebro-cerebellum with lower intensity (to prevent any major inhibitory behavior over M1 while still leading to improved motor learning). Meanwhile, a single cathode can be placed in an alternative location (such as FC1 or FC2).

Lastly, it should be said that the online BCI is perfectly apt for use in real-time applications, such as active therapies involving exoskeletons. In the future, we intend to use active tDCS and the online BCI to treat patients that have suffered a CVA accident. The idea is to improve their rehabilitation by engaging them in therapy where they have to control a lower limb exoskeleton in real time.

5. Conclusions

In this work, a new tDCS configuration intended to boost the cerebro-cerebellar pathway to improve the detection of lower limb MI via the use of a real-time BCI is tested. One anode is located over M1 and one cathode over the right cerebro-cerebellum. A single-blind experiment with duration of five days is completed using healthy subjects who are randomly separated into two groups: sham and active tDCS. The mental tasks they have to perform are: relax and pedaling MI. The online BCI designed is based on finding the power at an optimal frequency at each of nine carefully selected electrodes in the proximity of M1, S1, PM and SMA. From the very first day, the real-time detection accuracy achieved by the active tDCS group is over 60% and remains around 62.6% on average. However, the sham group needs three days of training to reach that same level of accuracy. This, along with other supporting evidence, indicates possibly that the tDCS has an immediate effect in activating the desired neural pathway, and shows the potential advantages in accelerating recovery of patients undergoing therapy. However, overall, the long term effects of tDCS seems to have been moderate at best. With this in mind, the stimulation montage could possibly be further improved to increase the effects of tDCS and hopefully justify its use. Lastly, the online BCI designed, with or without tDCS, is a desirable stepping stone in designing therapies that allow recovering patients' real-time control of lower limb exoskeletons, which is a future endeavor of interest.

Acknowledgments: This research was carried out in the framework of the project Associate titled: Decoding and stimulation of motor and sensory brain activity to support long term potentiation through Hebbian and paired associative stimulation during rehabilitation of gait (DPI2014-58431-C4-2-R), which is funded by the Spanish Ministry of Economy and Competitiveness and by the European Union through the European Regional Development Fund (ERDF), "A way to build Europe". M. Rodríguez-Ugarte wishes to thank Federico Fuentes for useful discussions and the careful revision of the manuscript. Lastly, the authors wish to thank the reviewers for the helpful comments that significantly improved the quality of the manuscript.

Author Contributions: M.d.l.S.R.-U. is responsible for the design, implementation, acquisition and data analysis. In addition, E.I. and M.O.-G. supervised the work and contributed with the revision process. J.M.A. actively contributed as director of the work.

Conflicts of Interest: The authors declare no conflict of interest.

Abbreviations

The following abbreviations are used in this manuscript:

CVA	Cerebrovascular accident
MI	Motor imagery
tDCS	Transcranial direct current stimulation
BCI	Brain–computer interface
M1	Primary motor cortex
S1	Primary somatosensory cortex
SMA	Supplementary motor area
PM	Premotor
EEG	Electroencephalographic
SVM	Support vector machine
ERD	Event-related desynchronization

References

1. Kumru, H.; Murillo, N.; Benito-Penalva, J.; Tormos, J.M.; Vidal, J. Transcranial direct current stimulation is not effective in the motor strength and gait recovery following motor incomplete spinal cord injury during Lokomat® gait training. *Neurosci. Lett.* **2016**, *620*, 143–147.
2. Flöel, A. tDCS-enhanced motor and cognitive function in neurological diseases. *Neuroimage* **2014**, *85*, 934–947.
3. Chieffo, R.; Ciocca, M.; Leocani, L.; Miranda, P.; Rothwell, J. 113. Short-term effect of different tdcs intensities on motor cortex excitability. *Clin. Neurophysiol.* **2015**, *126*, e26.
4. Batsikadze, G.; Moliadze, V.; Paulus, W.; Kuo, M.F.; Nitsche, M. Partially non-linear stimulation intensity-dependent effects of direct current stimulation on motor cortex excitability in humans. *J. Physiol.* **2013**, *591*, 1987–2000.
5. Bai, S.; Dokos, S.; Ho, K.A.; Loo, C. A computational modelling study of transcranial direct current stimulation montages used in depression. *Neuroimage* **2014**, *87*, 332–344.
6. Lee, S.J.; Chun, M.H. Combination transcranial direct current stimulation and virtual reality therapy for upper extremity training in patients with subacute stroke. *Arch. Phys. Med. Rehabil.* **2014**, *95*, 431–438.

7. Achilles, E.; Weiss-Blankenhorn, P.; Moos, K.; Hesse, M.; Sparing, R.; Fink, G. P649: Transcranial direct current stimulation (tDCS) of left parietal cortex facilitates gesture processing in healthy subjects. *Clin. Neurophysiol.* **2014**, *125*, S226–S227.

8. Wei, P.; He, W.; Zhou, Y.; Wang, L. Performance of motor imagery brain-computer interface based on anodal transcranial direct current stimulation modulation. *IEEE Trans. Neural Syst. Rehabil. Eng.* **2013**, *21*, 404–415.

9. Shah, B.; Nguyen, T.T.; Madhavan, S. Polarity independent effects of cerebellar tDCS on short term ankle visuomotor learning. *Brain Stimul.* **2013**, *6*, 966–968.

10. Fernandez, L.; Albein-Urios, N.; Kirkovski, M.; McGinley, J.L.; Murphy, A.T.; Hyde, C.; Stokes, M.A.; Rinehart, N.J.; Enticott, P.G. Cathodal transcranial direct current stimulation (tDCS) to the right cerebellar hemisphere affects motor adaptation during gait. *Cerebellum* **2017**, *16*, 168–177.

11. Cevallos, C.; Zarka, D.; Hoellinger, T.; Leroy, A.; Dan, B.; Cheron, G. Oscillations in the human brain during walking execution, imagination and observation. *Neuropsychologia* **2015**, *79*, 223–232.

12. Sahyoun, C.; Floyer-Lea, A.; Johansen-Berg, H.; Matthews, P. Towards an understanding of gait control: Brain activation during the anticipation, preparation and execution of foot movements. *Neuroimage* **2004**, *21*, 568–575.

13. Fukuyama, H.; Ouchi, Y.; Matsuzaki, S.; Nagahama, Y.; Yamauchi, H.; Ogawa, M.; Kimura, J.; Shibasaki, H. Brain functional activity during gait in normal subjects: A SPECT study. *Neurosci. Lett.* **1997**, *228*, 183–186.

14. Solodkin, A.; Hlustik, P.; Chen, E.E.; Small, S.L. Fine modulation in network activation during motor execution and motor imagery. *Cerebral Cortex* **2004**, *14*, 1246–1255.

15. Parsons, L.M.; Fox, P.T.; Downs, J.H.; Glass, T.; Hirsch, T.B.; Martin, C.C.; Jerabek, P.A.; Lancaster, J.L. Use of implicit motor imagery for visual shape discrimination as revealed by PET. *Nature* **1995**, *375*, 54–58.

16. D'Angelo, E.; Casali, S. Seeking a unified framework for cerebellar function and dysfunction: from circuit operations to cognition. *Front. Neural Circ.* **2012**, *6*, doi:10.3389/fncir.2012.00116.

17. Penhune, V.B.; Steele, C.J. Parallel contributions of cerebellar, striatal and M1 mechanisms to motor sequence learning. *Behav. Brain Res.* **2012**, *226*, 579–591.

18. Galea, J.M.; Jayaram, G.; Ajagbe, L.; Celnik, P. Modulation of cerebellar excitability by polarity-specific noninvasive direct current stimulation. *J. Neurosci.* **2009**, *29*, 9115–9122.

19. Block, H.J.; Celnik, P. Can cerebellar transcranial direct current stimulation become a valuable neurorehabilitation intervention? *Expert Rev. Neurother.* **2012**, *12*, 1275–1277.

20. Soekadar, S.R.; Witkowski, M.; Birbaumer, N.; Cohen, L.G. Enhancing Hebbian learning to control brain oscillatory activity. *Cerebral Cortex* **2014**, *25*, 2409–2415.

21. Cengiz, B.; Boran, H.E. The role of the cerebellum in motor imagery. *Neurosci. Lett.* **2016**, *617*, 156–159.

22. Wagner, T.; Fregni, F.; Fecteau, S.; Grodzinsky, A.; Zahn, M.; Pascual-Leone, A. Transcranial direct current stimulation: A computer-based human model study. *Neuroimage* **2007**, *35*, 1113–1124.

23. Sehm, B.; Kipping, J.; Schäfer, A.; Villringer, A.; Ragert, P. A comparison between uni-and bilateral tDCS effects on functional connectivity of the human motor cortex. *Front. Hum. Neurosci.* **2013**, *7*, doi:10.3389/fnhum.2013.00183.

24. Pfurtscheller, G.; Brunner, C.; Schlögl, A.; Da Silva, F.L. Mu rhythm (de) synchronization and EEG single-trial classification of different motor imagery tasks. *NeuroImage* **2006**, *31*, 153–159.

25. Lew, E.; Chavarriaga, R.; Silvoni, S.; Millán, J.d.R. Detection of self-paced reaching movement intention from EEG signals. *Front. Neuroeng.* **2012**, *5*, doi:10.3389/fneng.2012.00013.

26. Pfurtscheller, G.; Neuper, C. Motor imagery and direct brain-computer communication. *Proc. IEEE* **2001**, *89*, 1123–1134.

27. Cincotti, F.; Mattia, D.; Aloise, F.; Bufalari, S.; Schalk, G.; Oriolo, G.; Cherubini, A.; Marciani, M.G.; Babiloni, F. Non-invasive brain–computer interface system: Towards its application as assistive technology. *Brain Res. Bull.* **2008**, *75*, 796–803.

28. Müller-Putz, G.R.; Kaiser, V.; Solis-Escalante, T.; Pfurtscheller, G. Fast set-up asynchronous brain-switch based on detection of foot motor imagery in 1-channel EEG. *Med. Biol. Eng. Comput.* **2010**, *48*, 229–233.

29. Ang, K.K.; Guan, C.; Chua, K.S.G.; Ang, B.T.; Kuah, C.W.K.; Wang, C.; Phua, K.S.; Chin, Z.Y.; Zhang, H. A large clinical study on the ability of stroke patients to use an EEG-based motor imagery brain-computer interface. *Clin. EEG Neurosci.* **2011**, *42*, 253–258.

30. Liu, D.; Chen, W.; Lee, K.; Pei, Z.; Millán, J.d.R. An EEG-based brain-computer interface for gait training. In Proceedings of the 2017 29th Chinese Control And Decision Conference (CCDC), Chongqing, China, 28–30 May 2017; pp. 6755–6760.

31. Pfurtscheller, G.; Neuper, C. Motor imagery activates primary sensorimotor area in humans. *Neurosci. Lett.* **1997**, *239*, 65–68.

32. Naros, G.; Naros, I.; Grimm, F.; Ziemann, U.; Gharabaghi, A. Reinforcement learning of self regulated sensorimotor β-oscillations improves motor performance. *Neuroimage* **2016**, *134*, 142–152.

33. Zich, C.; Debener, S.; Kranczioch, C.; Bleichner, M.G.; Gutberlet, I.; De Vos, M. Real-time EEG feedback during simultaneous EEG–fMRI identifies the cortical signature of motor imagery. *Neuroimage* **2015**, *114*, 438–447.

34. Guger, C.; Ramoser, H.; Pfurtscheller, G. Real-time EEG analysis with subject-specific spatial patterns for a brain-computer interface (BCI). *IEEE Trans. Rehabil. Eng.* **2000**, *8*, 447–456.

35. Prasad, G.; Herman, P.; Coyle, D.; McDonough, S.; Crosbie, J. Applying a brain-computer interface to support motor imagery practice in people with stroke for upper limb recovery: A feasibility study. *J. Neuroeng. Rehabil.* **2010**, *7*, 60.

36. Yu, T.; Xiao, J.; Wang, F.; Zhang, R.; Gu, Z.; Cichocki, A.; Li, Y. Enhanced motor imagery training using a hybrid BCI with feedback. *IEEE Trans. Biomed. Eng.* **2015**, *62*, 1706–1717.

37. Horki, P.; Solis-Escalante, T.; Neuper, C.; Müller-Putz, G. Combined motor imagery and SSVEP based BCI control of a 2 DoF artificial upper limb. *Med. Biol. Eng. Comput.* **2011**, *49*, 567–577.

38. Bolognini, N.; Spandri, V.; Ferraro, F.; Salmaggi, A.; Molinari, A.C.; Fregni, F.; Maravita, A. Immediate and sustained effects of 5-day transcranial direct current stimulation of the motor cortex in phantom limb pain. *J. Pain* **2015**, *16*, 657–665.

39. Ferrucci, R.; Mameli, F.; Ruggiero, F.; Priori, A. Transcranial direct current stimulation as treatment for Parkinson's disease and other movement disorders. *Basal Ganglia* **2016**, *6*, 53–61.

40. Marangolo, P.; Marinelli, C.; Bonifazi, S.; Fiori, V.; Ceravolo, M.; Provinciali, L.; Tomaiuolo, F. Electrical stimulation over the left inferior frontal gyrus (IFG) determines long-term effects in the recovery of speech apraxia in three chronic aphasics. *Behav. Brain Res.* **2011**, *225*, 498–504.

41. Nitsche, M.A.; Paulus, W. Sustained excitability elevations induced by transcranial DC motor cortex stimulation in humans. *Neurology* **2001**, *57*, 1899–1901.

42. Thielscher, A.; Antunes, A.; Saturnino, G.B. Field modeling for transcranial magnetic stimulation: A useful tool to understand the physiological effects of TMS? In Proceedings of the 2015 37th Annual International Conference of the IEEE Engineering in Medicine and Biology Society (EMBC), Milan, Italy, 25–29 August 2015; pp. 222–225.

43. Angulo-Sherman, I.N.; Rodríguez-Ugarte, M.; Sciacca, N.; Iáñez, E.; Azorín, J.M. Effect of tDCS stimulation of motor cortex and cerebellum on EEG classification of motor imagery and sensorimotor band power. *J. Neuroeng. Rehabil.* **2017**, *14*, 31.

44. Antal, A.; Alekseichuk, I.; Bikson, M.; Brockmöller, J.; Brunoni, A.; Chen, R.; Cohen, L.; Dowthwaite, G.; Ellrich, J.; Flöel, A.; et al. Low intensity transcranial electric stimulation: Safety, ethical, legal regulatory and application guidelines. *Clin. Neurophysiol.* **2017**, *128*, 1774–1809.

45. Llinas, R.; Negrello, M.N. Cerebellum. *Scholarpedia* **2015**, *10*, 4606.

46. Rodríguez-Ugarte, M.; Costa, Á.; Iáñez, E.; Úbeda, A.; Azorín, J. Pseudo-online detection of intention of pedaling start cycle through EEG signals. In *Converging Clinical and Engineering Research on Neurorehabilitation II*; Springer: New York, NY, USA, 2017; pp. 1103–1107.

47. Hortal, E.; Úbeda, A.; Iáñez, E.; Azorín, J.M.; Fernández, E. EEG-Based Detection of Starting and Stopping During Gait Cycle. *Int. J. Neural Syst.* **2016**, *26*, 1650029.

48. McFarland, D.J.; McCane, L.M.; David, S.V.; Wolpaw, J.R. Spatial filter selection for EEG-based communication. *Electroencephalogr. Clin. Neurophysiol.* **1997**, *103*, 386–394.

49. Steinwart, I.; Christmann, A. *Support Vector Machines*; Springer Science & Business Media: New York, NY, USA, 2008.

50. Hamedi, M.; Salleh, S.H.; Noor, A.M.; Mohammad-Rezazadeh, I. Neural network-based three-class motor imagery classification using time-domain features for BCI applications. In Proceedings of the Region 10 Symposium, Kuala Lumpur, Malaysia, 14–16 April 2014; pp. 204–207.

51. Field, A. *Discovering Statistics Using IBM SPSS Statistics*; Sage: Thousand Oaks, CA, USA, 2013.

52. Nguyen-Vu, T.B.; Kimpo, R.R.; Rinaldi, J.M.; Kohli, A.; Zeng, H.; Deisseroth, K.; Raymond, J.L. Cerebellar Purkinje cell activity drives motor learning. *Nat. Neurosci.* **2013**, *16*, 1734–1736.

53. Priori, A.; Ciocca, M.; Parazzini, M.; Vergari, M.; Ferrucci, R. Transcranial cerebellar direct current stimulation and transcutaneous spinal cord direct current stimulation as innovative tools for neuroscientists. *J. Physiol.* **2014**, *592*, 3345–3369.

54. Grimaldi, G.; Manto, M. Anodal transcranial direct current stimulation (tDCS) decreases the amplitudes of long-latency stretch reflexes in cerebellar ataxia. *Ann. Biomed. Eng.* **2013**, *41*, 2437–2447.

sensors

MDPI

Article

A Novel Feature Optimization for Wearable Human-Computer Interfaces Using Surface Electromyography Sensors

Han Sun, Xiong Zhang *, Yacong Zhao, Yu Zhang, Xuefei Zhong and Zhaowen Fan

Department of Electronic Science and Engineering, Southeast University, Nanjing 210096, China; 230139593@seu.edu.cn (H.S.); 220161239@seu.edu.cn (Ya.Z.); 230129600@seu.edu.cn (Yu.Z.); zxf@seu.edu.cn (Xu.Z.); zhaowenfan@seu.edu.cn (Z.F.)
* Correspondence: zxbell@seu.edu.cn; Tel.: +86-138-0903-3793

Received: 4 January 2018; Accepted: 13 March 2018; Published: 15 March 2018

Abstract: The novel human-computer interface (HCI) using bioelectrical signals as input is a valuable tool to improve the lives of people with disabilities. In this paper, surface electromyography (sEMG) signals induced by four classes of wrist movements were acquired from four sites on the lower arm with our designed system. Forty-two features were extracted from the time, frequency and time-frequency domains. Optimal channels were determined from single-channel classification performance rank. The optimal-feature selection was according to a modified entropy criteria (EC) and Fisher discrimination (FD) criteria. The feature selection results were evaluated by four different classifiers, and compared with other conventional feature subsets. In online tests, the wearable system acquired real-time sEMG signals. The selected features and trained classifier model were used to control a telecar through four different paradigms in a designed environment with simple obstacles. Performance was evaluated based on travel time (TT) and recognition rate (RR). The results of hardware evaluation verified the feasibility of our acquisition systems, and ensured signal quality. Single-channel analysis results indicated that the channel located on the extensor carpi ulnaris (ECU) performed best with mean classification accuracy of 97.45% for all movement's pairs. Channels placed on ECU and the extensor carpi radialis (ECR) were selected according to the accuracy rank. Experimental results showed that the proposed FD method was better than other feature selection methods and single-type features. The combination of FD and random forest (RF) performed best in offline analysis, with 96.77% multi-class RR. Online results illustrated that the state-machine paradigm with a 125 ms window had the highest maneuverability and was closest to real-life control. Subjects could accomplish online sessions by three sEMG-based paradigms, with average times of 46.02, 49.06 and 48.08 s, respectively. These experiments validate the feasibility of proposed real-time wearable HCI system and algorithms, providing a potential assistive device interface for persons with disabilities.

Keywords: human-computer interface; surface electromyogram; channel selection; feature optimization; multi-class recognition; support vector machine

---

1. Introduction

Human-computer interfaces (HCI) for those with motor deficits based on bioelectrical signals have received increasing attention in the last decade. HCI provides communication and control channels between human subjects and the surrounding environment with the purpose of replacement or augmentation of muscle activity [1]. Common classes of bio-signals used to control assistive devices include electromyography (EMG) [2,3], electroencephalography (EEG) [4,5], electrooculography (EOG) [6,7], and fusions of these signals [8–11].

This work presents the performance results of an HCI system based on electrical currents generated in the contraction and relaxation phase of muscles [12] known as sEMG. Measurement of sEMG is noninvasive, can offer excellent signal-to-noise ratio compared to EEG, and provides real-time information about muscle intents [13].

Applications of sEMG include early disease detection [14], seizure [15,16] and fall detection [10,17], gesture [18,19], and sign language recognition [3,20]. These applications recorded activities of facial [21,22], upper [18,23] and lower-limb [17,24] muscles.

The sEMG-based HCIs are sensitive to electronic noise, movement artefacts and muscle fatigue [12]. To overcome the problem, wearable rigid printed circuit board (PCB)-based systems are often used [3,25,26]. However, these devices still need to be fully fixed and tethered. Another solution is development of wearable and flexible electronics. These sEMG sensors have been utilized to monitor electrocardiogram (ECG), EMG and body posture [27,28] signals as well as to detect arm gestures [29].

The sEMG-based HCIs require effective features to provide real-time efficiency from the data processing standpoint. Time-domain feature extraction is the most common method because these features provide high recognition rate and low computational cost. Root mean square (RMS) [30–33] and mean absolute value (MAV) [29,33,34] are most popular among these features. Time-frequency domain features which characterize varying frequency information at different time locations have received attention recently. Time-frequency domain analysis can efficiently eliminate the non-stationary noise in either time or frequency domain [12,35–38].

Feature optimization/selection techniques further enhance the performance. These techniques include filter, wrapper, and embedded methods [39]. Filter and wrapper methods have very low computational costs because they optimize features independent of the classification performance [40,41]. In contrast, embedded methods rely on criteria that are generated during the classifier training process. Examples of embedded techniques are the support vector machine (SVM)-based Recursive Feature Elimination (RFE) [42] and the linear discriminant analysis (LDA)-based Fisher's Discriminant (FD) function [17].

Channel selection is another simple feature selection/optimization technique. Within this method, channel optimization is based on the single-channel classification accuracy. A more sophisticated adaptation is about selecting important muscles/channels that highly contribute to the specific movements according to the power and frequency distribution of sEMG signals [31].

Classification is the next step to distinguish between different movements. Simple and linear classifiers are preferable because of their simplicity and ease of implementation. For example, Linear Discriminant Analysis (LDA) [29], k-Nearest Neighbor (kNN) [3] and Decision Tree (DT) [3,13] are feasible to separate small numbers of uncomplicated classes clearly. To increase expandability and performance in complicated systems, Support Vector Machine (SVM) [43], Artificial Neural Network (ANN) [44], Fuzzy Min-Max Neural Network (FMMNN) [40] and Random Forest (RF) [37] are suggested for classification. Neural Network classifiers can be used for both simple and complex cases due to their high performance. However, with too many hidden layers or hidden units, the classifiers need long training times and require large amounts of training data. Unsupervised learning methods including K-Means and Fuzzy C-Means have also provided high classification performance in sEMG recognition [32].

An important application of biopotential-based HCI is smart wheelchair control. The conventional paradigm of smart wheelchair control is through joysticks [45]. This paradigm is not applicable for patients with low or no control of their upper limb. The sEMG-based HCI is an advanced alternative controller [46,47], however with more sophisticated implementation than the conventional joystick control. A critical problem is the functionality requirement for accurate control. (1) The recognition for more motions or gestures can increase degree-of-freedom [37,42]. (2) A state-machine based control and a proportional control could meet the requirement [48,49]. In this method, a modality or a channel is used to switch control modes. (3) Fusion of two or more types of sensors can realize the high-dimensional control for assistive devices. The combination of sEMG and inertial measurement units (IMU) consisted of conventional accelerometers (ACC) and gyroscopes is frequently used because of the high performance in long-term control. Muscle fatigue and skin sweat can make sEMG signals drift, while they have no influence to IMU signals [50,51].

In this paper, we explore the performance of sEMG-based HCI in controlling a telecar. We present a wireless wearable sEMG system based on flexible printed circuit (FPC) with embedded dry metallic detecting electrodes that avoids channels connection and fixation challenges. We also present methodology to select sEMG feature subsets for the recognition of four movements. The novel entropy criteria (EC) and Fisher discrimination (FD) criteria are compared with the conventional RFE method. Dunn-Heriksen et al. [52] introduced EC in EEG channel selection. Here we adopt EC for sEMG feature optimization. Fisher's discriminant based separability measurement has been widely utilized in feature optimization [17,40]. However, we introduce a novel method to compute the ratio of between-class distances to within-class distances. Finally, subjects control a designed telecar based on different paradigms using the optimal channels and features.

The structure of this paper is as follows: Section 2 describes two types of sEMG acquisition systems. Experimental methods about anticipants, experiment design, sensors placement are discussed in Section 3. Signal processing techniques including preprocessing, feature extraction, selection, and classification are explained in Section 4. Section 5 introduces four different control paradigms, followed by the results of offline analysis and real-time control in Section 6. Finally, Sections 7 and 8 discuss and conclude with the strengths of the current work.

2. System Architecture

The entire circuit structure of two types of sEMG acquisition systems consists of four main parts: a power module, a signal-conditioning module, a signal-processing module, and a signal-transmission module. The power module provides required power and safety precaution for using in human recordings. The function of signal-conditioning module is to amplify and filter raw signals. Analog filtered sEMG signals are converted into digital signals in the signal-processing module. Finally, the signal-transmission module transmits these digital signals to PC. Our overall hardware design includes the requirements of low cost, low power, small size, human compatibility, and ease of programming and interfacing with standard computers.

2.1. Offline sEMG Acquisition System

We used disposable disc sensors in offline sEMG acquisition, as shown in Figure 1b. These sensors consist of Ag/AgCl electrodes, conductive gel, an adhesive area and a snap connection. Wet sensors with conductive gel ensure an easy conversion between ionic current and electron current, resulting in low electrode impedance up to few kilo-ohms [53]. Diameters of the Ag/AgCl electrodes, conductive gel, and disc sensors are 9, 15 and 34 mm, respectively. Two sensors, at a distance of roughly 30 mm, constitute a pair of bi-polar sEMG electrodes.

The sEMG signal has small amplitude and is severely distorted by electromagnetic interference. An approach to reduce the electrode-skin interference is to employ an amplifier with high input impedance. The acquisition system board presented in Figure 1a (32×22 mm^2) includes a high-performance voltage follower (AD8626, Analog Devices, Norwood, MA, USA) as well as a differential amplifier (INA128, TI, Dallas, TX, USA) with very large input impedance. Together they largely reduce common mode interference (CMI) as well as improve the common mode rejection ratio (CMRR) and signal to noise ratio (SNR). We further reduce differential mode interference (DMI) with anti-aliasing and on chip digital filters. In both hardware and software solutions, the high-pass filter may have a 3 dB cutoff frequency of 10–20 Hz and the low-pass filter a 3 dB cutoff frequency of 400–450 Hz to avoid loss of information from the sEMG signals [54]. Therefore, a pass-band filter between 10 Hz to 450 Hz is designed using the AD8626. The sEMG signals are further passed through a notch filter at 50 Hz (UAF42, TI).

For the signal-processing module, we used an ATmega8 low-power 8-bit microcontroller (ATMEL, Microchip Technology, Chandler, AZ, USA), as the central processor and analog to digital converter. The supply voltage and reference voltage are both 3.7 V. The amplified and filtered signals (in the range of −1.8 V to 1.8 V) are then transformed to unipolar signals in a dynamic range of 3.7 V, sent to the analog-to-digital converter, and finally transmitted to a PC with Bluetooth UART module (HC05). We receive the signals at 1000 Hz sampling rate. They are further filtered (bandpass 10 Hz to 450 Hz as well as notch 50 Hz and its harmonics) and then stored via MATLAB.

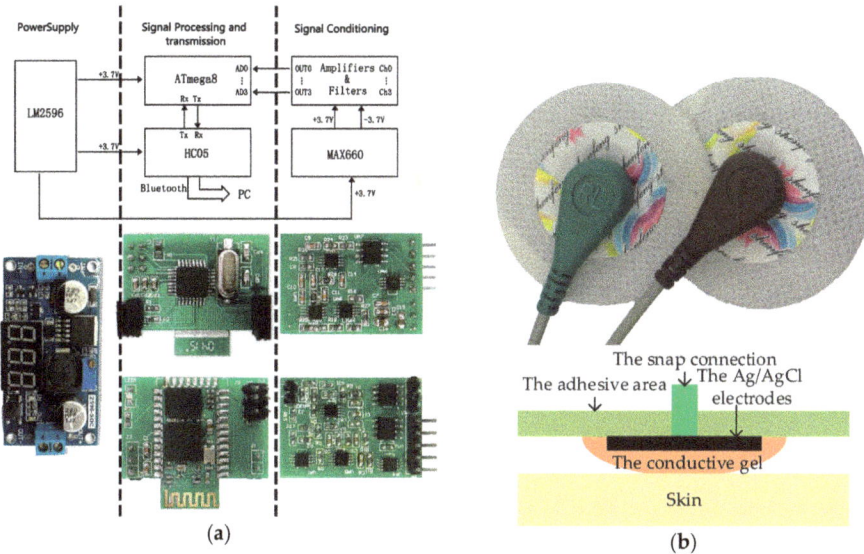

Figure 1. (**a**) The four-channel offline acquisition board; (**b**) The wet disc sensors and architecture.

The voltage convertor (LM2596, National Semiconductor, Santa Clara, CA, USA) including the thermal shutdown and current limit protection cells can provide the power of +3.7 V efficiently. The CMOS monolithic voltage converter chip (MAX660, Maxim, San Jose, CA, USA) produces a −3.7 V power to supply the negative voltage to dual-supply amplifiers.

2.2. Wearable sEMG Acquisition System

The wearable sEMG acquisition system, shown in Figure 2a, is almost same as the offline system, with small differences. First, the FPC-based real-time system is more flexible, small-sized, lightweight, and low-cost compared to the PCB-based system. Therefore, this design minimizes noise pickup in sEMG recording stations, and allows for recording without additional pre-amplification steps.

Second, in term of sensors materials, we designed two pairs of metallic dry sensors. This type of disc sensors is plated with copper on the top layer of FPC-based board. The next step is plating nickel and gold on the copper disc to stabilize contact impedance as shown in Figure 2b. Low electrode-skin impedance is critical for recording high-quality signals. The traditional solution is to gently exfoliate skin using abrasive gel or 75% alcohol. The diameter of each sensor is 3 mm and the fixed distance within a pair of sensors is 30 mm. The inter-electrode distance can be minimized further due to the smaller size of electrodes. Therefore, metallic dry sensors can fit an uneven skin for more precise applications.

Thirdly, the optimization of circuit structures is considered to satisfy demands of miniaturization and high-reliability for systems. The contact-impedance problem is a much more pressing problem for dry sensors compared with wet sensors [55]. Except for cleaning skins, another practice is to employ an amplifier with high input impedance. Therefore, we replaced the amplifier INA128 with a higher performance instrumentation amplifier INA2126 (TI). In order to minimize the system size, we replaced the AD8626 amplifier with a TLV4333 (TI). The TLV4333 contains a four-channel amplifier with an input filter to reduce both common mode interference (CMI) and differential mode interference (DMI). To combine the signal-processing and signal-transmission modules, the nRF51822 (Nordic, Oslo, Norway) is used as central processors because it integrates Cortex-M0 kernel, analog to digital converter (ADC), and Bluetooth 4.0 module. We used a 3 V button battery to power the real-time system.

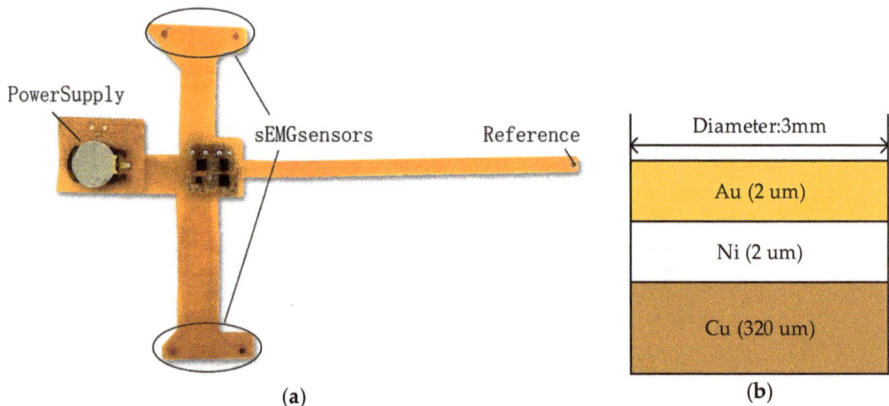

Figure 2. The two-channel real-time acquisition system. Subfigure (**a**) correspond to the acquisition board, and subfigure (**b**) correspond to the architecture of the metallic dry sensors.

Finally, wearable online systems are paired with a personal computer which performs more computationally intensive processing steps. These include independent component analysis (ICA), multiscale principle component analysis (MSPCA), specified features calculation, and prediction based on the trained model.

3. Methods

3.1. Subjects

Nigh able-bodied subjects (seven male and two female; mean age: 25.1; range: 21–31) participated in the data acquisition, all of whom had no prior experience with sEMG based HCIs and signed the consent form approved by the Academic Ethics Committee of Southeast University before experiments. All subjects are university-educated and non-smoking with no history of forearm muscle injuries and neurological disorders. Instructions for offline and online experiments were carefully explained and illustrated, and the first trial did not begin until subjects indicated full understanding of the required tasks. During all implementation processes, subjects sat motionless in a comfortable chair and rested their hands on a desktop.

3.2. Acquisition Setup

In offline experiments, we selected four different muscle groups: flexor carpi ulnaris (FCU), extensor carpi radialis (ECR), extensor carpi ulnaris (ECU), and abductor pollicis longus (APL); as shown in Figure 3a. The activity of these muscles were measured via pairs of bipolar sEMG sensors during four wrist movements including wrist extension (WE), wrist flexion (WF), make a fist (MF), and rest (REST). The reference sensor was placed on the upper arm far from recording channels.

Figure 3b shows the real-time system setup. Four controlling states namely forward, backward, clockwise rotation and anti-clockwise rotation were mapped with four motions. Subjects can control the vehicle easily because the direction and rhythm of wrist movements correspond to commands of vehicle motions. We performed channel selection technique on data pooled from all subjects and found that the location with highest classification accuracy are ECU and ECR.

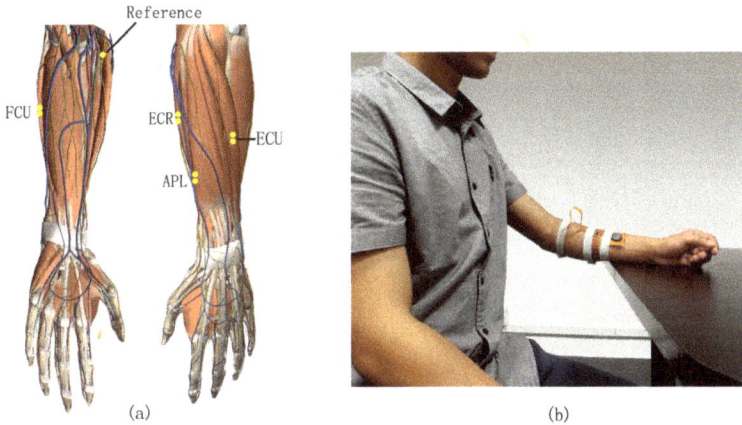

Figure 3. Sensor placement and implementation of the system. (**a**) Sensor placement in the offline analysis and the pairs of yellow points correspond to the bi-polar electrodes; (**b**) Real-time system setup for data acquisition.

3.3. Experiments Protocol

Each subject was connected and sat for one complete recording period. The experimental period was divided into five offline sessions (1 h) and eight real-time sessions (0.5 h). In offline sessions, subjects maintained 1.5 s movements according to cues on the screen. Signals were recorded continuously and saved separately for each session. Within each session, subjects performed 40 individual motion split evenly and randomly ordered among these four movements.

The same group of subjects attended real-time tests. We optimized channels, features, and classification parameters during offline sessions and utilized these optimal values for online processing. The telecar was controlled by four methods: the joystick paradigm, the fixed-moving sEMG paradigm, the channel-combination paradigm, and the state-machine paradigm. Each paradigm was repeated two times. The real-time analysis window was 125 ms with 20% overlap. The entire trajectory was a square.

4. Signal Processing and Pattern Recognition

All data processing was performed within MATLAB. The steps of signal analysis and the relationship between offline and online sessions are illustrated in the flowchart of Figure 4. In the offline phase, we used infinite impulse response (IIR) filters, ICA and MSPCA to de-noise sEMG signals. The feature extraction module includes time-domain, frequency-domain, and time-frequency-domain features computation. Feature selection refers to separability. The rank of single-channel accuracies selected optimal channels, and the EC and FD methods determined a uniform subset of features. Four machine learning algorithms (kNN, ANN, RF and SVM) were employed to classify features, and the best parameters and model were saved. In online sessions, same preprocessing approaches except a different segmentation method were adopted for signals from the selected channels. We then extracted the optimal feature subset according to offline sessions and utilized a classifier model to specify the control commands during online sessions.

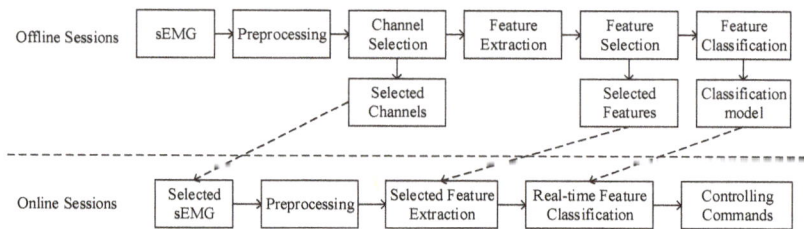

Figure 4. Flowchart of proposed processing framework.

4.1. Preprocessing

4.1.1. Data Filtering and Segmentation

We employed an IIR Butterworth bandpass filter (10–450 Hz, with order 16) according to the previous works [54,56]. An elliptic notch filter and several bandstop filters were used to eliminate the power line interference at 50 Hz and its harmonics. There are two main sEMG segmentation methods: disjoint and overlapped methods [57]. In offline sessions, subjects maintained each motion in a 1.5-s task time, which had precise onset and offset boundaries. The task period was further divided into 3 periods—a transient period of onset (0–0.25 s); a one-second execution period (0.25–1.25 s); and another transient period. Feature values were extracted only from the one-second execution window segmented by the disjoint method with a predefined length. For example, four feature vectors could be extracted from the execution period with a 250 ms analysis window. Real-time analysis adopted the overlapped segmentation method. The size of sliding window was predefined with 20% overlap.

4.1.2. Independent Component Analysis (ICA)

Subsequently, remaining artifact signals could be removed conveniently by applying ICA. For the general ICA model of sEMG, suppose we have N channels filtered sEMG signals $x_i, i = 1, \dots, N$. Each channel has N independent source signals $s_i, i = 1, \dots, N$ and records different mixture of s_i. Mathematically, the principle of mixing processes can be expressed as follows:

$$x = A \cdot s, \tag{1}$$

where A is the unknown mixing matrix, x and s represent the combination of x_i and s_i respectively. Then, the algorithm extracts a matrix with independent components (ICs) that recovers original sEMG signals when applied to the data set x according to:

$$u = W \cdot x, \tag{2}$$

where W called unmixing matrix equals A^{-1}, and u denotes the sources (ICs). After performing an ICA, clean sEMG signals used in future processing are obtained by removing the ICs with artifacts and neglecting the corresponding column of W.

4.1.3. Multiscale Principle Component Analysis (MSPCA)

The MSPCA algorithm was proposed by Bakshi [58] to merge the strengths of PCA with the benefits of the wavelet transform (WT). While PCA extracts the linear or nonlinear relationships among variables, WT extracts deterministic features and approximately removes the autocorrelation within measurements [59].

MSPCA has been applied for EMG [36] and ECG [60] signal modeling and de-noising. In terms of sEMG signals, the algorithm steps are: (1) The j-th column $x_j(t)$ of raw data are decomposed to its wavelet coefficients by WT. (2) The covariance matrix of each scale is computed along with the number of principle components separately from other scales. (3) The appropriate number of principle components is selected. (4) The combination of WT and PCA is used to reconstruct the de-noised signals.

4.2. Feature Extraction

Three main types of features—time, frequency and hybrid domains—have been used to classify sEMG signals for HCI [61]. These features are computed based on signals' amplitudes (time domain), estimated power spectrum density (frequency domain) and time-frequency transformation (hybrid domain). The time domain features are most popular because of their computational simplicity. Time-domain features include mean absolute value, modified mean absolute value with the weighting window function (function (1)) and the improved weighting window function (function (2)), root mean square, variance, waveform length, Willison amplitudes, simple square integral, zero crossing, slope sign change, and histogram of sEMG. Frequency-domain features are mostly used to study muscle fatigue and to recognize movements. Widely used frequency-domain features include auto-regressive coefficients, median and mean frequencies.

Time-frequency analysis, with its ability to represent time dependent frequency responses, has recently leveraged in the sEMG feature extraction. The most commonly used analysis method is discrete wavelet transform (DWT). We use the Daubechies 4 (DB4) wavelet because of higher classification accuracy and lower computation cost [62]. Average power of wavelet coefficients in each sub-band are extracted for evaluation of the frequency distribution. We also compute standard deviation of coefficients to evaluate changes in the distribution. Another time-frequency feature is power spectral density (PSD) of short time window Fourier transform.

Table 1 lists features used in this study, their abbreviation and dimensions. In offline sessions, all 42 listed features were extracted for each channel. These features were then cascaded into a final vector with a dimension of 168 for four channels.

Table 1. List of sEMG features and dimensions.

Feature Name	Abbreviation	Dimension
Mean Absolute Value	MAV	1
Modified Mean Absolute Value 1	MMAV1	1
Modified Mean Absolute Value 2	MMAV2	1
Root Mean Square	RMS	1
Variance	VAR	1
Waveform Length	WL	1
Zero Crossings	ZC	2
Slope Sign Change	SSC	1
Willison Amplitude	WAMP	5
Simple Square Integral	SSI	1
Histogram of sEMG	HEMG	6
Four-order Auto-Regressive Coefficients	AR	4
Median Frequency	MDF	1
Mean Frequency	MNF	1
Short Time Fourier Transform	STFT	1
Average Power of the Wavelet Coefficients	APWC	7
Standard Deviation of the Wavelet Coefficients	SDWC	7

4.3. Feature Selection

The dimension of features extracted in the last section should be reduced before sending them to classifiers. Efficient feature and channel selection algorithms improve the prediction performance and provide less computational complexity. The common approach to evaluate selected features is estimating the rank of classification rate or the separability criteria [63].

Here, the optimal-feature selection combined the rank of separability values (SV) of each feature with classification rates by SVM. First, we obtained a feature subset with the highest accuracy for individual channel. The mean number of optimal feature subsets across all channels and subjects represented the size of features in real-time algorithms. Next step was to confirm the detailed common features. We calculated the summation of each feature's SV via EC and FD methods. We then used the RFE method to compute and rank the frequency each feature was contained in optimal subsets. The selected features were determined by specific criteria according to these different methods. Ideally, different algorithms should obtain almost same optimal feature subsets.

4.3.1. Entropy Criterion Based Feature Selection (EC)

A modified entropy based method is used to calculate separability values. The variance of features among different classes can provide classification information, and the entropy of features' variance is a measurement of uncertainty. When the variance of different classes is close, it means that the specific class has little classification information and vice versa. Therefore, the entropy of variance measures separability of each feature. The definition is as follows:

$$J_i = -\sum_{k=1}^{n} V_k^i \cdot \ln(V_k^i), \tag{3}$$

where J_i denotes SV of the i-th feature, and V_k^i denotes normalized variance of the i-th feature for the k-th class (totally n class). Within this method, smaller J_i corresponds to the feature with larger variance entropy.

4.3.2. Fisher Discrimination Based Feature Selection (FD)

The ratio of between-class and within-class distance could evaluate the extracted features' separability numerically. The principle of this method is similar with the Fisher linear discriminant analysis (LDA) algorithm [64]. The SV is calculated by:

$$J_i = D^2(a^i, b^i)/(D^2(a^i, a^i) + D^2(b^i, b^i)), \tag{4}$$

where a^i and b^i denote the i-th feature in class a and b respectively, and the function $D^2(a,b)$ is the mean Euclidean distance between all combination of different trials of two groups. The separability improves when the ratio increases. In multi-class (totally n class) separability analysis, we separate the problem into n two-class problems according to one-versus-all strategy [65]. The average ratio of these two-class problems is computed as the multi-class SV for each feature.

4.3.3. Recursive Feature Elimination (RFE)

Compared with the mentioned methods offering the numerical evaluation of features, the RFE algorithm based on SVM outputs a list of ranked features. In detail, the RFE algorithm mainly contains following steps [66]. (1) Features and class labels are combined. (2) Training the model of SVM. (3) Computing the weight vector and rank criteria. (4) The feature with the smallest rank criteria is eliminated. (5) Steps (2)–(4) are repeated until only one feature is left. Finally, the algorithm outputs the feature rank list. The rank criteria is the squared coefficients w^2 [67]. Importance of a feature is determined by the loss of the margin between classification boundaries when the feature is removed. The rank criteria is defined as:

$$J_i = \min_i \left| w - w^{(-i)} \right|, \tag{5}$$

where w is the inverse of margins which means $w = \|w\|_2$, and $w^{(-i)}$ represents the w without the i-th feature at this SVM iteration.

4.4. Classification

Although we quantify the separability of various sEMG features, it is still unclear whether they interact well with the classification process. Therefore, it is important to choose the best classifier for recognizing sEMG patterns. Here, four widely-used classifiers—kNN, ANN, RF and SVM—were Considered. We optimized classification parameters based on offline data (train) and then applied the optimal classifier to real-time data (test).

4.4.1. k-Nearest Neighbor (kNN)

The kNN is one of the simplest learning methods that divides data into two or more classes. The kNN is frequently used for small training datasets, because it is easy to implement and has low computational cost. Inputs consist of the k closest training samples in the feature space. In the sEMG classification, the distance of k nearest neighbors from one another determines the label of test samples. Performance of the kNN depends on the selection of parameter k. Wan et al. tested the relationship between k and ten-fold CV accuracy [68]. When k is in the range of 3 to 10, the difference of accuracy is not huge. In this work, six nearest neighbors were selected to evaluate accuracies.

4.4.2. Artificial Neural Network (ANN)

The ANN follows a learning method with self-learning capability [69]. However, because the network contains numerous parameters its training process is time-consuming. These parameters including thresholds of hidden layers and connection weights between layers. In the classification of four-motion sEMG signals, the ANN structure consisted of one input layer, one hidden layer and one output layer. The dimension of input feature vectors was n. The neurons of input, hidden and output layers were n, $2n$ and 4, respectively. The activation function was a sigmoid. We estimated parameters by the back-propagation algorithm to reduce the cost function and gradient [70]. Because of long training time, we used a five-fold CV to validate classification of sEMG data.

4.4.3. Random Forest (RF)

The RF is a type of ensemble learning method. Although the design and computation are easy, it works better than other high-performance classifiers, such as SVM and ANN, in some applications [71]. In order to ensure performance of the RF, each base learner should have high precision. Simultaneously, to improve generalization ability of the RF, the diversity of base learners is guaranteed by two methods. The first method is to sample training data randomly as the input of each base learner. The second method is to choose the best decision feature from a subset of features (dimension: s) instead of from all features (dimension: d) for each node. The output is a final class voted by all base learners. In this study, the optimum number of base decision tree was 30 according to the research results of Gokgoz and Subasi [12]. The optimal feature subsets for nodes were determined as follows [72]:

$$s = \log_2 d. \tag{6}$$

4.4.4. Support Vector Machine (SVM)

The SVM has high speed in calibration and classification of high dimensional sEMG features. The goal of this algorithm is to solve classification problem by finding maximal margin hyper-plane (w,b) to separate training data with a given set of labels. Briefly, a positive real constant α is computed by training data to determine parameters w and b. When using the test feature f, a label is assigned according to the decision boundary function, which is:

$$g(f) = \text{sign}(\langle w,f \rangle + b) = \text{sign}(\sum_{j=1}^{m} \alpha_j y^j K(f,f^j) + b), \tag{7}$$

where f^j denotes the j-th trial (totally m trials) in training data with a corresponding label y^j, and K is a kernel function including a high dimensional model. In this work, we chose the radial basis kernel function in LIBSVM [73]. Despite more than two movements present, the binary SVM was still used with one-versus-all technique. In offline sessions, ten-fold cross validation (CV) was used to assess classification accuracies and F-score.

5. Control Methods

Subjects can control a wheelchair—the final aim of sEMG-based HCIs—only when they achieve high performance in the telecar control with a pre-defined path. In real-time sessions, subjects controlled the designed toy vehicle with the wearable sEMG system to finish two loops in a square-loop environment with some simple obstacles. The length of each side was 40 cm. The vehicle was randomly positioned at any corner after the obstacle localization was completed. The moving and rotating speeds are set to a constant value of 12 cm/s and 0.25π rad/s, respectively. Figure 5a shows an ideal route to finish the loops and simple obstacle map. The differential distance of two loops in the figure is only for clear visualization.

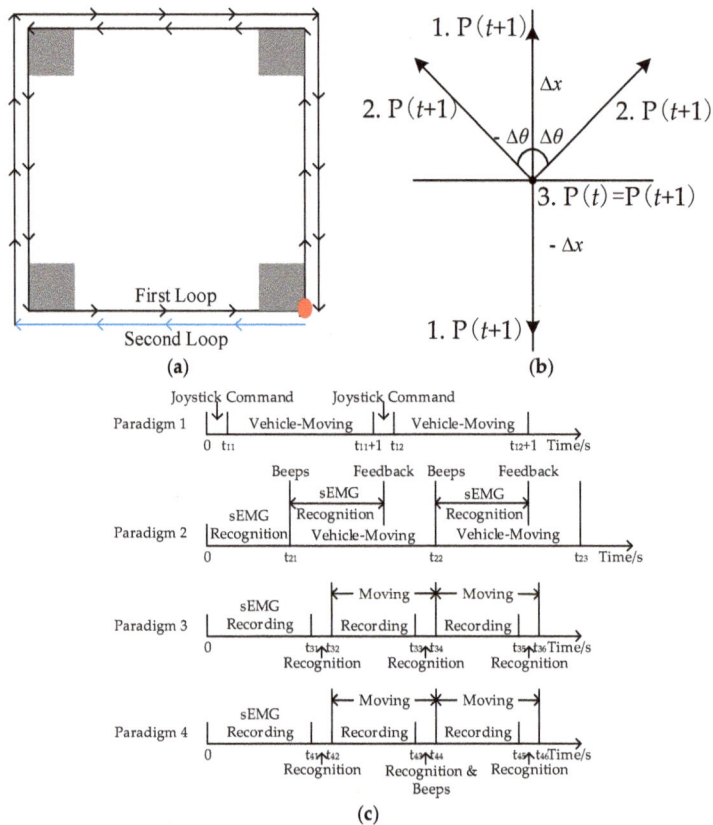

Figure 5. (a) Environment and obstacle maps. It contains the forward information (black vectors), backward information (blue vectors), starting point (red circle) and obstacle location (gray areas); (b) Three possibilities of the location update for Paradigm 1 and 2 ('1' denotes the straight-line movements, '2' denotes the rotation movements, and '3' is no movements); (c) The protocol for each paradigm. The onset of each real-time session is presented as the reference time point of 0 s.

This section introduced four control methods. The response time of a real-time system should not introduce a delay that was perceivable by users, and the threshold was generally regarded to be roughly 300 ms [74]. Therefore, the real-time control in this work adopted a 125 ms window. There are two controlling rules: First, only one motion was classified at a time during the implementation. Second, in Paradigm 2, the classification result of newly acquired sEMG features combined with the last motion to confirm whether two consecutive windows were same. The protocol for each paradigm is shown in Figure 5c, and introduced as follows:

Paradigm 1: The vehicle moved to a direction according to the joystick command. The maneuverability is best for healthy subjects.

Paradigm 2: Considering safety and a continuous control, the fixed-moving paradigm was introduced. In this paradigm, the next motion was determined during vehicle moving. Once the vehicle moved, subjects were prompted to start a motion with an auditory signal (beeps). After our system identified the motion, subjects received another auditory feedback and prepared for the next move. The epoch of sEMG data processing was 125 ms, and the same results of two consecutive epochs were considered as a valid control command. The HCI system would translate predicted classes into corresponding actions as described in Table 2. Suppose $P(t)$ is the location at the time window t, and Δx and $\Delta \theta$ denote the position change in straight-line and veer directions. Then, the position at the next time window $t + 1$ could be updated as:

$$P(t + 1) = P(t) \pm \Delta x, \; P(t + 1) = P(t) \pm \Delta \theta, \; \text{or} \; P(t + 1) = P(t). \tag{8}$$

The equation reveals that the position at $t + 1$ has three possibilities as shown in Figure 5b: (1) Fixed Δx determines the vehicle moves forward or backward by 12 cm. (2) Fixed $\Delta \theta$ influences veer movements, which means the vehicle rotates 45 degrees clockwise or anti-clockwise. (3) If there is no movement commands, the vehicle waits and then stops.

Because of the high separability of sEMG signals among different movements as well as the efficient auditory cues and feedbacks, this paradigm had a high degree of maneuverability in continuous HCIs.

Table 2. Definition of muscular movements and corresponding commands to vehicle.

Muscular Movements	Vehicle Motions
Make a Fist (MF)	Move forward by the length of 12 cm
Keep Relaxed (REST)	Move backward by the length of 12 cm
Wrist Extension (WE)	Clockwise rotation of 45 degrees
Wrist Flexion (WF)	Anti-clockwise rotation of 45 degrees
No continuous same results come	Stop

Paradigm 3: Although the second paradigm has a good performance, some problems still exist. The main problem is about fixed moving periods, which leads to challenges for paths that frequently change directions. Another problem is that when no continuous same results come, the delay may still exists.

To overcome these problems, we proposed the channel-combination paradigm with 125 ms window. The same results from both selected channels determined moving directions during recognition periods (i.e., t_{31}–t_{32}, t_{33}–t_{34} and t_{35}–t_{36} in the figure). Table 3 shows control methods in this paradigm. The sEMG recording for the next process was synchronous with the vehicle moved. For example, the recording time was from t_{32} to t_{33} and the vehicle-moving period was from t_{32} to t_{34}. The respond time is short in this paradigm, but subjects could not stop by their autonomous motions in the four-state control, which is a hidden danger for patients.

Paradigm 4: The state-machine-based control paradigm could increase the functionality [48]. Five-dimension control could be achieved by four motions in this paradigm as shown in Table 4. The REST state was a switching of straight-line and rotational movements. The protocol was similar with Paradigm 3 in Figure 5c. The initial state was straight-line movements, and the detailed control method was as following. Recognition results from t_{41}–t_{42} determined moving states in t_{42}–t_{44}. Upon the REST state appeared (e.g., the period of t_{43}–t_{44}), an auditory beeps was offered and the mode was switched to rotational movements. Consequently, another REST state set back the mode to straight-line state. A critical control rule was the mode could not be changed unless at least one motion was implemented. Subjects can stop control by keeping fisting.

Table 3. Definition of muscular movements and control method in Paradigm 3 and 4.

Muscular Movements	Control Method	Vehicle Motions
Make a Fist (MF)	Ch1 = Ch2 = 1	Forward
Keep Relaxed (REST)	Ch1 = Ch2 = 2	Backward
Wrist Extension (WE)	Ch1 = Ch2 = 3	Clockwise rotation
Wrist Flexion (WF)	Ch1 = Ch2 = 4	Anti-clockwise rotation
Others	Ch1 ≠ Ch2	Stop

Table 4. Definition of muscular movements and control method in Paradigm 5 and 6.

Muscular Movements	Vehicle Motions
Keep Relaxed (REST)	Switching: on/off
Wrist Flexion (WF)	On: Forward Off: Anti-clockwise rotation
Wrist Extension (WE)	On: Backward Off: Clockwise rotation
Make a Fist (MF)	Stop

6. Results

Our objective is to choose and use effective features for the sEMG-controlled vehicle with wearable HCI designed by our group. We present four steps: first, we validate the feasibility and performance of our proposed hardware and filters. Second, the rank of classification accuracy picks two channels. Thirdly, optimal feature subsets and multi-class recognition rates are computed by the proposed feature selection algorithms, and compared with the RFE method. Finally, we generalize findings through comparing different control paradigms, and investigate whether the selected common channels and features are applicable to online sessions.

6.1. Acquisition System Testing

Tests of this part are to verify the feasibility of acquisition systems and preprocessing methods. The results show that real-time high-quality signals can be transmitted to computers and saved within the Bluetooth communication distance.

6.1.1. Hardware Evaluation

Signal amplifiers and filters are the main components in acquisition systems. Two stages of amplifiers were used to avoid effects on the signals' bandwidth when the gain of one stage amplifier was too large. The first stage amplifier has a large input impedance, and the gain is 51. The second stage is an inverting amplifier with high gain (−35.7), gain bandwidth product (GBP) and CMRR. Therefore, the total gain for amplification of raw sEMG signals is −1821 as shown in Figure 6a.

The acquisition systems contain a low-pass filter of 450 Hz, a high-pass filter of 10 Hz and a notch filter of 50 Hz. We validated designs of filters and circuit components in the FilterPro (TI) instead of via manual derivation process. The low-pass and high-pass filters are four-order and two-order Butterworth structure with the Sallen-Key topology, respectively. In addition, a 50 Hz notch filter is integrated in the UAF42.

Frequency responses of filters were measured by applying a 1 Vpp sinusoidal signal logarithmically generated by a function generator. Figure 6 depicts the amplitude-versus-frequency curves. The frequency range of testing signals is from 1 Hz to 480 Hz. From Figure 6b,c, acquisition systems show a flat operation on the edge of frequencies of interest (10–450 Hz). The selection and parameter errors of resistors and capacitors resulted in the real cut-off frequency range of on-chip filters is from 6 Hz (f_L) to 451 Hz (fc). In Figure 6d, we show effects of the notch filer, the interference of 50 Hz has been reduced.

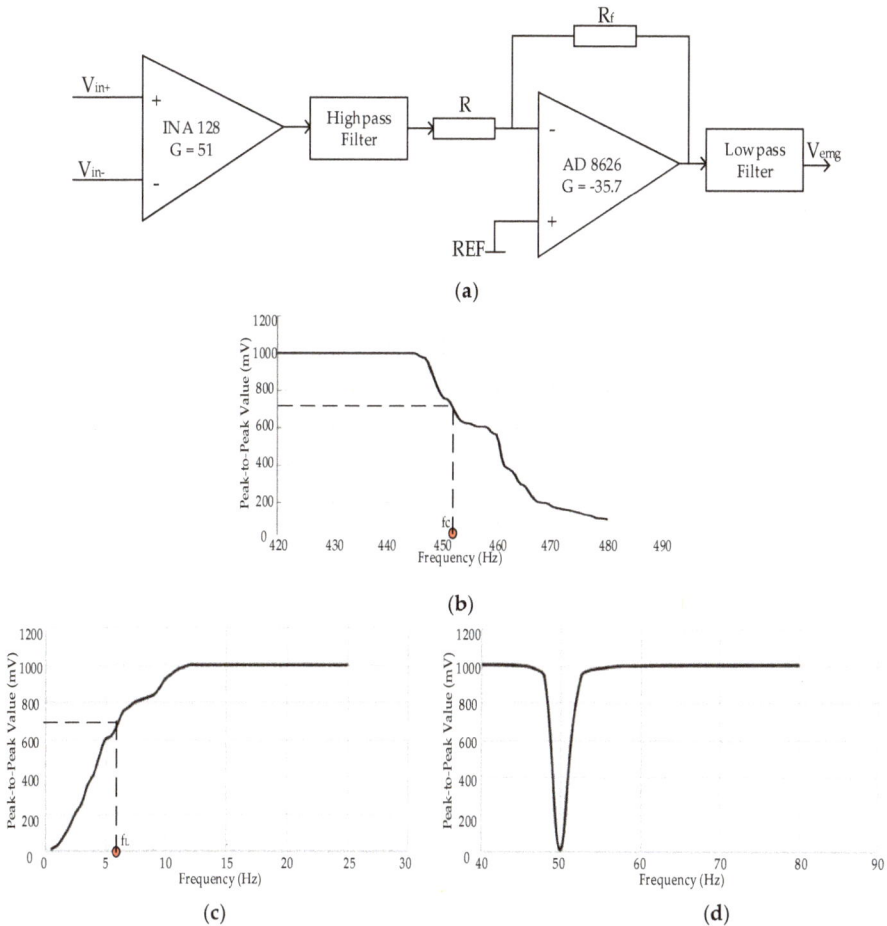

(a)

(b)

(c)

(d)

Figure 6. (a) is the system electrical schema of amplifiers and filters, (**b–d**) are the amplitude-versus-frequency curves of low-pass filter with the cut-off frequency fc, high-pass filter with the cut-off frequency f_L, and notch filter of 50 Hz, respectively.

In order to evaluate performance of the recording hardware, we computed the SNR as follows:

$$SNR = 10\log_{10}\left(\frac{A_m^2}{A_r^2}\right),\tag{9}$$

where A_m is the maximum RMS amplitude of continuous strained muscles signals and A_r is the maximum RMS value of noise when a muscle is not activated. SNR of the offline and online acquisition systems are 47.42 dB and 54.09 dB, respectively. The slightly higher SNR in the wearable system can be attributed to the optimized circuit structure and selection of high-performance chips.

6.1.2. Test on the Digital Signal Preprocessing

To improve SNR of systems and quality of sEMG signals, the preprocessing module contains several digital filters, including band-pass filter in the frequency range of interest (10–450 Hz). When adopting dry sensors, significant 50 Hz noise pickup interfered with signals. In order to eliminate this power line noise, we used a combination of analog and digital filters including the elliptic notch and band-stop filters.

The features in time and frequency domains were extracted. Therefore, this section presents the time-domain and frequency-domain verification. Figure 7a shows time-domain signals at the APL after preprocessing and normalization. An increased amplitude appears after zero to one second in the last three movements, because subjects keep rest in this period. Because of the normalization of each state, amplitudes of signals within 0–1 s in the last three plots are not similar to amplitudes of the REST state. The differences among different movements are clear. The REST and WE states have respectively the smallest and largest amplitudes. The time-series for the WF and MF tasks closely resemble each other. In detail, the MF state has slightly higher amplitudes than the WF state.

The power spectral density (PSD) was estimated for each trial. Averaged PSD zoomed in the range of 0–5 is depicted in Figure 7b for the time span from 1 to 2.5 s. The MF state has the highest mean PSD, followed by the WF, WE and REST states. In detail, the MF and the WF states have the highest PSD in sub-bands of 10–105 Hz and 105–195 Hz, respectively. Relevant frequencies of all movements are between approximately 10 Hz and 450 Hz. Neural information plateaus around 40–95 Hz, and the range of interest extends to 40–195 Hz for the WF state. Then, the PSD diminishes slowly as the frequency increased to 450 Hz. Mentioned notch and bandstop filters can eliminate the 50 Hz and its harmonic. SNR is improved to 61.47 dB and 68.91 dB for these two acquisition systems, respectively.

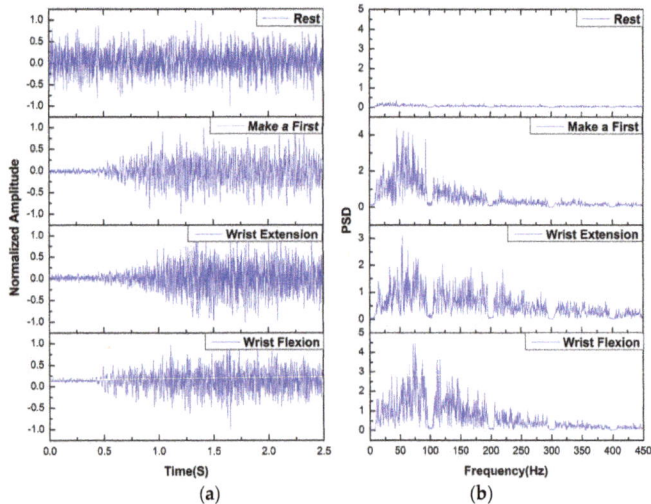

Figure 7. Time-frequency verification of different movements: (**a**) Normalized amplitudes in time domain; (**b**) PSD in frequency domain. (The names of four movements were represented to their abbreviations in the following analysis)

6.2. Channel and Feature Selection

According to previous studies, longer time windows would not have significantly improved prediction accuracy [43,74]. All sEMG signals during training were analyzed in non-overlapping windows as mentioned in Section 4.1.1. Each motion modality could extract 160 feature vectors with 250 ms window. If the window length decreased to 125 ms, training sessions contained 1280 feature vectors. To ensure classification performance and reduce training complexity, the window length was 250 ms in the channel and feature selection.

6.2.1. Channel Selection

Although we use only four pairs of sEMG channels, it is still necessary to minimize the number of channels to make systems more mobile and easier to maintain. We rank single-channel classification accuracies using all features. Table 5 shows accuracies of each channel across all nine subjects.

Table 5. Classification accuracy of each channel for different pairs of movements.

Motion Pairs	APL	ECR	ECU	FCU
REST vs. MF	99.32%	99.42%	99.56%	99.47%
REST vs. WE	96.17%	98.99%	97.98%	99.30%
REST vs. WF	98.86%	99.13%	98.64%	99.34%
MF vs. WE	90.62%	95.16%	93.90%	94.96%
MF vs. WF	84.95%	91.08%	96.86%	91.31%
WE vs. WF	82.32%	95.52%	97.76%	85.62%
Average	92.04%	96.55%	97.45%	95.00%

We then select channels located on the ECU and ECR according to Table 5. Features from the channel ECU achieve the best classification accuracy equal to 97.45%, followed by the channel ECR and FCU reaching 96.55% and 95.00%, respectively. The channel FCU is best for recognizing between the REST and motions states (i.e., the first three pairs in Table 5). The channel ECU provides the best accuracies compared among different motion states (i.e., the last three pairs in Table 5). When comparing all motion pairs, first three pairs have higher accuracies than last three pairs. The REST and MF pair obtains the best performance for all channels. Furthermore, the MF and WE pair has the highest distinction among last three pairs.

For each subject, we divided sEMG features from these two channels into training and testing sets by ten-fold CV to estimate mean classification accuracies of different pairs of movements as shown in Figure 8. Accuracies of one subject are lower, which are marked as outliers in the boxplot. Mean accuracies of the first three pairs are 99.56%, 98.99% and 99.12%, respectively. Classification results of the last three pairs are more than 97%. Especially for the third and sixth pairs, median accuracies reach 100%. Above all, compared with single-channel analysis, the selected-channel performance is not significantly improved in recognizing the rest state with other movements. However, channel selection improves accuracies of the last three pairs by 3.75%, 6.33% and 2.80%, respectively.

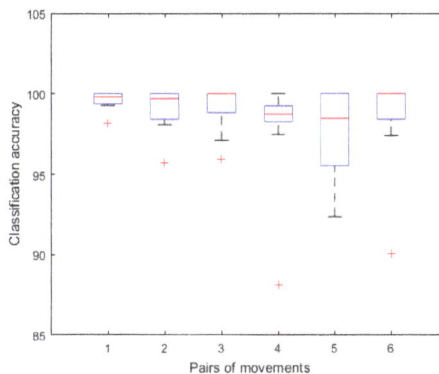

Figure 8. Boxplot for EMG tasks classification accuracy with ECU and ECR. The horizontal axis represents the different pairs of movements (1: REST vs. MF, 2: REST vs. WE; 3: REST vs. WF, 4: MF vs. WE, 5: MF vs. WF, and 6: WE vs. WF).

6.2.2. Classification Performance between Each Two Motions

This part tests the single-channel separabilty between each two movements, and investigates the feasibility of feature selection methods. Proposed EC and FD methods ranked features between two motions. Classification rates were performed as features increased from 1 to 42 to determine the best combination of feature space. Mean classification accuracies (MCA) with optimal feature numbers (OFN) across all subjects were evaluated by LIBSVM in the proposed two situations, and compared with the RFE method.

Table 6 lists MCA with OFN at the channel ECU. Classification accuracies are above 97% for all pairs except for the MF and WE pair. On average, accuracies are slightly higher with lower numbers of features opting for the FD and RFE based feature subsets. When comparing classification performance among different pairs, first three pairs obtain larger MCA with less OFN. Almost same accuracies between the REST and motions states reach about 98% by three feature selection methods. For last three pairs, average accuracies in the range of 94.72–98.16% and 94.68–98.15% are achieved by the FD and RFE methods compared with the range of 94.52–98.06% by the EC method.

Table 6. Mean optimal feature numbers and classification accuracies of the channel ECU.

Motion Pairs	OFN with EC	MCA with EC	OFN with FD	MCA with FD	OFN with RFE	MCA with RFE
REST vs. MF	19.3	99.73%	3.1	99.73%	1.4	99.73%
REST vs. WE	24.8	98.16%	5.8	98.24%	5.8	98.36%
REST vs. WF	15.2	98.69%	1.8	98.79%	2.4	98.89%
MF vs. WE	28.8	94.52%	26.4	94.72%	23.1	94.68%
MF vs. WF	22.2	97.35%	20.8	97.25%	21.3	97.11%
WE vs. WF	29.9	98.06%	26.1	98.16%	27.8	98.15%
Average	23.4	97.75%	14.0	97.82%	13.6	97.82%

One-way analyses of variance (ANOVA) are used for statistical analysis. The factors for analysis are six pairs of motions and three feature selection algorithms. (1) According to Table 6, the OFN is influenced by different pairs ($F(5,156) = 14.416$, $p < 0.001$) as well as three algorithms ($F(2,159) = 7.641$, $p = 0.001$). Post hoc tests of the influence of pairs show that first three pairs use significantly small feature subsets compared to last three pairs, but no differences are found within these two groups. Post hoc tests also show that the FD and RFE methods differ significantly from the EC method ($p = 0.002$ and $p = 0.003$, respectively), indicating the EC method uses more features to reach the optimal accuracy. There are no differences between the FD and RFE methods ($p = 0.991$), because these two algorithms are both based on classifiers learning. (2) The MCA is also affected by pairs ($F(5,156) = 8.716$, $p < 0.001$), but do not show reliable relationship with algorithms ($F(2,159) = 0.006$, $p = 0.994$). Post hoc comparisons reveal that the MF and WE pair has significantly lower accuracy than other pairs. All other pairs have no differences within each other except for comparing the first and fifth pairs ($p = 0.039$).

Furthermore, Figure 9 shows classification accuracies of three pairs of motions (i.e., the last three pairs in Table 6) as features increases by the EC and RFE ranks. The classification accuracy increases as the feature space increases. We assert that this is due to insufficient information provided with small feature subsets. However, when the feature size exceeds OFN, the accuracy remains high and then begins to decrease due to over-fitting. It illustrates one reason why feature selection is necessary. When number of selected features is less than 25, the RFE method performs better than the EC method. Then, performance of these two method reaches the same level. Figure 9a plots the best recognition rates of subject S6 could be improved to 89.38%, 95.63% and 97.5% with 36, 23 and 31 features picked by the RFE method for mentioned three pairs, respectively. As shown in Figure 9b, subject S7 uses the EC method to select 33 features to yield 96.25% accuracy comparing the MF and WE states, to select 37 features to yield 98.13% accuracy comparing the MF and WE states, and to select 33 features to yield 98.44% accuracy comparing the WE and WF states.

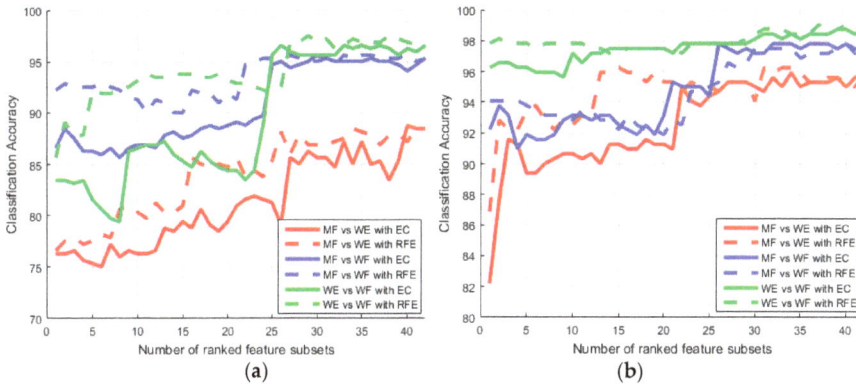

Figure 9. Comparisons of the RFE and EC methods. Evolution of classification accuracies was computed with feature subsets increasing according to different separability ranks: (**a**) classification accuracies of three pairs of motions for subject S6; (**b**) Same results for subject S7.

6.2.3. Feature Selection

The feature selection analysis is as follows. First, the dimension of optimal features is determined by single-channel analysis of ECU and ECR. This step combines feature selection and classification processes. Mean optimal feature numbers across all channels and subjects are 31, 23.3, 32.8 for the EC, FD and RFE methods, respectively. Then, the focus in this section is to identify common features from these subjects for future application. The advantage is that under repeated use, limitation to specified features reduces training and processing times. Two weighting methods were used to select specified features derived from the SV rank of different feature selection algorithms.

The weighting methods based on proposed EC and FD methods belong to quantitative weighting methods because they have detailed numerical evaluation. The single-channel SV of each feature is normalized to the range of 0 to 1, and averaged across all subjects. Then the separability criteria is the summation of all single-channel SV. Detailed features are determined by considering the 1st–31st and 1st–23rd features of EC and FD methods respectively according to their SV ranks. Table 7 shows total selected features from mentioned two feature selection algorithms, as well as the separability value of each feature. For the FD method, improper amplitude thresholds lead to the exclusion of WAMP1 and entire HEMG features except for HEMG1. All frequency features and the STFT are also neglected. Removal of APWC_D4 to APWC_A6 and SDWC_D5 to SDWC_A6 indicates a low effect of low-frequency components. For the EC method, features in the time domain including ZC1, WAMP4, WAMP5, HEMG1 and HEMG2, as well as all frequency features are eliminated.

The RFE method can indicate whether an individual feature is within the optimal subset with 1st–33rd features for a particular subject. The frequency with which each feature occurred among the top 33 features, across all subjects and channels, is considered. These frequencies are sorted in a descending order. Table 7 presents the top 33 features and their frequencies marked as T_{32}. Because we select two channels for nine subjects, the highest time should be 18. The results show that seven features form the best feature combination for this method. These features include variance, the first and third thresholds for Willison amplitude, and average power of the wavelet coefficients in the 1st–4th sub-bands. In contrast, two dimensions of the AR coefficients never enter top 33.

From the table, the optimal feature subsets with qualitative and quantitative weighting analysis indicate that time-domain and time-frequency-domain features lead to a better separability performance than frequency-domain features.

Table 7. The selected features and SV rank.

Rank	EC Method		FD Method		RFE Method	
	Feature	Aver. SV	Feature	Aver. SV	Feature	Aver. T_{32}
1	APWC_D3	0.2296	WAMP5	1.9005	VAR	18
2	APWC_D2	0.3066	WAMP4	1.7716	WAMP1	18
3	APWC_D4	0.3447	ZC2	1.5245	WAMP3	18
4	APWC_D1	0.3657	ZC1	1.4595	APWC_D1	18
5	STFT	0.3665	WAMP3	1.4290	APWC_D2	18
6	APWC_D5	0.3723	SSC	1.2673	APWC_D3	18
7	VAR	0.3882	SDWC_D1	1.2508	APWC_D4	18
8	SSI	0.3883	SDWC_D3	1.2128	WAMP2	17
9	APWC_D6	0.5859	SDWC_D2	1.2079	SSI	17
10	APWC_A6	0.6015	APWC_D1	1.1998	HEMG4	17
11	WAMP1	0.8886	HEMG1	1.1997	HEMG5	17
12	WAMP2	0.8907	HEMG2	1.1963	STFT	17
13	SDWC_D3	0.9152	MAV	1.1721	APWC_A6	17
14	MMAV2	0.9875	RMS	1.1626	APWC_D5	17
15	SDWC_D4	0.9898	MMAV1	1.1620	APWC_D6	17
16	MMAV1	0.9905	APWC_D3	1.1595	SDWC_D3	17
17	MAV	0.9908	WAMP3	1.1477	MAV	16
18	SDWC_D2	0.9915	APWC_D2	1.1458	MMAV1	16
19	RMS	0.9972	SDWC_D4	1.0873	MMAV2	16
20	SDWC_D5	1.0365	MMAV2	1.0643	RMS	16
21	SDWC_D1	1.0448	WL	1.0626	SDWC_A6	16
22	HEMG4	1.0605	SSI	1.0454	WAMP4	15
23	HEMG5	1.0714	VAR	1.0454	WAMP5	15
24	SDWC_A6	1.2020			SDWC_D1	15
25	SDWC_D6	1.2354			SDWC_D2	15
26	WAMP3	1.2441			SDWC_D4	15
27	WL	1.2672			SDWC_D6	15
28	SSC	1.3096			ZC2	14
29	HEMG3	1.4682			SDWC_D5	14
30	ZC2	1.4725			HEMG3	13
31	HEMG6	1.5413			HEMG6	13
32					HEMG1	11
33					HEMG2	11

6.3. Classification Performance

Four amplifiers (kNN, ANN, RF and SVM) and three analysis windows (125, 250 and 500 ms) were compared in this section.

6.3.1. Comparisons of Feature Subsets and Classifiers

Table 8 summarizes classification results for different feature combinations. In this study, seven different feature subsets are classified by four different algorithms such as kNN, ANN, RF and SVM. Each classifier is trained and tested with data from the same subject. Bold numbers in Table 8 indicate the best classifier for each feature subset. RF and ANN classifiers perform better for all subsets. RF with FD-based features ranks first at 96.77%, followed by ANN at 96.67%, SVM at 95.40% and kNN at 94.41%. In classification of EC-based features, ANN provides the superior accuracy with 96.74%, and RF ranks second with 96.66%. SVM gives 95.37% and kNN is with 94.73% ACC. All classifiers deliver above 94% accuracies after feature selection.

As shown in Table 8, classification performance of EC-based and FD-based features almost have no differences. Both of them are slightly better than RFE-based features and single-type features. Compared among single-type features (RMS, MAV, APWC and SDWC), the wavelet coefficients have better classification accuracies than RMS and MAV. The reason is that features of sEMG signals after time-frequency preprocessing offer a better classification precession [75]. Above all, smart combinations by feature selection methods provide more accurate features.

Table 8. Classification performance of machine learning algorithms. ACC is the abbreviation of Accuracy.

Features	kNN		ANN		RF		SVM	
	ACC	F-Score	ACC	F-Score	ACC	F-Score	ACC	F-Score
EC	94.73%	0.9473	**96.74%**	0.9672	96.66%	0.9656	95.37%	0.9533
FD	94.41%	0.9441	96.67%	0.9667	**96.77%**	0.9669	95.40%	0.9529
RFE	94.37%	0.9436	**95.82%**	0.9678	95.60%	0.9651	95.01%	0.9489
RMS	86.36%	0.8630	87.75%	0.8723	**89.43%**	0.8918	89.14%	0.8755
MAV	87.23%	0.8719	87.21%	0.8664	**89.12%**	0.8885	88.20%	0.8773
APWC	86.08%	0.8585	90.20%	0.9007	**91.63%**	0.9150	87.07%	0.8678
SDWC	85.36%	0.8535	88.90%	0.8861	**91.60%**	0.9150	82.85%	0.8004

F-Score is another index to evaluate classification performance calculated by the formula:

$$F\text{-Score} = \frac{2 \times TP}{2 \times TP + FP + FN}, \tag{10}$$

where TP, FP and FN are the numbers of true positives, false positives and false negatives in the confusion matrix, respectively. ACC and F-Score are close to each other, which indicates all classifiers achieve reliable performance on these feature subsets. With EC-based features, RF obtains 96.66% ACC and 0.9656 F-Score. The F-Score of FD-based features classified by RF is 0.9669, which is coincident with 96.77% ACC. It is also the case for other classifiers.

For statistical analysis of classification accuracy, different feature subsets and classifiers are the factors. The ANOVA reveals significant effect of feature subsets ($F(6,245) = 14.323$, $p < 0.001$) and classifiers ($F(3,248) = 2.990$, $p = 0.032$). However, the two factors interacted missed the 5% criteria ($p = 0.677$). (1) From post hoc analysis, the feature subsets could be divided into two groups. The first group contains feature space selected by algorithms. The second group is all single-type features. The two groups differ significantly ($p < 0.001$), but no differences appear within each group ($p > 0.5$). The results demonstrate that performance of feature selection algorithms is significantly better than single-type features. (2) In view of different classifiers, RF is significantly better than kNN ($p = 0.036$) and marginally better than SVM ($p = 0.109$). Furthermore, ANN has almost similar performance with RF ($p = 0.796$).

6.3.2. Comparisons of Analysis Window

Figure 10 shows the effects of analysis window length and accuracies. The mean accuracies are calculated by RF and SVM classifiers with the FD-based feature subset. RF performs better with these three epochs. Mean classification accuracies are 96.29%, 96.77% and 97.09% for the 125, 250, and 500 ms windows, respectively. The difference is non-significant ($p = 0.744$). Statistical analysis implicates that when shortening window length to 125 ms, the accuracy is not deteriorated. The advantages of adopting shorter windows are low computational cost and little storage space. Moreover, it is important with regard to the real-time classifier.

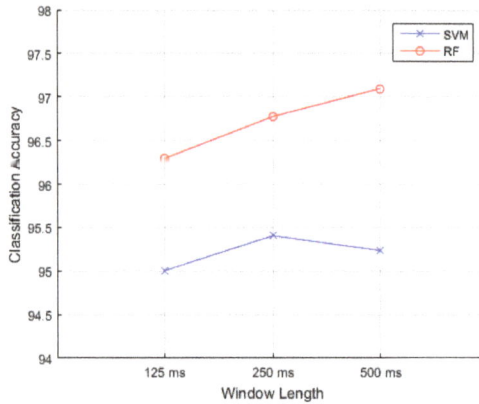

Figure 10. Effect of window length on the classification accuracy. The classification accuracies are averaged over all nine subjects.

6.3.3. Comparisons of Confusion Matrices

Each motion indicates a detailed command in the online system. Therefore, we structure the confusion matrix of each modality to investigate results of parameters and model selection. Figure 11 shows the recognition performance of FD-based, RFE-based and APWC feature subsets, respectively.

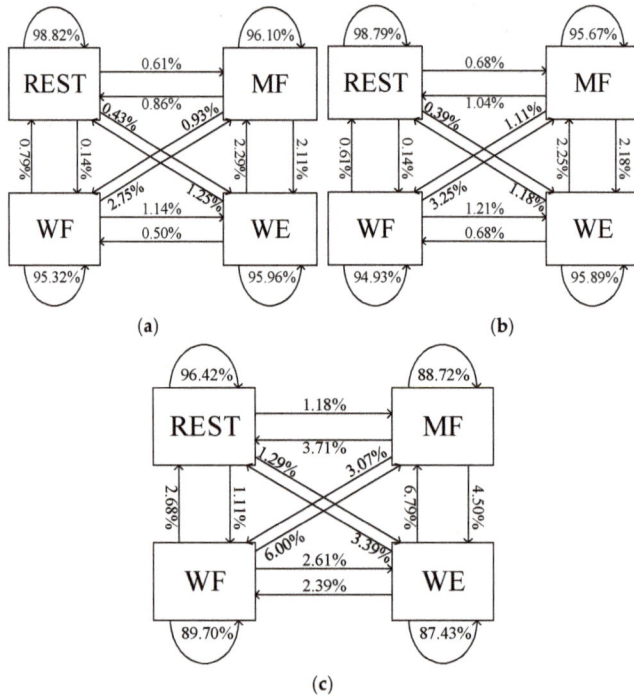

Figure 11. Recognition rate of the RF and confusion graph of different feature subsets: (**a**) the FD-based features; (**b**) the RFE-based features; (**c**) the APWC features. The numbers denote the percentage of samples in the class (arrow tail) classified as the class (arrowhead).

These features are extracted from 125 ms windows, and classified by RF. The FD method (REST: 98.82%, MF: 96.10%, WE: 95.96% and WF: 95.32%) is slightly better than the RFE method (REST: 98.79%, MF: 95.67%, WE: 95.89% and WF: 94.93%). The feature selection process is helpful to classify all states compared to single-type APWC features (REST: 96.42%, MF: 88.72%, WE: 87.43% and WF: 89.70%). The REST modality achieves the best recognition performance as sEMG amplitudes of keeping rest and moving have large differences shown in Figure 7. Whereas, the MF, WE and WF modalities are misclassified to others, especially for the MF and WE modalities. These observations are in line with the results in previous sections. They indicate that assessment of multi-class recognition is feasible with proposed feature selection methods. Our proposed FD method can improve the prediction performance and reduce the feature numbers compared with the conventional RFE method and single-type features.

6.4. Online Evaluation by Wearable EMG-based HCI

Each subject performs eight online sessions according to four separate paradigms. The selected features extracted from optimal channels and the trained RF model are opted for online sessions. Averaged recognition rates (RR) and travel time (TT) of each paradigm for all subjects are recorded and referred to Table 9. In Paradigm 2, moving directions offer labels for classification. Therefore, the RR equals to classification accuracies. However, in Paradigm 3 and 4, subjects control the vehicle according to their thoughts. Here, the RR in Paradigm 3 denotes success rates of recognizing classification results of both channels are same. Since it is hard to define which motion is right in Paradigm 4, the RR is not presented in Table 9 as a criterion.

Table 9. The recognition rate and travel time for different online paradigms.

Subject	Paradigm 1		Paradigm 2		Paradigm 3		Paradigm 4	
	TT (s)	RR	TT (s)	RR	TT (s)	RR	TT (s)	RR
S1	44.98	–	44.88	99.25%	47.21	96.88%	46.93	–
S2	45.86	–	44.82	98.38%	50.31	89.61%	46.42	–
S3	44.44	–	44.94	97.66%	45.07	98.03%	46.29	–
S4	45.08	–	46.93	95.75%	48.87	95.77%	47.08	–
S5	44.73	–	46.60	96.00%	47.10	92.50%	49.61	–
S6	45.55	–	46.86	92.19%	51.43	84.45%	47.81	–
S7	44.82	–	45.52	95.91%	48.87	96.09%	46.97	–
S8	45.70	–	47.04	86.16%	52.03	85.08%	51.37	–
S9	45.46	–	46.59	91.77%	50.67	88.54%	50.26	–
Mean	45.18	–	46.02	94.79%	49.06	91.77%	48.08	–

The results illustrate all subjects are able to complete these online paradigms with acceptable accuracies and travel time. The TT in last three paradigms is close to the joystick paradigm (Paradigm 1). Averaged time of Paradigm 1 to finish the loops is 45.18 s. In the fixed-moving paradigm (Paradigm 2), subjects can accomplish two sessions within the mean time of 46.02 s with 95.01% RR. Paradigm 3 and 4 increase the control time to 49.06 s and 48.08 s, respectively. The increment is achieved that when two channels have different classification results in Paradigm 3, the vehicle stops and waits. Transience pauses also happen in mode switches during Paradigm 4. Although Paradigm 3 and 4 are more sophisticated, these paradigms are closer to daily life. In Paradigm 2, only one command could make the vehicle move a fixed distance. However, in Paradigm 3 and 4, subjects decide each minor motion by their own ideas. Compared these two complex paradigms, subjects use less travel time in Paradigm 4. The reason is that success rates are lower to ensure both channels have the same class for Paradigm 3. S3 completes all sEMG-based sessions using the shortest time with the highest RR. On the contrary, S8 performs the worst.

The rough relationship between the TT and RR in Paradigm 2 and 3 is that more TT uses, less RR obtains. However, there are some special situations in detail. For example, the TT of S9 is similar with S5 (46.59 s vs. 46.60 s in Paradigm 2), but S5 has higher RR (96.00% vs. 91.77% in Paradigm 2). S9 can send out control commands before the end of vehicle-moving periods, although he misclassifies some motions. The performance decreases sharply controlling Paradigm 3 for a small number of subjects. The RR reduces 9.37% and 7.74% for S2 and S6, respectively. According to the offline and online analysis, performances of two selected-channels have some differences. For S2, accuracies of FCR and FCU are 90.76% and 96.80% in the offline analysis. The problem is solved in Paradigm 4 to a certain extent because of using combined features from both channels. The real-time RR obtained in Paradigm 2 is slightly lower than offline sessions. A main reason is the states with high accuracies are less than offline experiments. For example, Paradigm 2 needs only four backward commands controlled by the REST state which has 98.82% offline RR.

The results of statistical analysis illustrate that the travel time shows a significant effect of paradigms ($F(3,68) = 14.149$, $p < 0.001$). Post hoc tests reveal the TT of Paradigm 1 is significantly shorter than Paradigm 3 and 4 ($p < 0.001$), but it has no difference with Paradigm 2 ($p = 0.607$). Subjects use slightly less time in Paradigm 4 compared with Paradigm 3 ($p = 0.548$).

The route tracking performance of two subjects for the rectangular route is provided in Figure 12. Position measurements are taken when the vehicle reaches a certain position as green circles in the figure. During the online implementation, rotational movements are more difficult than straight-line movements for most subjects. The difference between S3 and S8 is related to two control methods facing a turn. A group of subjects turns a degree, moves forward and adjusts the direction for the next straight-line motion, as shown in Figure 12a. The other group of subjects moves forward for an enough distance and makes an approximate 90 degree turn at the corner, as plotted in Figure 12b. The first group needs short path length, but also needs to change modes three times. After turning, the vehicle could move in a straight line without too many fluctuations.

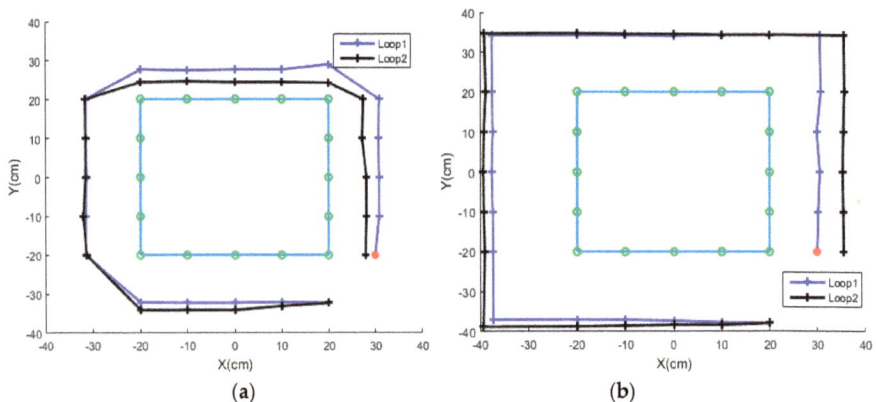

Figure 12. Rectangular route results for sEMG based control by different subjects: (**a**) path traced by S3; (**b**) path traced by S8.

7. Discussion

The purpose of this work was to design and demonstrate a type of sEMG-based HCI. The optimal combination of sEMG feature selection and classification methods is found and applied for online telecar control with the wearable acquisition system. The results demonstrate that the system with selected channels and features could achieve the classification accuracy and F-score above 90% in both offline and online experiments. This study provides potentials that patients with little motor ability could control the actual wheelchair with our system and algorithms.

7.1. Wearable EMG-Based HCI System Design

sEMG monitoring systems are suitable for wearable wireless applications that require small size, excellent mobility, low power consumption, and high transmission rates [76]. The most common systems were based on rigid PCBs [44,50,77]. The work studied by Kundu [50] proposed an EMG acquisition system equipped with a 7.4 V Li-ion battery, and then data were transmitted to computers via USB. Youn et al. [77] proposed a wireless sEMG system, whose size was 37 × 17 mm², with a Bluetooth transmission module. In the design of another system, a 3.7 V Li-ion battery provided power to the system with a size of 34 × 25 mm² [44]. Data were sent to a PC through a wireless module pair. Our proposed PCB-based system had a slightly smaller size (32 × 22 mm²) and the power was ±3.7 V. The main problem of these systems is mobility. Because the PCB-based systems still have connection wires between systems and sensors, they require complex fixation. Another problem is the sensor material. Although disposable sensors are convenient, they did not provide good performance in the accurate control on uneven skins due to the large distance between electrode pairs.

To overcome these problems, we implemented all systems on the FPC with embedded metallic sensors. Flexible dry sensors based on the FPC substrate achieved comparative performance with standard wet Ag/AgCl sensors [78], and were approved for clinical applications. FPC lines connected the signal-conditioning and signal-processing modules designed on the PCB to transmit signals and power [79]. Here, dry gold-plated copper sensors were used and the inter-pair sensor spacing was set at 12 mm. The fixation distance between sensors pairs was 30 mm and could be adjusted as needed.

SNR of a system could influence the signal quality. Phinyomark et al. [80] demonstrated the relationship between classification accuracies and SNR. Different white Gaussian noises were added to make the SNR varied from 20 to 0 dB. When the level of SNR noise reached 20 dB, accuracies were close to clean signals. SNR of standard-wet and FPC-based sensors were 18.1 dB and 20.2 dB [78]. SNR of the system designed in Youn's work was 59.06 dB [77]. SNR of our proposed PCB-based and FPC-based systems were 61.47 dB and 68.91 dB, respectively.

7.2. Feature Selection and Classification

Other studies have recognized several sEMG patterns to different applications such as motions/hand gesture recognition, prosthesis control and diagnostic decision. To allow comparison of our findings with these literatures, we list methods, classification results and applications in Table 10. The averaged ACC of our paper is best.

Table 10. Comparison of results and methods with other studies using sEMG-based signals.

Year	Features	Feature Selction Classification	ACC (%)	Applications	Classes	Channels
2015 [12]	DWT	RF	96.67	Diagnostic Decision	3	5
2017 [26]	MAV	SVM	>90	Prosthetic Hands	4	1
2017 [29]	RMS & WL	LDA	90%	Prosthetic Hands	6	4
2015 [32]	RMS & DRMS	Fuzzy C-Means	89.15	Motions Recognition	4	16
2012 [35]	AR & DWT	ANFIS [1]	95	Diagnostic Decision	3	1
2014 [36]	MUSIC [2]	SVM	92.55	Diagnostic Decision	3	5
2017 [37]	WPD	RF	92.1	Motions Recognition	10	8
2017 [40]	178	Fisher Criterion + SVM	84.01	Gait Phase Recognition	4	4
2017 [42]	21 TD [3] & 6 FD [4]	RFE + Extreme Learning Machine	>90	Motions Recognition	17	12
2017 [43]	15	SVM	63–99	Prosthetic Hands	6	1
2017 [44]	MAV	ANN	94	Virtual Trackpad	10	4
2017 [69]	DFT & MAV	SVM	70.2	Motions Recognition	14	8
2017 [81]	22	RES [5] + LBN [6]	93.25	Speech Recognition	11	5
This work	46	FD + RF	96.77	Motions Recognition	4	2

[1] ANFIS: the Adaptive Neuro-Fuzzy Inference System. [2] MUSIC: feature extraction using Multiple Signal Classification. [3] TD: Time Domain. [4] FD: Frequency Domain. [5] RES: ratio of Euclidian distance and standard deviation. [6] LBN: Linear Bayes Normal classifier.

Efficient features selection algorithms could exclude many irrelevant and redundant features to provide higher performance. Nevertheless, the methods were not mentioned in some studies in Table 10. Fang et al. [32] just mentioned RMS was one of the most important sEMG features because of lower computational cost and decent performance. Another study used MAV for the same reason [44]. The optimal features should be extracted by some criteria. Tosin et al. [42] demonstrated that RFE was a powerful feature selection algorithm. However, the output was a list of ranks in separability without detailed values. Then, quantitative feature selection methods including the Davies-Bouldin index [82], RES [81] and Fisher Criterion [40] were introduced. There are two problems in these methods. First, success rates of these methods are not high enough (Table 10). The second problem involves the optimal number of features. Huang et al. [82] used a feature subset selected by the Davies-Bouldin index to obtain 85% and 71% classification rates in offline and online tests, respectively. In Lee's work, the authors tested classification performance fixing the numbers of feature subsets to 100 and 150 [40]. In others, the feature selection process including CV of classifiers was complicated [81]. In our paper, the modified EC method was proposed because of low computational cost. We also combined Euclidian distances with the Fisher's discriminant to obtain the modified FD method. The accuracies of these two methods were 96.66% and 96.77%, which were slightly better than the conventional RFE method and other feature selection methods in the table. The average number of features to receive the best accuracy for each selected channel across all subjects was defined as the optimal feature number. This method was more reasonable than Lee's work [40], and easier than Srisuwan's work [81].

7.3. The Online Performance

Since the final target of our systems is for a wheelchair control, the performance of smart wheelchairs is compared and discussed in this section. Delicate motions of the upper limb controlled joysticks-based smart wheelchairs [83], but they are not capable for patients with complete or partial loss of muscle activities. The EEG-based [83] and EOG-based [7] wheelchairs with automated navigation systems were proposed. In Huang's work [7], subjects could control the wheelchair to finish all tasks within 227 s and 277 s by joysticks and EOG signals, respectively. The recognition rate for healthy subjects was 91.7%. The main challenge was to decrease misclassification rates of unwanted blinks or rotational motions of eyes. In Zhang's work [84], the destination selection was fast, but the critical problem was subjects needed 4.5 s to stop control.

The sEMG-based control method was considered in this work, because sEMG signals could achieve higher accuracies and use in long-term applications. The fixed-moving paradigm could improve safety. The average time was 46.02 s, which means each vehicle motion including the sEMG-recognition and vehicle-moving periods cost 1.05 s. The waiting time was much shorter than the same method in an EEG-based wheelchair [85].

According to Kucukyildiz's work [86], the fixed-moving control paradigm had challenges for paths with frequently directional changes. Their work used very short analysis window (50 ms) for the sEMG control. Englehart et al. analyzed the effects of analysis window length upon classification accuracy [74]. The results showed that the best performance is with 32 ms analysis window with a majority vote decision. There was no differences when the window length ranged from 32 ms to 256 ms. However, in single-window analysis, the accuracy degraded rapidly with decreasing analysis window length. According to this work, the real-time processing window is 125 ms in our work.

To improve control's continuity, the channel-combination and state-machine paradigms were introduced. The travel time of controlling by joysticks was 45.18 s. Subjects used 49.06 s and 48.08 s by these two continuous sEMG-based paradigms. The accuracy of Paradigm 2 was 94.79%. The recognition rate of motions was 91.77% in Paradigm 3. The same comparisons were shown in Kundu's work [50]. The travel time of a designed wheelchair was 67.18 s and 72.88 s for joysticks and sEMG signals. The real time recognition accuracy was 90.58%. Despite the moving speed was lower and the path length was shorter in our work, the real-time results were acceptable.

The trend of sEMG-based HCIs is to increase the degree-of-freedoms. Maeda et al. designed an omnidirectional wheelchair with four-channel sEMG signals [87]. They adopted amplitude combinations of different channels during straining muscles. The similar method was defined as proportional control [48]. The performance in classifying 10 functions with a linear discriminant classifier, reaching 94%, 93% and 87% at 16, 8 and 4 channels, respectively. In Ishii's work [49], the combinations of different motions corresponded to eight control commands based on the state machines.

In our work, Paradigm 3 and 4 were similar with the proportional and state-machine control. Our vehicle could move to five directions with two channels. The travel time of Paradigm 4 was slightly shorter than Paradigm 3. These two paradigms were more sophisticated than Paradigm 2, but they were closer to the real-life control method. However, these paradigms were hard to remember or implement for some subjects, especially for the high-dimensional control.

7.4. Limitations and Future Work

There are several basic limitations associated with this study that need further development to provide the wearable sEMG system for clinical purposes. (1) Although we obtained a stable acquisition system, Balouchestni et al. [76] designed a system to recover the original bio-signals with good level of accuracy and SNR greater than 95.8 dB. Therefore, the circuit architecture optimization are still needed. (2) From Table 10, the next step of research should extend the motion pool. (3) The current research study recorded and analyzed the sEMG data performed only by healthy subjects. (4) In the future, minimization of analysis windows and improvement of single-window performance are main works for real-time algorithms. (5) In this study, subjects controlled the designed telecar in a laboratory environment. We are combining our system with a smart wheelchair. To control it in a complicated real scenario, efforts still need to be made.

8. Conclusions

Two wearable sEMG acquisition systems are designed and implemented successfully in this work. The PCB-based prototype can capture four-channel sEMG signals simultaneously from different forearm muscles, and the FPC-based system with two channels are utilized for online control. The system could communicate with a laptop wirelessly through Bluetooth. The high SNR of 61.47 dB and 68.91 dB for these systems ensure the signal quality. Temporal and frequency responses indicate that the system can remove noise and are stable during all motions.

The ECU and ECR channels are selected with 97.45% and 96.55% mean classification accuracies across all pairs of motions and subjects. In single-channel analysis, the FD and RFE methods achieve the optimal accuracy with significantly less features than the EC method ($p = 0.002$ and $p = 0.003$ respectively). For the channel ECU, the average accuracy increases to 97.82% with only 14 features. Accuracies above 98% are achieved comparing the REST state with other states. The FD method produces recognition rates in the range of 94.72% to 98.16% comparing among three motions.

Detailed features are selected according to the level of feature separability provided by the EC, FD and RFE methods. According to qualitative and quantitative weighting analysis, these three methods opt for 31, 23 and 33 features, respectively. The feature selection results also prove that time-domain and time-frequency-domain features provide more discriminative information than frequency-domain features. The FD-based feature subset with RF classifier achieves 96.77% accuracy, which is better than other methods and single-type features referred in some references.

Furthermore, to validate the feasibility of proposed methods, we invited same group of subjects to control the designed toy vehicle using four different paradigms. Subjects can accomplish the online task by joysticks with averaged 45.18 s. For the fixed-moving paradigm, the mean travel time is 46.02 s with 94.79% recognition rate. The results of Paradigm 3 and 4 reveal that these paradigms can improve the maneuverability and provide potentials in more sophisticated paths.

Therefore, all mentioned results suggest that our proposed acquisition systems and algorithms can be used in the HCI research. The future work focuses on recording and discerning more motions to realize the accurate implementation of smart wheelchairs.

Acknowledgments: The authors appreciate anonymous reviewers for their valuable comments, and F. Bahari and B. Gluckman for their technical and editorial assistance. This work was supported by the National Key Research and Development Program of China (No. 2016YFB0401201), the National Science Foundation of China (No. 61505028), and Enterprises, Universities and Research Institutes Prospective Joint Research Program of Jiangsu Province (No. BY2016076-07).

Author Contributions: Xi.Z. and H.S. conceived and designed the experiments; Ya.Z. and Z.F. designed the acquisition systems; Xu Z. performed the statistical analysis; Yu.Z. and H.S. proposed the algorithms and analyzed the data; H.S. wrote the paper; Xi.Z. reviewed the paper.

Conflicts of Interest: The authors declare no conflict of interest.

References

1. Skavhaug, I.M.; Lyons, K.R.; Nemchuk, A.; Muroff, S.D.; Joshi, S.S. Learning to modulate the partial powers of a single sEMG power spectrum through a novel human-computer interface. *Hum. Mov. Sci.* **2016**, *47*, 60–69. [CrossRef] [PubMed]

2. Fukuda, O.; Tsuji, T.; Kaneko, M.; Otsuka, A. A human-assisting manipulator teleoperated by EMG signals and arm motions. *IEEE Trans. Robot. Autom.* **2003**, *19*, 210–222. [CrossRef]

3. Wu, J.; Sun, L.; Jafari, R. A wearable system for recognizing American sign language in real-time using IMU and surface EMG sensors. *IEEE J. Biomed. Health Inform.* **2016**, *20*, 1281–1290. [CrossRef] [PubMed]

4. Galán, F.; Nuttin, M.; Lew, E.; Ferrez, P.W.; Vanacker, G.; Philips, J.; Millán, J.D.R. A brain-actuated wheelchair: Asynchronous and non-invasive brain–computer interfaces for continuous control of robots. *Clin. Neurophysiol.* **2008**, *119*, 2159–2169. [CrossRef] [PubMed]

5. Ang, K.K.; Guan, C. EEG-based strategies to detect motor imagery for control and rehabilitation. *IEEE Trans. Neural Syst. Rehabil. Eng.* **2017**, *25*, 392–401. [CrossRef] [PubMed]

6. Barea, R.; Boquete, L.; Mazo, M.; López, E. Wheelchair guidance strategies using EOG. *J. Intell. Robot. Syst.* **2002**, *34*, 279–299. [CrossRef]

7. Huang, Q.; He, S.; Wang, Q.; Gu, Z.; Peng, N.; Li, K.; Zhang, Y.; Shao, M.; Li, Y. An EOG-based human-machine interface for wheelchair control. *IEEE Trans. Biomed. Eng.* **2017**. [CrossRef] [PubMed]

8. Pfurtscheller, G.; Allison, B.Z.; Brunner, C.; Bauernfeind, G.; Solis-Escalante, T.; Scherer, R.; Zander, T.O.; Mueller-Putz, G.; Neuper, C.; Birbaumer, N. The hybrid BCI. *Front. Neurosci.* **2010**, *4*, 42. [CrossRef] [PubMed]

9. Kawase, T.; Sakurada, T.; Koike, Y.; Kansaku, K. A hybrid BMI-based exoskeleton for paresis: EMG control for assisting arm movements. *J. Neural Eng.* **2017**, *14*, 016015. [CrossRef] [PubMed]

10. De Venuto, D.; Annese, V.F.; de Tommaso, M.; Vecchio, E.; Vincentelli, A.L. Combining EEG and EMG signals in a wireless system for preventing fall in neurodegenerative diseases. In *Ambient Assisted Living*; Andò, B., Siciliano, P., Marletta, V., Monteriù, A., Eds.; Springer International Publishing: Cham, Switzerland, 2015; Volume 11, pp. 317–327.

11. Lin, K.; Cinetto, A.; Wang, Y.; Chen, X.; Gao, S.; Gao, X. An online hybrid BCI system based on SSVEP and EMG. *J. Neural Eng.* **2016**, *13*, 026020. [CrossRef] [PubMed]

12. Gokgoz, E.; Subasi, A. Comparison of decision tree algorithms for EMG signal classification using DWT. *Biomed. Signal Process. Control* **2015**, *18*, 138–144. [CrossRef]

13. Stepp, C.E. Surface electromyography for speech and swallowing systems: Measurement, analysis, and interpretation. *J. Speech Lang. Hear. Res.* **2012**, *55*, 1232–1246. [CrossRef]

14. Mishra, V.K.; Bajaj, V.; Kumar, A. Classification of normal, ALS, and myopathy EMG signals using ELM classifier. In Proceedings of the 2016 2nd International Conference on Advances in Electrical, Electronics, Information, Communication and Bio-Informatics (AEEICB), Chennai, India, 27–28 February 2016; pp. 455–459.

15. Beniczky, S.; Conradsen, I.; Henning, O.; Fabricius, M.; Wolf, P. Automated real-time detection of tonic-clonic seizures using a wearable EMG device. *Neurology* **2018**, *90*, e1–e7. [CrossRef] [PubMed]

16. Szabó, C.Á.; Morgan, L.C.; Karkar, K.M.; Leary, L.D.; Lie, O.V.; Girouard, M.; Cavazos, J.E. Electromyography-based seizure detector: Preliminary results comparing a generalized tonic-clonic seizure detection algorithm to video-EEG recordings. *Epilepsia* **2015**, *56*, 1432–1437. [CrossRef] [PubMed]

17. Xi, X.; Tang, M.; Miran, S.M.; Luo, Z. Evaluation of feature extraction and recognition for activity monitoring and fall detection based on wearable sEMG sensors. *Sensors* **2017**, *17*, 1229. [CrossRef] [PubMed]

18. Du, Y.; Jin, W.; Wei, W.; Hu, Y.; Geng, W. Surface EMG-based inter-session gesture recognition enhanced by deep domain adaptation. *Sensors* **2017**, *17*, 458. [CrossRef] [PubMed]
19. Chen, Z.; Zhang, N.; Wang, Z.; Zhou, Z.; Hu, D. Hand gestures recognition from multi-channel forearm EMG signals. In Proceeding of the International Conference on Cognitive Systems and Signal Processing, Beijing, China, 19–23 November 2016; pp. 119–125.
20. Cheng, J.; Chen, X.; Liu, A.; Peng, H. A novel phonology-and radical-coded Chinese sign language recognition framework using accelerometer and surface electromyography sensors. *Sensors* **2015**, *15*, 23303–23324. [CrossRef] [PubMed]
21. Kang, H.; Wang, W.; Li, X.; Du, H.; Li, L.; Ma, J. Multichannel s-EMG system of masticatory muscles: Design and clinical application in diagnosis of dysfunction in stomatognathic system. *Biomed. Eng. Appl. Basis Commun.* **2015**, *27*, 1550008. [CrossRef]
22. Yamaguchi, T.; Mikami, S.; Saito, M.; Okada, K.; Gotouda, A. A newly developed ultraminiature wearable electromyogram system useful for analyses of masseteric activity during the whole day. *J. Prosthodont. Res.* **2018**, *62*, 110–115. [CrossRef] [PubMed]
23. Hu, X.; Wang, Z.Z.; Ren, X.M. Classification of forearm action surface EMG signals based on fractal dimension. *J. Southeast Univ.* **2005**, *21*, 324–329.
24. Michieletto, S.; Tonin, L.; Antonello, M.; Bortoletto, R.; Spolaor, F.; Pagello, E.; Menegatti, E. GMM-based single-joint angle estimation using EMG signals. In Proceedings of the 13th International Conference of Intelligent Autonomous Systems, Padova, Italy, 15–18 July 2014; pp. 1173–1184.
25. Chen, X.; Wang, Z.J. Pattern recognition of number gestures based on a wireless surface EMG system. *Biomed. Signal Process. Control* **2013**, *8*, 184–192. [CrossRef]
26. Tavakoli, M.; Benussi, C.; Lourenco, J.L. Single channel surface EMG control of advanced prosthetic hands: A simple, low cost and efficient approach. *Expert Syst. Appl.* **2017**, *79*, 322–332. [CrossRef]
27. Pantelopoulos, A.; Bourbakis, N.G. A survey on wearable sensor-based systems for health monitoring and prognosis. *IEEE Trans. Syst. Man Cybern. C* **2010**, *40*, 1–12. [CrossRef]
28. Roland, T.; Baumgartner, W.; Amsuess, S.; Russold, M.F. Signal evaluation of capacitive EMG for upper limb prostheses control using an ultra-low-power microcontroller. In Proceedings of the 2016 IEEE EMBS Conference on Biomedical Engineering and Sciences (IECBES), Kuala Lumpur, Malaysia, 4–8 December 2016; pp. 317–320.
29. Xiong, P.; Wu, C.; Zhou, H.; Song, A.; Hu, L.; Liu, X.P. Design of an accurate end-of-arm force display system based on wearable arm gesture sensors and EMG sensors. *Inform. Fusion* **2018**, *39*, 178–185. [CrossRef]
30. Mejić, L.; Došen, S.; Ilić, V.; Stanišić, D.; Jorgovanović, N. An implementation of movement classification for prosthesis control using custom-made EMG system. *Serbian J. Electr. Eng.* **2017**, *14*, 13–22. [CrossRef]
31. Costa, Á.; Itkonen, M.; Yamasaki, H.; Alnajjar, F.S.; Shimoda, S. Importance of muscle selection for EMG signal analysis during upper limb rehabilitation of stroke patients. In Proceedings of the 2017 39th Annual International Conference of the IEEE on Engineering in Medicine and Biology Society (EMBC), Seogwipo, Korea, 11–15 July 2017; pp. 2510–2513.
32. Fang, Y.; Liu, H.; Li, G.; Zhu, X. A multichannel surface EMG system for hand motion recognition. *Int. J. Hum. Robot.* **2015**, *12*, 1550011. [CrossRef]
33. Chen, H.; Zhang, Y.; Zhang, Z.; Fang, Y.; Liu, H.; Yao, C. Exploring the relation between EMG sampling frequency and hand motion recognition accuracy. In Proceedings of the 2017 IEEE International Conference on Systems, Man and Cybernetics (SMC), Banff, AB, Canada, 5–8 October 2017.
34. Alkan, A.; Günay, M. Identification of EMG signals using discriminant analysis and SVM classifier. *Expert Syst. Appl.* **2012**, *39*, 44–47. [CrossRef]
35. Subasi, A. Classification of EMG signals using combined features and soft computing techniques. *Appl. Soft Comput.* **2012**, *12*, 2188–2198. [CrossRef]
36. Gokgoz, E.; Subasi, A. Effect of multiscale PCA de-noising on EMG signal classification for diagnosis of neuromuscular disorders. *J. Med. Syst.* **2014**, *38*, 31. [CrossRef] [PubMed]
37. Abdullah, A.A.; Subasi, A.; Qaisar, S.M. Surface EMG signal classification by using WPD and ensemble tree classifiers. In Proceedings of the International Conference on Medical and Biological Engineering 2017, Sarajevo, Bosnia and Herzegovina, 16–18 March 2017; pp. 475–481.
38. Wang, G.; Wang, Z.; Chen, W.; Zhuang, J. Classification of surface EMG signals using optimal wavelet packet method based on Davies-Bouldin criterion. *Med. Biol. Eng. Comput.* **2006**, *44*, 865–872. [CrossRef] [PubMed]

39. Lai, C.; Guo, S.; Cheng, L.; Wang, W. A comparative study of feature selection methods for the discriminative Analysis of Temporal Lobe Epilepsy. *Front. Neurol.* **2017**, *8*, 1–13. [CrossRef] [PubMed]

40. Lee, S.W.; Yi, T.; Jung, J.W.; Bien, Z. Design of a gait phase recognition system that can cope with EMG electrode location variation. *IEEE Trans. Autom. Sci. Eng.* **2017**, *14*, 1429–1439. [CrossRef]

41. Wu, J.; Tian, Z.; Sun, L.; Estevez, L.; Jafari, R. Real-time American sign language recognition using wrist-worn motion and surface EMG sensors. In Proceedings of the 2015 IEEE 12th International Conference on Wearable and Implantable Body Sensor Networks, Cambridge, MA, USA, 9–12 June 2015; pp. 1–6.

42. Tosin, M.C.; Majolo, M.; Chedid, R.; Cene, V.H.; Balbinot, A. sEMG feature selection and classification using SVM-RFE. In Proceedings of the 2017 39th Annual International Conference of the IEEE Engineering in Medicine and Biology Society, Seogwipo, Korea, 11–15 July 2017; pp. 390–393.

43. Gailey, A.; Artemiadis, P.; Santello, M. Proof of concept of an online EMG-based decoding of hand postures and individual digit forces for prosthetic hand control. *Front. Neurol.* **2017**, *8*, 7. [CrossRef] [PubMed]

44. Liu, X.; Sacks, J.; Zhang, M.; Richardson, A.G.; Lucas, T.H.; Van der Spiegel, J. The virtual trackpad: An electromyography-based, wireless, real-time, low-power, embedded hand-gesture-recognition system using an event-driven artificial neural network. *IEEE Trans. Circuits Syst. II Express Briefs* **2017**, *64*, 1257–1261. [CrossRef]

45. Simpson, R.C. Smart wheelchairs: A literature review. *J. Rehabil. Res. Dev.* **2005**, *42*, 423–436. [CrossRef] [PubMed]

46. Lung, C.W.; Chen, C.L.; Jan, Y.K.; Chao, L.F.; Chen, W.F.; Liau, B.Y. Activation sequence patterns of forearm muscles for driving a power wheelchair. In Proceedings of the International Conference on Applied Human Factors and Ergonomics, Los Angeles, CA, USA, 17–21 July 2017; pp. 141–147.

47. Al-Okby, M.F.R.; Neubert, S.; Stoll, N.; Thurow, K. Development and testing of intelligent low-cost wheelchair controller for quadriplegics and paralysis patients. In Proceedings of the 2017 2nd International Conference on Bio-engineering for Smart Technologies (BioSMART), Paris, France, 30 August–1 September 2017; pp. 1–4.

48. Parker, P.; Englehart, K.; Hudgins, B. Myoelectric signal processing for control of powered limb prostheses. *J. Electromyogr. Kinesiol.* **2006**, *16*, 541–548. [CrossRef] [PubMed]

49. Ishii, C.; Konishi, R. A control of electric wheelchair using an emg based on degree of muscular activity. In Proceedings of the 2016 Euromicro Conference on Digital System Design (DSD), Limassol, Cyprus, 31 August–2 September 2016; pp. 567–574.

50. Kundu, A.S.; Mazumder, O.; Lenka, P.K.; Bhaumik, S. Hand gesture recognition based omnidirectional wheelchair control using IMU and EMG sensors. *J. Intell. Robot. Syst.* **2017**, 1–13. [CrossRef]

51. Shi, J.; Ren, X.; Liu, Z.; Chen, Z.; Duan, F. The development of a wheelchair control method based on sEMG signals. In Proceedings of the 2017 Chinese Intelligent Systems Conference, Mudanjiang, China, 14–15 October 2017; pp. 429–438.

52. Duun-Henriksen, J.; Kjaer, T.W.; Madsen, R.E.; Remvig, L.S.; Thomsen, C.E.; Sorensen, H.B.D. Channel selection for automatic seizure detection. *Clin. Neurophysiol.* **2012**, *123*, 84–92. [CrossRef] [PubMed]

53. Xu, J.; Mitra, S.; Van Hoof, C.; Yazicioglu, R.F.; Makinwa, K.A. Active electrodes for wearable EEG acquisition: Review and electronics design methodology. *IEEE Rev. Biomed. Eng.* **2017**, *10*, 187–198. [CrossRef] [PubMed]

54. Merletti, R.; Botter, A.; Troiano, A.; Merlo, E.; Minetto, M.A. Technology and instrumentation for detection and conditioning of the surface electromyographic signal: State of the art. *Clin. Biomech.* **2009**, *24*, 122–134. [CrossRef] [PubMed]

55. Chi, Y.M.; Jung, T.P.; Cauwenberghs, G. Dry-contact and noncontact biopotential electrodes: Methodological review. *IEEE Rev. Biomed. Eng.* **2010**, *3*, 106–119. [CrossRef] [PubMed]

56. Merletti, R.; Hermens, H. Detection and conditioning of the surface EMG signal. In *Electromyography: Physiology, Engineering and Noninvasive Applications*; Merletti, R., Parker, P.A., Eds.; IEEE Press and John Wiley & Sons: Hoboken, NJ, USA, 2004; pp. 107–132. ISBN 0-471-67580-6.

57. Oskoei, M.A.; Hu, H. Support vector machine-based classification scheme for myoelectric control applied to upper limb. *IEEE Trans. Biomed. Eng.* **2008**, *55*, 1956–1965. [CrossRef] [PubMed]

58. Bakshi, B.R. Multiscale PCA with application to multivariate statistical process monitoring. *AIChE J.* **1998**, *44*, 1596–1610. [CrossRef]

59. Sheriff, M.Z.; Mansouri, M.; Karim, M.N.; Nounou, M. Fault detection using multiscale PCA-based moving window GLRT. *J. Process Control* **2017**, *54*, 47–64. [CrossRef]

60. Alickovic, E.; Subasi, A. Effect of multiscale PCA de-noising in ECG beat classification for diagnosis of cardiovascular diseases. *Circuits Syst. Signal Process.* **2015**, *34*, 513–533. [CrossRef]

61. Zhang, M.; Sawchuk, A.A. Human daily activity recognition with sparse representation using wearable sensors. *IEEE J. Biomed. Health Inform.* **2013**, *17*, 553–560. [CrossRef] [PubMed]

62. Xu, Q.; Zhou, H.; Wang, Y.; Huang, J. Fuzzy support vector machine for classification of EEG signals using wavelet based features. *Med. Eng. Phys.* **2009**, *31*, 858–865. [CrossRef] [PubMed]

63. Guyon, I.; Elisseeff, A. An introduction to variable and feature selection. *J. Mach. Learn. Res.* **2003**, *3*, 1157–1182.

64. Schiff, S.J.; Sauer, T.; Kumar, R.; Weinstein, S.L. Neuronal spatiotemporal pattern discrimination: The dynamical evolution of seizures. *Neuroimage* **2005**, *28*, 1043–1055. [CrossRef] [PubMed]

65. Akata, Z.; Perronnin, F.; Harchaoui, Z.; Schmid, C. Good practice in large-scale learning for image classification. *IEEE Trans. Pattern Anal. Mach. Intell.* **2014**, *36*, 507–520. [CrossRef] [PubMed]

66. Yin, Z.; Wang, Y.; Liu, L.; Zhang, W.; Zhang, J. Cross-subject EEG feature selection for emotion recognition using transfer recursive feature elimination. *Front. Neurorobot.* **2017**, *11*, 19. [CrossRef] [PubMed]

67. Zhou, X.; Tuck, D.P. MSVM-RFE: Extensions of SVM-RFE for multiclass gene selection on DNA microarray data. *Bioinformatics* **2007**, *23*, 1106–1114. [CrossRef] [PubMed]

68. Wan, B.; Wu, R.; Zhang, K.; Liu, L. A new subtle hand gestures recognition algorithm based on EMG and FSR. In Proceedings of the 2017 IEEE 21st International Conference on Computer Supported Cooperative Work in Design (CSCWD), Wellington, New Zealand, 26–28 April 2017; pp. 127–132.

69. Zhang, H.; Yang, D.; Shi, C.; Jiang, L.; Liu, H. Robust EMG pattern recognition with electrode donning/doffing and multiple confounding factors. In Proceedings of the International Conference on Intelligent Robotics and Applications, Wuhan, China, 16–18 August 2017; pp. 413–424.

70. Kondo, G.; Kato, R.; Yokoi, H.; Arai, T. Classification of individual finger motions hybridizing electromyogram in transient and converged states. In Proceedings of the 2010 IEEE International Conference on Robotics and Automation (ICRA), Anchorage, AK, USA, 3–7 May 2010; pp. 2909–2915.

71. Liaw, A.; Wiener, M. Classification and regression by random forest. *R News* **2002**, *2*, 18–22.

72. Breiman, L. Random forests. *Mach. Learn.* **2001**, *45*, 5–32. [CrossRef]

73. Chang, C.C.; Lin, C.J. LIBSVM: A library for support vector machines. *ACM Trans. Intell. Syst. Technol.* **2011**, *2*, 1–27. [CrossRef]

74. Englehart, K.; Hudgins, B. A robust, real-time control scheme for multifunction myoelectric control. *IEEE Trans. Biomed. Eng.* **2003**, *50*, 848–854. [CrossRef] [PubMed]

75. Dobrowolski, A.P.; Wierzbowski, M.; Tomczykiewicz, K. Multiresolution MUAPs decomposition and SVM-based analysis in the classification of neuromuscular disorders. *Comput. Methods Programs Biomed.* **2011**, *107*, 393–403. [CrossRef] [PubMed]

76. Balouchestani, M.; Krishnan, S. Effective low-power wearable wireless surface EMG sensor design based on analog-compressed sensing. *Sensors* **2014**, *14*, 24305–24328. [CrossRef] [PubMed]

77. Youn, W.; Kim, J. Development of a compact-size and wireless surface EMG measurement system. In Proceedings of the ICROS-SICE International Joint Conference 2009, Fukuoka, Japan, 18–21 August 2009; pp. 1625–1628.

78. Lin, K.; Wang, X.; Zhang, X.; Wang, B.; Huang, J.; Huang, F. An FPC based flexible dry electrode with stacked double-micro-domes array for wearable biopotential recording system. *Microsyst. Technol.* **2017**, *23*, 1443–1451. [CrossRef]

79. Guo, W.; Yao, P.; Sheng, X.; Liu, H.; Zhu, X. A wireless wearable sEMG and NIRS acquisition system for an enhanced human-computer interface. In Proceedings of the 2014 IEEE International Conference on Systems, Man and Cybernetics, San Diego, CA, USA, 5–8 October 2014; pp. 2192–2197.

80. Phinyomark, A.; Phukpattaranont, P.; Limsakul, C. Feature reduction and selection for EMG signal classification. *Expert Syst. Appl.* **2012**, *39*, 7420–7431. [CrossRef]

81. Srisuwan, N.; Phukpattaranont, P.; Limsakul, C. Comparison of feature evaluation criteria for speech recognition based on electromyography. *Med. Biol. Eng. Comput.* **2017**, 1–11. [CrossRef] [PubMed]

82. Huang, H.P.; Liu, Y.H.; Liu, L.W.; Wong, C.S. EMG classification for prehensile postures using cascaded architecture of neural networks with self-organizing maps. In Proceedings of the 2003 IEEE International Conference on Robotics & Automation, Taipei, Taiwan, 14–19 September 2003; Volume 1, pp. 1497–1502.

83. Cler, M.J.; Stepp, C.E. Discrete Versus Continuous Mapping of Facial Electromyography for Human–Machine Interface Control: Performance and Training Effects. *IEEE Trans. Neural Syst. Rehabil. Eng.* **2015**, *23*, 572–580. [CrossRef] [PubMed]
84. Zhang, R.; Li, Y.; Yan, Y.; Zhang, H.; Wu, S.; Yu, T.; Gu, Z. Control of a wheelchair in an indoor environment based on a brain–computer interface and automated navigation. *IEEE Trans. Neural Syst. Rehabil. Eng.* **2016**, *24*, 128–139. [CrossRef] [PubMed]
85. Chin, Z.Y.; Ang, K.K.; Wang, C.; Guan, C. Navigation in a virtual environment using multiclass motor imagery Brain-Computer Interface. In Proceedings of the 2013 IEEE Symposium on the Computational Intelligence, Cognitive Algorithms, Mind, and Brain (CCMB), Singapore, 16–19 April 2013; pp. 152–157.
86. Kucukyildiz, G.; Ocak, H.; Karakaya, S.; Sayli, O. Design and implementation of a multi sensor based brain computer interface for a robotic wheelchair. *J. Intell. Robot. Syst.* **2017**, *87*, 247–263. [CrossRef]
87. Maeda, Y.; Ishibashi, S. Operating instruction method based on EMG for omnidirectional wheelchair robot. In Proceedings of the 2017 Joint 17th World Congress of International Fuzzy Systems Association and 9th International Conference on Soft Computing and Intelligent Systems (IFSA-SCIS), Otsu, Japan, 27–30 June 2017; pp. 1–5.

sensors

[MDPI]

Article

Evaluating the Influence of Chromatic and Luminance Stimuli on SSVEPs from Behind-the-Ears and Occipital Areas

Alan Floriano [1,†], Pablo F. Diez [2,3] and Teodiano Freire Bastos-Filho [1,*]

[1] Postgraduate Program in Electrical Engineering, Federal University of Espirito Santo, Vitoria 29075-910, Brazil; afloriano.ufes@gmail.com

[2] Gabinete de Tecnologia Medica (GATEME), Facultad de Ingenieria, Universidad Nacional de San Juan, San Juan J5400ARL, Argentina; pdiez@gateme.unsj.edu.ar

[3] Consejo Nacional de Investigaciones Científicas y Técnicas (CONICET), San Juan C1425FBQ, Argentina

* Correspondence: teodiano.bastos@ufes.br; Tel.: +55-27-4009-2077

† Current address: Av. Fernando Ferrari, 514, Goiabeiras, Vitoria CEP 29075-910, Brazil.

Received: 30 November 2017; Accepted: 14 February 2018; Published: 17 February 2018

Abstract: This work presents a study of chromatic and luminance stimuli in low-, medium-, and high-frequency stimulation to evoke steady-state visual evoked potential (SSVEP) in the behind-the-ears area. Twelve healthy subjects participated in this study. The electroencephalogram (EEG) was measured on occipital (Oz) and left and right temporal (TP9 and TP10) areas. The SSVEP was evaluated in terms of amplitude, signal-to-noise ratio (SNR), and detection accuracy using power spectral density analysis (PSDA), canonical correlation analysis (CCA), and temporally local multivariate synchronization index (TMSI) methods. It was found that stimuli based on suitable color and luminance elicited stronger SSVEP in the behind-the-ears area, and that the response of the SSVEP was related to the flickering frequency and the color of the stimuli. Thus, green-red stimulus elicited the highest SSVEP in medium-frequency range, and green-blue stimulus elicited the highest SSVEP in high-frequency range, reaching detection accuracy rates higher than 80%. These findings will aid in the development of more comfortable, accurate and stable BCIs with electrodes positioned on the behind-the-ears (hairless) areas.

Keywords: SSVEP; visual stimuli; BCI; hairless area

---

1. Introduction

The elicited response in the visual cortex by light stimuli flickering at a constant frequency is known as steady-state visual evoked potential (SSVEP) [1]. In the electroencephalogram (EEG), these potentials manifest as an oscillatory component in the signal, with the same frequency (and/or its harmonics) of the visual stimulation [1]. SSVEP can normally be evoked up to 90 Hz [2], and three stimuli bands can be identified: low (up to 12 Hz), medium (12–30 Hz), and high-frequency ($\geq$30 Hz) [3–5].

SSVEPs have been used for studies concerning the brain, vision, and the development of brain–computer interfaces (BCIs) [6]. People with severe disabilities can use BCIs as an alternative channel for interaction and communication with the environment around them, only using brain activity [7]. In SSVEP-based BCIs, each stimulus flickers at a specified frequency [8]. Thus, when a person gazes at one of the stimuli, an SSVEP is evoked in the brain [6,9], which can be detected in the EEG signal, and later translated into a control command [10].

Researches have used the SSVEP response to develop assistive technologies, such as robotic wheelchairs [11,12] and robotic exoskeletons [13], as well as for rehabilitation [14,15], communication [16,17], mobile robot control [18], cursor control for computer interaction [4,19], and entertainment [8,20,21].

The SSVEP response is generally maximum on the occipital area of the scalp, and consequently, SSVEPs are strongly detected in the electrodes located at this area [6]. Hence, most of existing SSVEP-based BCIs use electrodes located at O1, O2, and Oz positions. However, this area is generally covered by hair, which causes some complications in the electrode contact with the skin [22,23]. This represents an important drawback in BCI implementation due to the loss of contact between electrode and skin, drying of the gel, especially in long-term operation. In contrast, in hairless regions, it is possible to use different kinds of electrodes, and these drawbacks may be mitigated. Thus, more comfortable BCIs can be designed.

Different studies based on magnetoencephalography (MEG), positron emission tomography (PET), functional magnetic resonance imaging (fMRI), and EEG have demonstrated that SSVEPs can be found in other brain areas, such as parietal, temporal, frontal, and prefrontal areas [9,24–30]. Thus, using EEG signals from hairless regions to develop a SSVEP-based BCI is a possible option. For example, a system measured SSVEP right in-the-ear using stimulation at low-frequency range (8–11 Hz) [31]. In other work, an electrode was positioned behind-the-ear to acquire the SSVEP [32]. Hsu et al. measured the EEG on the forehead, employing medium-frequency stimuli [33]. In another work, electrodes placed at three hairless areas (behind-the-ears, neck, and face) were used to detect SSVEP [34], in which the authors concluded that electrodes positioned behind-the-ears are the best candidates to build an SSVEP-based BCI in a hairless area. These works used stimuli based only on luminance modulation.

However, visual stimuli that use colors (green-blue or green-red) and luminance combination can increase the evoked response [35–39]. Moreover, it was found that color information is mediated by specialized neurons that are clustered within the temporal areas [40]. Besides, there are color-selective neurons in the inferotemporal cortex [41]. The inferotemporal cortex receives projections from the primary visual cortex (ventral pathways) [42,43], which are both color-sense-associated and object recognition pathways that detect luminance and color. Thus, colored stimuli can enhance the detection of SSVEP in behind-the-ears regions, due to its proximity to the temporal area. Our hypothesis is that chromatic and luminance stimuli can evoke a better response than luminance stimuli in this area.

Therefore, in our work, we present a comparative study of chromatic and luminance stimuli flickering at low-, medium-, and high-frequencies to evoke SSVEP responses in behind-the-ears areas. Basically, this study aims to answer three questions: (1) What is the influence of chromatic and luminance stimuli on SSVEP from behind-the-ears? (2) What is the best combination (green-blue or green-red—note that we did not use the blue-red combination, as these colors are the worst case for photosensitive epilepsy, especially at 15 Hz [44])? (3) How is the SSVEP response evoked by these stimuli in low-, medium-, and high-frequency bands? Therefore, the results of the current work will help in the development of more accurate and comfortable BCI systems.

2. Materials and Methods

2.1. Data Acquisition

Twelve healthy subjects (ages 26.1 ± 4.1; 6 F and 6 M) with normal or corrected-to-normal vision participated in this study. The EEG recordings were conducted in a laboratory with low background noise and dim luminance. Previous to participation in this study, all volunteers read an information sheet and provided written consent to participate. Ethical approval was granted by the institutional ethics committee. The subjects did not receive any financial reward for their participation.

The EEG was measured over occipital (Oz) and left and right temporal (TP9 and TP10) areas (see Figure 1). The ground electrode was placed at A2. The EEG signals were acquired with a Grass 15LT amplifier system, and digitalized with a NI-DAQ-Pad6015 (sampling frequency: 256 Hz).

The cut-off frequencies of the analogical pass-band filter were set to 1 and 100 Hz. Additionally, a notch filter for 50 Hz line interference was applied.

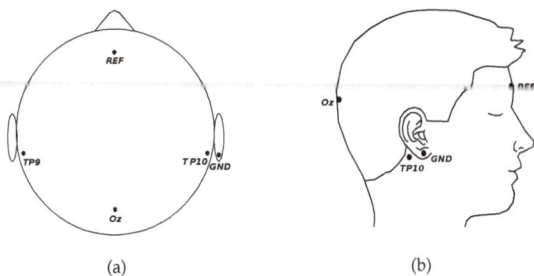

(a)   (b)

Figure 1. Positions on the scalp where the electrodes were located. (**a**) Top view of positions; (**b**) Side view of the positions. Oz: occipital area; TP9: left temporal area; TP10: right temporal area; REF: reference electrode; GND: ground electrode.

2.2. Visual Stimulation

The visual stimulation was performed by light-emitting diodes (LEDs) that illuminated a diffusion board of 4 cm × 4 cm. The LEDs were red, blue, green, and white. Each LED could flicker at different frequencies from 5 Hz to 65 Hz with an interval of 5 Hz. Therefore, the stimulation range comprised the three SSVEP bands (low-, medium-, and high-frequency). The frequency of the LEDs was precisely controlled with an Xilinx Spartan3E field-programmable gate array (FPGA) on a Nexys board (Digilent Inc., Pullman, USA). The 50 Hz frequency was not used as a stimulation frequency, because this is the Argentinian power line frequency. The light intensity of the the green and white LEDs was 750 mcd and 250 mcd for blue and red LEDs.

The setup consisted of three different stimuli (see Figure 2). The first stimulus was white (W) LED for the luminance condition. The W stimulus was configured with 90% of contrast between off and on state, as done by [33]. The other two stimuli were green-red (G-R) stimulus and green-blue (G-B) stimulus for the chromatic and luminance conditions. In this case, the contrast was configured with 50%, such as done by [39]. Figure 2 shows the transition of the two states of the visual stimuli. Each state remained activated for half of the period of the stimulation frequency ($f = 1/T$, where T is the period). For the luminance stimulus (W), the two states represented the light on and off. For the G-R and G-B stimulus, the two states were green-red and green-blue, respectively.

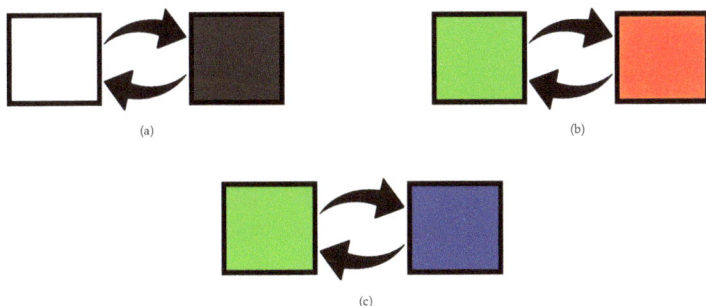

Figure 2. Visual stimulation used for the experiments: (**a**) Luminance stimulus (white, W); (**b**) green-red (G-R) stimulus; (**c**) green-blue (G-B) stimulus.

2.3. Experimental Protocol

Each subject sat in a chair at 60 cm from the stimulus. The experiment was divided into five runs (Figure 3a). At each run, the three possible stimuli (G-R, G-B, and W) were showed to the volunteer (Figure 3b). At each colored stimulus, the 12 frequencies were presented. Thus, each stimulus comprised 12 trials, and each trial lasted 7 s (Figure 3c). Thus, 12 trials (one per frequency) of the same colored stimulus were presented to the volunteer. Later, the process was repeated for the other two colored stimuli, which comprised a run. Finally, the run was repeated five times. The stimulation frequencies and the colored stimuli were randomly presented to each volunteer. In order to avoid expectation effects, a variable separation time (2–4 s) between trials was used. The trial began with a beep (at t = 0 s), and 2 s later the stimulus was turned on. The stimulus stayed on until the end of the trial at t = 7 s. At this moment, a feedback was presented to the volunteer indicating whether the SSVEP was detected or not. The volunteer could relax for 2–5 min.

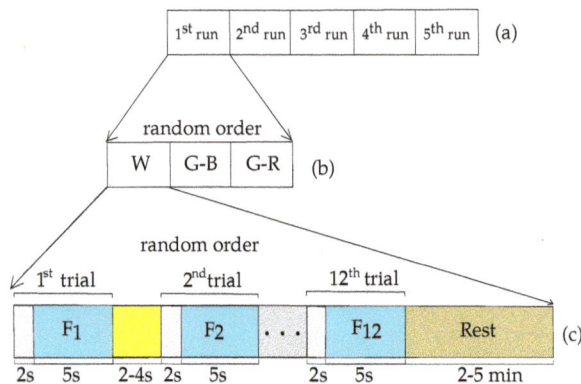

Figure 3. Protocol of the experiment: (**a**) experiment divided into five runs; (**b**) three colored stimuli presented in random order to each volunteer; (**c**) 12 frequencies randomly presented for each colored stimulus.

2.4. EEG Signal Processing

The EEG was preprocessed using a Butterworth filter, order 6, with cut-off frequencies set at 3 and 70 Hz. Later, the EEG between t = 2 s and t = 7 s was extracted for analyzing in the next step. Then, the magnitude of the frequency components of the signal was calculated based on the discrete Fourier transform (DFT) of the signal $x[n]$ defined as:

$$F(f) = \left| \sum_{n=0}^{N-1} x[n] e^{-j2\pi f n T_s} \right|, \qquad (1)$$

where $F(f)$ is the magnitude of the signal, T_s is the sampling period, N is the total number of samples of the signal, and f is the frequency.

The signal-to-noise ratio (SNR) measurement was computed based on the values extracted from Equation (1). The SNR of SSVEP at a single channel is defined as the ratio of $F(f)$ to the mean amplitude of the K neighboring frequencies [34,45]:

$$SNR = \frac{K \times F(f)}{\sum_{n=1}^{K/2} [F(f + n\Delta f) + F(f - n\Delta f)]}, \qquad (2)$$

where Δf is the frequency resolution (0.2 Hz in this study), and K was set to 8 (i.e., four frequencies on each side) [46].

2.5. Statistical Evaluation

For the statistical analysis of the results, the Friedman test for simultaneous comparison of more than two groups was used. Post-hoc pairwise comparisons using Wilcoxon signed-rank test were also conducted, in which a level of $p < 0.05$ was selected as the threshold for statistical significance.

3. Results

This section is divided in three parts, where results about amplitude, SNR, and a simulation of SSVEP classification are presented.

3.1. Amplitude

Figure 4 shows the average amplitudes of the elicited SSVEP of all volunteers for the three visual stimuli. The frequencies marked with an asterisk show statistical significance (p-value < 0.05) using the Friedman test. The amplitudes were calculated according to Equation (1). Then, for each volunteer, the amplitude value was obtained at each frequency $F(f)$, and the average was computed across the volunteers.

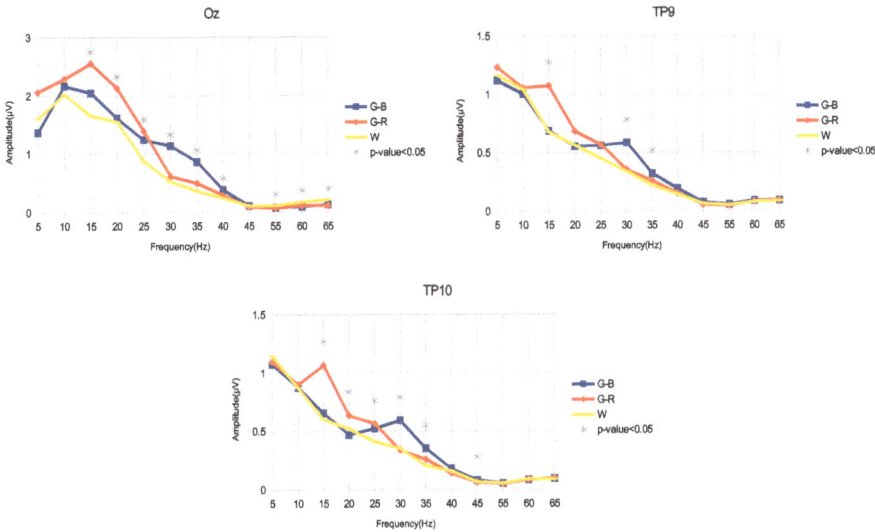

Figure 4. Average of the steady-state visual evoked potential (SSVEP) amplitudes of all volunteers for the Oz, TP9, and TP10 channels using three different stimuli. The frequencies with statistical significance (p-value < 0.05) based on the Friedman test are marked with an asterisk.

At the occipital region, the G-R stimulus showed a higher response when compared with the W stimulus in the medium-frequency range (15–25 Hz, with p-value < 0.05). In contrast, in the high-frequency range (30–40 Hz, p-value < 0.05) the G-B stimulus presented a better response when compared with the W stimulus. In the 55–65 Hz interval, the W stimulus achieved a better response than G-R and G-B stimuli.

In the temporal region (TP9 and TP10), a similar behavior was observed; i.e., in the medium-frequency range, the G-R stimulus achieved higher amplitudes (TP9: 15 Hz, with p-value < 0.05; TP10: 15–25 Hz, with p-value < 0.05) than W and G-B stimuli. In the high-frequency range, the G-B stimulus showed a better response (30–35 Hz, with p-value < 0.05) than the other stimuli.

3.2. SNR

Figure 5 shows the average SNR of the SSVEP of all volunteers for the three stimuli. The frequencies marked with an asterisk show statistical significance (*p*-value < 0.05) using the Friedman test.

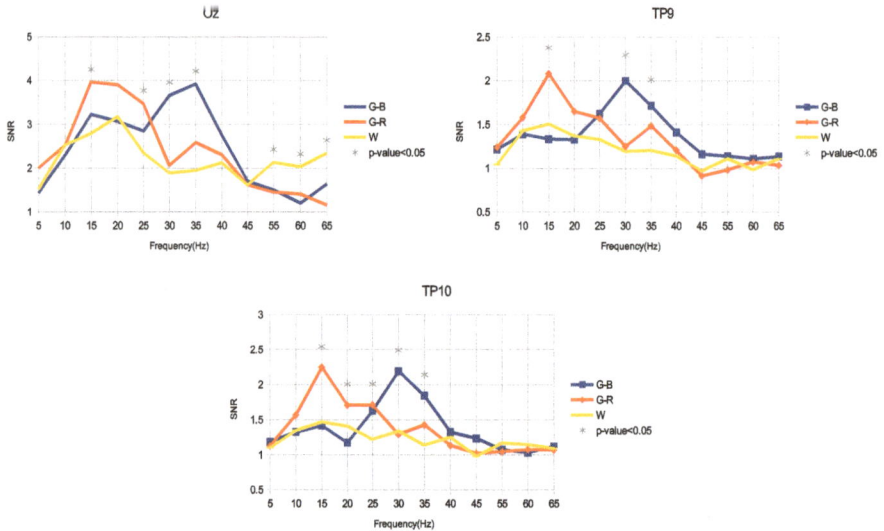

Figure 5. Average of the SSVEP SNR of all volunteers for the Oz, TP9, and TP10 channels using the three different stimulus configurations. The frequencies with statistical significance (*p*-value < 0.05) based on the Friedman test are marked with an asterisk.

At the occipital region, the G-R stimulus showed a higher response than the W stimulus in the medium-frequency range (15–25 Hz). In the high-frequency range (30–40 Hz), the G-B stimulus showed a better response (30–35 Hz, *p*-value < 0.05) than the W stimulus. In the 55–65 Hz range, the luminance stimulus (W) presented a better response than the G-R and G-B stimuli.

Again, a similar behavior was observed in the temporal region (TP9 and TP10); i.e., in the medium-frequency range, there was a higher response of the G-R stimulus (TP9: 15 Hz, with *p*-value < 0.05; TP10: 15–25 Hz, with *p*-value < 0.05) compared with the luminance stimulus (W). In the high-frequency range the G-B stimulus showed a better response (30–35 Hz, with *p*-value < 0.05) when compared with the W stimulus.

3.3. Simulated SSVEP Classification

Aiming to provide an overview of how the different SNR of SSVEP would impact the design of a future BCI system, a simulated online analysis was performed. Hence, the accuracy of the SSVEP detection [47,48] and information transfer rate (ITR) [9] were used. To emulate an online detection process, the EEG signal was segmented in 4 s windows (as done by [34]). Then, power spectral density analysis (PSDA) [45], canonical correlation analysis (CCA) [49] and temporally local multivariate synchronization index (TMSI) [50] methods, with criterion of maximum value were used for data classification. For this test, two frequencies were chosen in order to simulate a binary BCI. Thus, 15 and 20 Hz were chosen within the medium-frequency range, and 30 and 35 Hz were chosen for the high-frequency range, as these frequencies presented the best SNR (see Figure 5).

Figures 6 and 7 present the average accuracy of the classification of all volunteers for the three stimuli in medium- and high-frequency ranges.

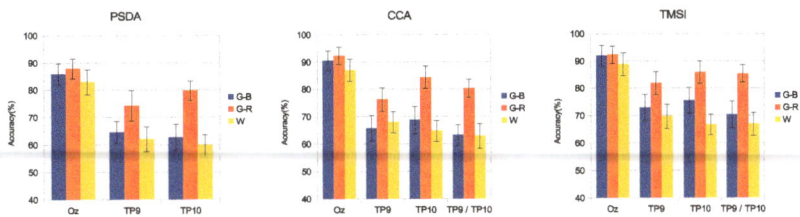

Figure 6. Average accuracy of all volunteers for the three stimuli in medium-frequency range. Error bars indicate standard errors. CCA: canonical correlation analysis; TMSI: temporally local multivariate synchronization index.

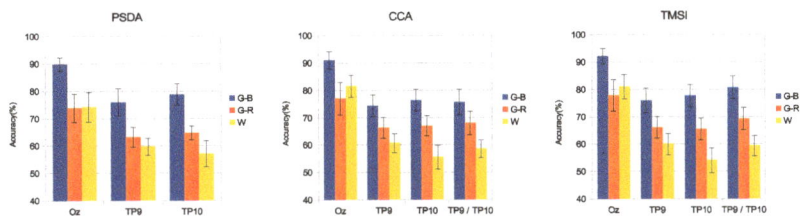

Figure 7. Average accuracy of all volunteers for the three stimuli in high-frequency range. Error bars indicate standard errors.

Figures 8 and 9 present the average ITR of all volunteers for the three stimuli in medium- and high-frequency ranges.

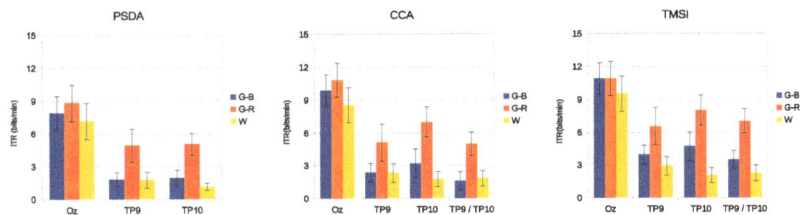

Figure 8. Average information transfer rate (ITR) of all volunteers for the three stimuli for the medium-frequency range. Error bars indicate standard errors.

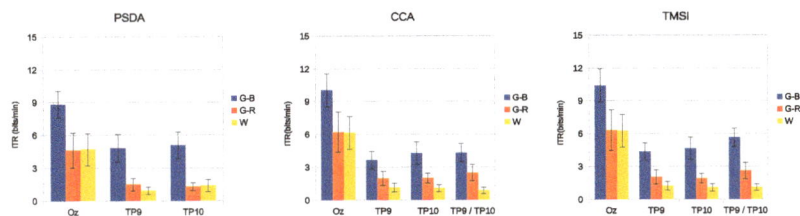

Figure 9. Average ITR of all volunteers for the three stimuli for the high-frequency range. Error bars indicate standard errors.

4. Discussion

The literature reports that the visual stimuli that combine colors and luminance can increase the evoked response of P300 potentials [35,36]. Similarly, it was demonstrated that the combination of G-B and luminance can evoke a better response in the SSVEP from the occipital region [37,38]. In other work [39], G-R stimuli combined with luminance changes obtained a better response at a modulated frequency of 15 Hz. These works measured the EEG on occipital and parietal regions of the scalp.

Currently, the BCI community is looking at how to transfer these systems from the lab to the patient's home. Thus, more accurate and comfortable BCI systems must be designed. This way, measuring the EEG from hairless positions presents advantages to the user, and recently, these kinds of BCI systems have been reported in the literature [31,32,34]. These studies demonstrated that it is possible to develop a BCI based on EEG measured from hairless regions; however, concerning the wide frequency range and types of stimulation (color and luminance), the question about the best frequency and type of stimulation remains unclear.

In the current work, SSVEP from behind-the-ears areas (TP9 and TP10) was elicited by three stimuli (G-B, G-R, and W) flickering at low-, medium-, and high-frequency. The aim of this work was to analyze how these stimuli influence the SSVEP response from the behind-the-ears areas. Higher amplitude (Figure 4) and SNR (Figure 5) of the SSVEP were observed when stimuli that combined color and luminance (G-R and G-B) were applied. Particularly, the best response in the medium-frequency band (15–25 Hz) was obtained with G-R stimulation. On the other hand, G-B stimulation showed the best response in the high-frequency range (30–40 Hz). Therefore, a suitable color and luminance stimulation allows the achievement of higher amplitudes and higher SNR from behind-the-ears areas, and consequently, an accurate and comfortable BCI may be designed.

At the occipital electrode, G-R presented the best response in the medium-frequency range, in accordance with [39]. In the current work, the Oz electrode was used as the standard measure in SSVEP and to corroborate the results with TP9 and TP10 electrodes.

The results of the simulated online analysis showed that the combination of colors and luminance also improved the SSVEP detection accuracy and ITR. When stimuli in the medium-frequency range were used, G-R presented better results (see Figures 6 and 8). Using PSDA in TP9 and TP10, the detection accuracy was increased by 12% and 19%, respectively, in comparison with the luminance stimulus (W). Similar behavior was observed with CCA and TMSI methods, with G-R stimulus obtaining better results than other stimuli. CCA and TMSI achieved higher accuracy detection values than PSDA. Specifically, CCA achieved at Oz: $92.3 \pm 3.0\%$, at TP9: $76.1 \pm 4.8\%$, at TP10: $84.3 \pm 3.7\%$, and at TP9/TP10: $80.2 \pm 2.8\%$. On the other hand, TMSI achieved at Oz: $92.3 \pm 3.2\%$, at TP9: $81.8 \pm 4.2\%$, at TP10: $85.8 \pm 4.2\%$, and at TP9/TP10: $85.3 \pm 3.3\%$. As a counterpart, the 15–25 Hz stimulation range can provoke epileptic seizures [1,51].

The high-frequency band is known for its low-amplitude SSVEP, making difficult to implement a BCI [52]. However, when using G-B stimuli, the detection accuracy and ITR were increased on both the occipital and temporal regions (see Figures 7 and 9). The G-B stimulus presented the higher accuracy when compared with G-R and W stimuli using PSDA (Oz: $89.8 \pm 2.5\%$, TP9: $76.0 \pm 5.0\%$ TP10: $78.8 \pm 4.0\%$), CCA (Oz: $91.0 \pm 3.1\%$, TP9: $74.3 \pm 4.0\%$, TP10: $76.3 \pm 4.0\%$, TP9/TP10: $75.7 \pm 4.6\%$), and TMSI (Oz: $92.2 \pm 2.8\%$, TP9: $76.0 \pm 4.5\%$, TP10: $77.7 \pm 4.0\%$, TP9/TP10: $80.7 \pm 4.2\%$). In addition, a previous work [44] reports that the green-blue chromatic flicker is the safest stimulus for human visual photosensitivity.

Moreover, the high-frequency stimulation band can provide a reduction of epileptic seizures [51], false positives due to alpha rhythm (8–13 Hz) [1,53], migraine headaches [54], and visual fatigue [54]. Hence, the development of more comfortable and stable BCIs are possible [53]. In the current work, we showed that using G-B stimulus at high frequencies allowed an accuracy rate close to 80% to be obtained in a hairless area.

5. Conclusions

This work presented a study of chromatic and luminance stimuli to evoke SSVEPs on behind-the-ears areas. The SSVEP was elicited at low-, medium-, and high-frequency stimulation, and the EEG was measured at left and right temporal (TP9 and TP10) areas. It was found that stimuli based on a suitable color and luminance elicited stronger SSVEP on the behind-the-ears areas. Interestingly, we found a different response of SSVEP related to frequency and color of the stimuli. G-R stimulus elicited the highest SSVEP in the medium-frequency range (15–25 Hz), and G-B stimulus elicited the highest SSVEP at high-frequencies (30–40 Hz). Moreover, detection accuracies of around 80% using PSDA and even more than 80% using CCA or TMSI on a hairless area were obtained. These findings allow the development of more comfortable, accurate, and stable BCIs with electrodes positioned on the behind-the-ears (hairless) areas.

6. Future Works

Currently, we are developing an online BCI based on the results of this research to command a robotic wheelchair. Other possible applications for this BCI are spellers for communication, command of telepresence robots, command of autonomous car or for turning on/off home appliances in domotics, and other applications required by people with severe disabilities. A recent work of Chien et al. applied a new paradigm to elicit SSVEPs with little or no flickering sensation and using color stimuli on a screen [48]. However, that work did not use its paradigm on non-hair positions, which can be evaluated using our methodology.

Acknowledgments: The authors acknowledge the financial support from Capes (Brazil) and CONICET (Argentina) and technical support from the Federal University of Espirito Santo (Brazil) and the National University of San Juan (Argentina).

Author Contributions: A.F. performed the experiments, analyzed the data and drafted the manuscript. P.D. contributed to designing the experimental protocol, analyzed the data and drafted the manuscript. T.B. contributed with materials and analysis tools, and contributed to the data analysis and interpretation of the findings. All authors read, edited, and approved the final manuscript.

Conflicts of Interest: The authors declare no conflict of interest.

Abbreviations

The following abbreviations are used in this manuscript:

SSVEP   Steady-State Visual Evoked Potential
BCI   Brain-Computer Interface
SNR   Signal-to-Noise Ratio
REF   Reference Electrode
GND   Ground Electrode
G-R   Green-Red Stimulus
G-B   Green-Blue Stimulus
W   White Stimulus

References

1. Zhu, D.; Bieger, J.; Molina, G.G.; Aarts, R.M. A survey of stimulation methods used in SSVEP-based BCIs. *Comput. Intell. Neurosci.* **2010**, *2010*, 1–12.
2. Herrmann, C. Human EEG responses to 1–100 Hz flicker: Resonance phenomena in visual cortex and their potential correlation to cognitive phenomena. *Exp. Brain Res.* **2001**, *137*, 346–353.
3. Regan, D. *Human Brain Electrophysiology: Evoked Potentials and Evoked Magnetic Fields in Science and Medicine*; Elsevier: New York, NY, USA, 1989.
4. Diez, P.F.; Mut, V.A.; Perona, E.M.A.; Leber, E.L. Asynchronous BCI control using high-frequency SSVEP. *J. Neuroeng. Rehabil.* **2011**, *8*, 39.

5. Chabuda, A.; Durka, P.; Żygierewicz, J. High frequency SSVEP-BCI with hardware stimuli control and phase-synchronized comb filter. *IEEE Trans. Neural Syst. Rehabil. Eng.* **2017**, *26*, 344–352.

6. Norcia, A.M.; Appelbaum, L.G.; Ales, J.M.; Cottereau, B.R.; Rossion, B. The steady-state visual evoked potential in vision research: A review. *J. Vis.* **2015**, *15*, 4.

7. Nicolas-Alonso, L.F.; Gomez-Gil, J. Brain computer interfaces, a review. *Sensors* **2012**, *12*, 1211–1279.

8. Chumerin, N.; Manyakov, N.V.; van Vliet, M.; Robben, A.; Combaz, A.; Van Hulle, M. Steady state visual evoked potential-based computer gaming on a consumer-grade eeg device. *IEEE Trans. Comput. Intell. AI Games* **2013**, *5*, 100–110.

9. Vialatte, F.B.; Maurice, M.; Dauwels, J.; Cichocki, A. Steady-state visually evoked potentials: Focus on essential paradigms and future perspectives. *Prog. Neurobiol.* **2010**, *90*, 418–438.

10. Allison, B.; Lüth, T.; Valbuena, D.; Teymourian, A.; Volosyak, I.; Gräser, A. BCI demographics: How many (and what kinds of) people can use an SSVEP BCI? *IEEE Trans. Neural Syst. Rehabil. Eng.* **2010**, *18*, 107–116.

11. Müller, S.T.; Celeste, W.C.; Bastos-Filho, T.F.; Sarcinelli-Filho, M. Brain-computer interface based on visual evoked potentials to command autonomous robotic wheelchair. *J. Med. Biol. Eng.* **2010**, *30*, 407–415.

12. Diez, P.F.; Müller, S.M.T.; Mut, V.A.; Laciar, E.; Avila, E.; Bastos-Filho, T.F.; Sarcinelli-Filho, M. Commanding a robotic wheelchair with a high-frequency steady-state visual evoked potential based brain–computer interface. *Med. Eng. Phys.* **2013**, *35*, 1155–1164.

13. Kwak, N.S.; Müller, K.R.; Lee, S.W. A lower limb exoskeleton control system based on steady state visual evoked potentials. *J. Neural Eng.* **2015**, *12*, 056009.

14. Zeng, X.; Zhu, G.; Yue, L.; Zhang, M.; Xie, S. A Feasibility Study of SSVEP-Based Passive Training on an Ankle Rehabilitation Robot. *J. Healthc. Eng.* **2017**, *2017*, 6819056.

15. Xie, S.; Meng, W. SSVEP-Based BCI for Lower Limb Rehabilitation. In *Biomechatronics in Medical Rehabilitation*; Springer: Cham, Switzerland, 2017; pp. 69–86.

16. Nakanishi, M.; Wang, Y.; Wang, Y.T.; Mitsukura, Y.; Jung, T.P. A high-speed brain speller using steady-state visual evoked potentials. *Int. J. Neural Syst.* **2014**, *24*, 1450019.

17. Won, D.O.; Hwang, H.J.; Dähne, S.; Müller, K.R.; Lee, S.W. Effect of higher frequency on the classification of steady-state visual evoked potentials. *J. Neural Eng.* **2015**, *13*, 016014.

18. Diez, P.F.; Mut, V.A.; Laciar, E.; Perona, E.M.A. Mobile robot navigation with a self-paced brain-computer interface based on high-frequency SSVEP. *Robotica* **2014**, *32*, 695–709.

19. Wu, C.H.; Chang, H.C.; Lee, P.L.; Li, K.S.; Sie, J.J.; Sun, C.W.; Yang, C.Y.; Li, P.H.; Deng, H.T.; Shyu, K.K. Frequency recognition in an SSVEP-based brain computer interface using empirical mode decomposition and refined generalized zero-crossing. *J. Neurosci. Methods* **2011**, *196*, 170–181.

20. Lalor, E.C.; Kelly, S.P.; Finucane, C.; Burke, R.; Smith, R.; Reilly, R.B.; Mcdarby, G. Steady-state VEP-based brain-computer interface control in an immersive 3D gaming environment. *EURASIP J. Appl. Signal Process.* **2005**, *2005*, 706906.

21. Shyu, K.K.; Lee, P.L.; Lee, M.H.; Lin, M.H.; Lai, R.J.; Chiu, Y.J. Development of a low-cost FPGA-based SSVEP BCI multimedia control system. *IEEE Trans. Biomed. Circuits Syst.* **2010**, *4*, 125–132.

22. Wang, Y.T.; Wang, Y.; Cheng, C.K.; Jung, T.P. Measuring steady-state visual evoked potentials from non-hair-bearing areas. In Proceedings of the 2012 Annual International Conference of the IEEE Engineering in Medicine and Biology Society (EMBC), San Diego, CA, USA, 28 August–1 September 2012; pp. 1806–1809.

23. Wei, C.S.; Wang, Y.T.; Lin, C.T.; Jung, T.P. Toward non-hair-bearing brain-computer interfaces for neurocognitive lapse detection. In Proceedings of the 2015 37th Annual International Conference of the IEEE Engineering in Medicine and Biology Society (EMBC), Milan, Italy, 25–29 August 2015; pp. 6638–6641.

24. Di Russo, F.; Pitzalis, S.; Aprile, T.; Spitoni, G.; Patria, F.; Stella, A.; Spinelli, D.; Hillyard, S.A. Spatiotemporal analysis of the cortical sources of the steady-state visual evoked potential. *Hum. Brain Mapp.* **2007**, *28*, 323–334.

25. Pastor, M.A.; Artieda, J.; Arbizu, J.; Valencia, M.; Masdeu, J.C. Human cerebral activation during steady-state visual-evoked responses. *J. Neurosci.* **2003**, *23*, 11621–11627.

26. Srinivasan, R.; Fornari, E.; Knyazeva, M.G.; Meuli, R.; Maeder, P. fMRI responses in medial frontal cortex that depend on the temporal frequency of visual input. *Exp. Brain Res.* **2007**, *180*, 677–691.

27. Pastor, M.A.; Valencia, M.; Artieda, J.; Alegre, M.; Masdeu, J. Topography of cortical activation differs for fundamental and harmonic frequencies of the steady-state visual-evoked responses. An EEG and PET H215O study. *Cereb. Cortex* **2007**, *17*, 1899–1905.

28. Srinivasan, R.; Bibi, F.A.; Nunez, P.L. Steady-state visual evoked potentials: Distributed local sources and wave-like dynamics are sensitive to flicker frequency. *Brain Topogr.* **2006**, *18*, 167–187.
29. Sammer, G.; Blecker, C.; Gebhardt, H.; Kirsch, P.; Stark, R.; Vaitl, D. Acquisition of typical EEG waveforms during fMRI: SSVEP, LRP, and frontal theta. *Neuroimage* **2005**, *24*, 1012–1024.
30. Fawcett, I.P.; Barnes, G.R.; Hillebrand, A.; Singh, K.D. The temporal frequency tuning of human visual cortex investigated using synthetic aperture magnetometry. *Neuroimage* **2004**, *21*, 1542–1553.
31. Wang, Y.T.; Nakanishi, M.; Kappel, S.L.; Kidmose, P.; Mandic, D.P.; Wang, Y.; Cheng, C.K.; Jung, T.P. Developing an online steady-state visual evoked potential-based brain-computer interface system using EarEEG. In Proceedings of the 2015 37th Annual International Conference of the IEEE Engineering in Medicine and Biology Society (EMBC), Milan, Italy, 25–29 August 2015; pp. 2271–2274.
32. Norton, J.J.; Lee, D.S.; Lee, J.W.; Lee, W.; Kwon, O.; Won, P.; Jung, S.Y.; Cheng, H.; Jeong, J.W.; Akce, A.; et al. Soft, curved electrode systems capable of integration on the auricle as a persistent brain–computer interface. *Proc. Natl. Acad. Sci. USA* **2015**, *112*, 3920–3925.
33. Hsu, H.T.; Lee, I.H.; Tsai, H.T.; Chang, H.C.; Shyu, K.K.; Hsu, C.C.; Chang, H.H.; Yeh, T.K.; Chang, C.Y.; Lee, P.L. Evaluate the Feasibility of Using Frontal SSVEP to Implement an SSVEP-Based BCI in Young, Elderly and ALS Groups. *IEEE Trans. Neural Syst. Rehabil. Eng.* **2016**, *24*, 603–615.
34. Wang, Y.T.; Nakanishi, M.; Wang, Y.; Wei, C.S.; Cheng, C.K.; Jung, T.P. An Online Brain-Computer Interface Based on SSVEPs Measured From Non-Hair-Bearing Areas. *IEEE Trans. Neural Syst. Rehabil. Eng.* **2017**, *25*, 14–21.
35. Takano, K.; Komatsu, T.; Hata, N.; Nakajima, Y.; Kansaku, K. Visual stimuli for the P300 brain-computer interface: A comparison of white/gray and green/blue flicker matrices. *Clin. Neurophysiol.* **2009**, *120*, 1562–1566.
36. Ikegami, S.; Takano, K.; Wada, M.; Saeki, N.; Kansaku, K. Effect of the green/blue flicker matrix for P300-based brain–computer interface: An EEG–fMRI study. *Front. Neurol.* **2012**, *3*, 1–10.
37. Aminaka, D.; Makino, S.; Rutkowski, T.M. Chromatic ssvep bci paradigm targeting the higher frequency eeg responses. In Proceedings of the 2014 Asia-Pacific Signal and Information Processing Association Annual Summit and Conference (APSIPA), Siem Reap, Cambodia, 9–12 December 2014; pp. 1–7.
38. Sakurada, T.; Kawase, T.; Komatsu, T.; Kansaku, K. Use of high-frequency visual stimuli above the critical flicker frequency in a SSVEP-based BMI. *Clin. Neurophysiol.* **2015**, *126*, 1972–1978.
39. Chen, X.; Wang, Y.; Zhang, S.; Gao, S.; Hu, Y.; Gao, X. A novel stimulation method for multi-class SSVEP-BCI using intermodulation frequencies. *J. Neural Eng.* **2017**, *14*, 026013.
40. Conway, B.R.; Moeller, S.; Tsao, D.Y. Specialized color modules in macaque extrastriate cortex. *Neuron* **2007**, *56*, 560–573.
41. Koida, K.; Komatsu, H. Effects of task demands on the responses of color-selective neurons in the inferior temporal cortex. *Nat. Neurosci.* **2007**, *10*, 108–116.
42. Mishkin, M.; Ungerleider, L.G. Contribution of striate inputs to the visuospatial functions of parieto-preoccipital cortex in monkeys. *Behav. Brain Res.* **1982**, *6*, 57–77.
43. Goodale, M.A.; Milner, A.D. Separate visual pathways for perception and action. *Trends Neurosci.* **1992**, *15*, 20–25.
44. Parra, J.; Lopes da Silva, F.H.; Stroink, H.; Kalitzin, S. Is colour modulation an independent factor in human visual photosensitivity? *Brain* **2007**, *130*, 1679–1689.
45. Wang, Y.; Wang, R.; Gao, X.; Hong, B.; Gao, S. A practical VEP-based brain-computer interface. *IEEE Trans. Neural Syst. Rehabil. Eng.* **2006**, *14*, 234–240.
46. Chen, X.; Chen, Z.; Gao, S.; Gao, X. A high-itr ssvep-based bci speller. *Brain-Comput. Interfaces* **2014**, *1*, 181–191.
47. Oikonomou, V.P.; Liaros, G.; Georgiadis, K.; Chatzilari, E.; Adam, K.; Nikolopoulos, S.; Kompatsiaris, I. Comparative evaluation of state-of-the-art algorithms for SSVEP-based BCIs. *arXiv Preprint* **2016**. arXiv:1602.00904.
48. Chien, Y.Y.; Lin, F.C.; Zao, J.K.; Chou, C.C.; Huang, Y.P.; Kuo, H.Y.; Wang, Y.; Jung, T.P.; Shieh, H.P.D. Polychromatic SSVEP stimuli with subtle flickering adapted to brain-display interactions. *J. Neural Eng.* **2017**, *14*, 016018.
49. Lin, Z.; Zhang, C.; Wu, W.; Gao, X. Frequency recognition based on canonical correlation analysis for SSVEP-based BCIs. *IEEE Trans. Biomed. Eng.* **2007**, *54*, 1172–1176.

50. Zhang, Y.; Guo, D.; Xu, P.; Zhang, Y.; Yao, D. Robust frequency recognition for SSVEP-based BCI with temporally local multivariate synchronization index. *Cogn. Neurodynamics* **2016**, *10*, 505–511.

51. Fisher, R.S.; Harding, G.; Erba, G.; Barkley, G.L.; Wilkins, A. Photic-and Pattern-induced Seizures: A Review for the Epilepsy Foundation of America Working Group. *Epilepsia* **2005**, *46*, 1426–1441.

52. Volosyak, I.; Valbuena, D.; Lüth, T.; Malechka, T.; Gräser, A. BCI demographics II: How many (and what kinds of) people can use a high-frequency SSVEP BCI? *IEEE Trans. Neural Syst. Rehabil. Eng.* **2011**, *19*, 232–239.

53. Yijun, W.; Ruiping, W.; Xiaorong, G.; Shangkai, G. Brain-computer interface based on the high-frequency steady-state visual evoked potential. In Proceeding of the IEEE 2005 First International Conference on Neural Interface and Control, Wuhan, China, 26–28 May 2005; pp. 37–39.

54. Lin, F.C.; Chien, Y.Y.; Zao, J.K.; Huang, Y.P.; Ko, L.W.; Shieh, H.P.D.; Wang, Y.; Jung, T.P. High-frequency polychromatic visual stimuli for new interactive display systems. *SPIE Newsroom* **2015**, 1–4. doi:10.1117/2.1201504.005851.

sensors

MDPI

Article

Disturbance-Estimated Adaptive Backstepping Sliding Mode Control of a Pneumatic Muscles-Driven Ankle Rehabilitation Robot

Qingsong Ai [1,2], Chengxiang Zhu [1,2], Jie Zuo [1,2], Wei Meng [1,2,3,*], Quan Liu [1,2], Sheng Q. Xie [1,3] and Ming Yang [4]

[1] School of Information Engineering, Wuhan University of Technology, Wuhan 430070, China; qingsongai@whut.edu.cn (Q.A.); zchengx0508@whut.edu.cn (C.Z.); zuojie@whut.edu.cn (J.Z.); quanliu@whut.edu.cn (Q.L.); s.q.xie@leeds.ac.uk (S.X.)
[2] Key Laboratory of Fiber Optic Sensing Technology and Information Processing, Ministry of Education, Wuhan University of Technology, Wuhan 430070, China
[3] School of Electronic and Electrical Engineering, University of Leeds, Leeds LS2 9JT, UK
[4] Faculty of Engineering, Environment and Computing, Coventry University, Coventry CV1 5FB, UK; ab2032@coventry.ac.uk
* Correspondence: weimeng@whut.edu.cn; Tel.: +86-131-6331-2360

Received: 1 November 2017; Accepted: 26 December 2017; Published: 28 December 2017

Abstract: A rehabilitation robot plays an important role in relieving the therapists' burden and helping patients with ankle injuries to perform more accurate and effective rehabilitation training. However, a majority of current ankle rehabilitation robots are rigid and have drawbacks in terms of complex structure, poor flexibility and lack of safety. Taking advantages of pneumatic muscles' good flexibility and light weight, we developed a novel two degrees of freedom (2-DOF) parallel compliant ankle rehabilitation robot actuated by pneumatic muscles (PMs). To solve the PM's nonlinear characteristics during operation and to tackle the human-robot uncertainties in rehabilitation, an adaptive backstepping sliding mode control (ABS-SMC) method is proposed in this paper. The human-robot external disturbance can be estimated by an observer, who is then used to adjust the robot output to accommodate external changes. The system stability is guaranteed by the Lyapunov stability theorem. Experimental results on the compliant ankle rehabilitation robot show that the proposed ABS-SMC is able to estimate the external disturbance online and adjust the control output in real time during operation, resulting in a higher trajectory tracking accuracy and better response performance especially in dynamic conditions.

Keywords: parallel robot; ankle rehabilitation; pneumatic muscles; disturbance estimation; adaptive sliding mode control

1. Introduction

The ankle joint plays a key role in maintaining balance during walking [1–3]. Recently, there have been an increasing number of people suffering from ankle injuries caused by diseases and accidents. In the US, more than 23,000 cases of ankle sprain injuries happen every day [4]. The postoperative recovery from ankle injury is slow and ineffective while the application of rehabilitation robots is supposed to be possible to solve this problem. Rehabilitation robots can help patients accomplish repetitive training tasks more accurately and effectively without physical therapists' excessive participation [5–7]. Increasing attention has been paid to the robotic rehabilitation that is appropriate to perform repetitive exercises for the recovery from neuromuscular injuries [8].

In the perspective of ankle rehabilitation, parallel robots can produce greater torque as well as achieve multiple movement degrees of freedom (DOFs) [9]. A series of parallel platform-based ankle rehabilitation robots have been developed [10]. Liu et al. [11], Alireza et al. [12], and Mozafar et al. [13] all proposed a 6-DOF ankle rehabilitation robot based on the Stewart platform. However, these robots utilized rigid actuators, such as electric motors or cylinders [14] that cannot achieve soft and compliant interaction with the patients. To overcome the limitations, some researchers started to use pneumatic muscles (PMs) as actuators to drive the ankle rehabilitation robot. PMs have inner compliance, high power/weight ratio [15] and can drive the robot in a safer way, so they have become increasingly popular in the rehabilitation robots [16]. Xie et al. [17,18] designed a four PMs-driven 3-DOF ankle rehabilitation robot with large workspace and good flexibility. Park et al. [19] in Harvard University designed a PMs-driven ankle rehabilitation robot by simulating the human muscle-tendon-ligament model, in which the PMs directly drove the foot to complete dorsiflexion/plantarflexion and inversion/eversion movements. Sawicki et al. [20] also used multiple PMs to provide dorsiflexion and plantar flexion torque for the ankle movement. Patrick et al. [21] designed a 2-DOFs ankle rehabilitation robot driven by three PMs to help patients achieve plantarflexion/dorsiflexion and inversion/eversion movements.

PMs have strong non-linearity and time-varying properties [22], which may cause difficulties in implementing precise control [23]. In order to solve these problems, a variety of control approaches have been developed. Zhao et al. [24] used neural network to adjust the parameters of PID controller. However, the method has the problems of long response time, poor tracking on desired trajectory and low tracking accuracy in the step response experiment. Zhang et al. [25] proposed a hybrid fuzzy controller to control the elbow exoskeleton robot actuated by PMs. However, this method cannot estimate the external disturbance when chattering happens, resulting in a large overshoot of step response. For the safety of human-robot interaction, Choi et al. [26] proposed a new approach to control the compliance and associated position independently. However, when an external disturbance occurs suddenly, the control method cannot detect the external disturbance quickly and it takes a long time to re-track the desired trajectory. Meng et al. [9] proposed an iterative feedback tuning control method for the repetitive training. However, the actual trajectory changed in a ladder shape because the external disturbance cannot be estimated. Jiang et al. also [27] proposed an adaptive fuzzy control algorithm based on neural network optimization to control the humanoid lower limb device driven by pneumatic muscles. However, this method cannot achieve high-accuracy tracking control and the error would significantly increase when the external load changes.

During the operation of rehabilitation robot, external disturbances are usually inevitable [28]. To obtain good control performance, the applied disturbance needs to be known exactly. However, external disturbances are often difficult to get accurately [29]. Therefore, one of the reasons why the above control method cannot achieve better control accuracy is that the external disturbance cannot be estimated. It has been recently accepted that the disturbance observer is a good choice to solve this problem [30]. Yang et al. [31] designed an error-feedback controller based on extended state observer to estimate the external disturbances and improve the trajectory tracking accuracy of a PMs-driven robot. Zhu et al. [32] presented an adaptive robust controller based on a pressure observer to control a three PMs-driven robot without pressure sensors. Wu et al. [33] proposed a novel nonlinear disturbance observer-based dynamic surface control (NDOBDSC) and can solve the friction and unknown external disturbances existing in the PM-driven device. Youssif et al. [34] designed a nonlinear disturbance observer (NDO) to estimate the lumped disturbance. Zhang et al. [35] proposed an active disturbance rejection controller for a PM actuator to achieve angle tracking precisely under varying load conditions. Plenty of studies have implied that external disturbance observer can reduce the error and improve the control accuracy effectively.

On the other hand, since the parallel robot actuated by PMs is a complex high-order nonlinear system, it would be increasingly difficult to develop an accurate control scheme for the system [36]. The backstepping sliding mode control (BS-SMC) can decompose a high-order nonlinear system into several lower order subsystems and design an intermediate virtual controller for each subsystem, which can improve the control performance [37]. In recent years, BS-SMC has attracted the interest of many researchers. Petit et al. [38] used backstepping sliding mode method to control a robot with variable stiffness and achieved satisfactory tracking performance. However, the tracking error would obviously increase if external disturbance occurred. Taheri et al. [39] designed a backstepping sliding mode controller for pneumatic cylinders suitable for wearable robots. The force and stiffness tracking performance were better than the previous pneumatic force-stiffness sliding mode controllers. However, the overshoot of this control scheme was still large and there was no experiment with variable loads. Esmaeili et al. [40] used a backstepping sliding mode controller to achieve balancing and trajectory tracking of Two Wheeled Balancing Mobile Robots (TWBMRs).

As concluded from the previous studies, there will be excessive overshoots or significantly increased errors when the external disturbance happens. The main reason is that the above methods cannot estimate the external disturbance, and as a result the control output cannot be adjusted in real time. This paper will propose an adaptive backstepping sliding mode control (ABS-SMC) with the capacity to estimate the external disturbance during operation, thus improving the robustness and accuracy of the control method. The ABS-SMC method is applied to a new 2-DOF parallel ankle rehabilitation robot which has been recently developed by us using pneumatic muscles. The controller can also deal with the nonlinearities and uncertainties of the robot system. The rest of this paper is arranged as follows: Section 2 presents mechanism design of the ankle robot. The control strategy is described in Section 3. In Section 4, experiments are carried out to verify the performance of the controller. Section 5 draws conclusion of the paper.

2. The Ankle Rehabilitation Robot

The complete system of the 2-DOF ankle rehabilitation robot and its hardware configuration are shown in Figures 1 and 2, respectively. The robot consists of a fixed platform, a moving platform, and three pneumatic muscle actuators. The moving platform is equipped with two angle sensors (GONIOMETER SG110) to measure its real-time orientation angle around the X and Y axis. Each pneumatic muscle (FESTO MAS-20-400N) is controlled by an air pressure proportional valve (ITV 2050-212N). The position information of each pneumatic muscle is collected by displacement transducers (MLO-POT-225-TLF). A force/toque sensor (ATI Mini85) is mounted between the platform and the footplate to measure the applied ankle torque. Through the data acquisition card, the sensing data are gathered by robRIO and then transmitted to the host computer. After the D/A conversion of the data, the control signals are input to the corresponding proportional valves to control pneumatic muscles, thus driving the upper platform to move. The ABS-SMC is implemented in the host computer and closed-loop control is realized on LabVIEW.

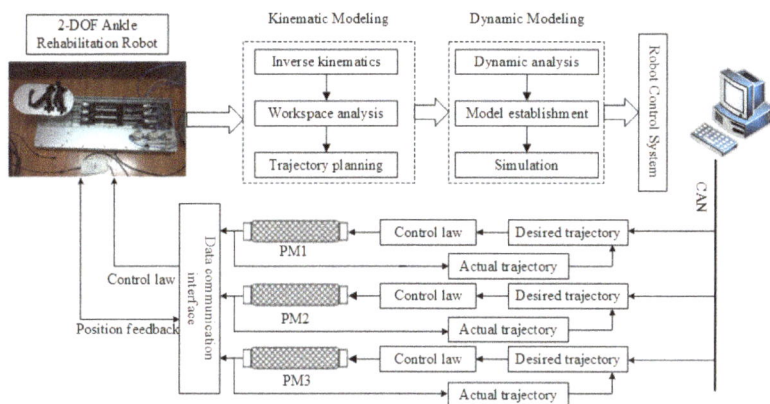

Figure 1. System structure of the ankle rehabilitation robot.

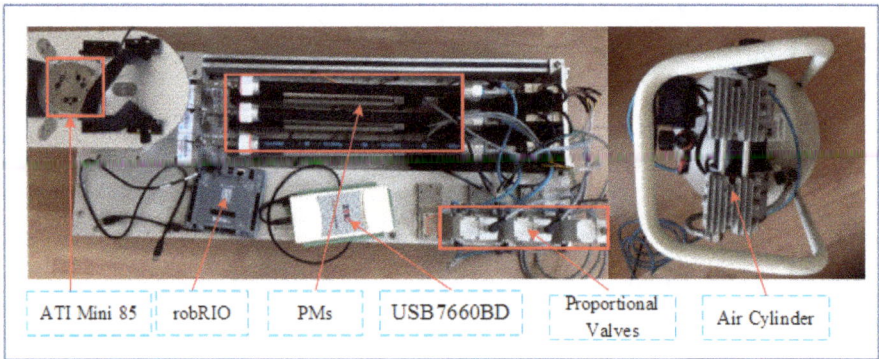

Figure 2. The developed ankle rehabilitation robot driven by PMs.

Figure 3a,b shows the simplified structure and geometrical model of the designed ankle rehabilitation robot. Since the PM can only provide pulling force, the robot must have a redundant actuation mechanism [41]. So the 2-DOF ankle rehabilitation robot is actuated by three pneumatic muscles. The lower fixed platform has three fixed holes, and the wires pass through the holes on the fixed platform. A strut is fixed between the fixed platform and the moving platform (end-effector). The Hooke joints between these two platforms guarantee that the robot can only move at two orientations. When the muscles' lengths change, the platform can be controlled to work on two orientations. In order to reduce the height of the robot and make it easier for human usage, three PMs are placed in the horizontal direction, using three fixed pulleys to change the direction of actuating forces. In this case, the overall height of the robot is only 0.3 m.

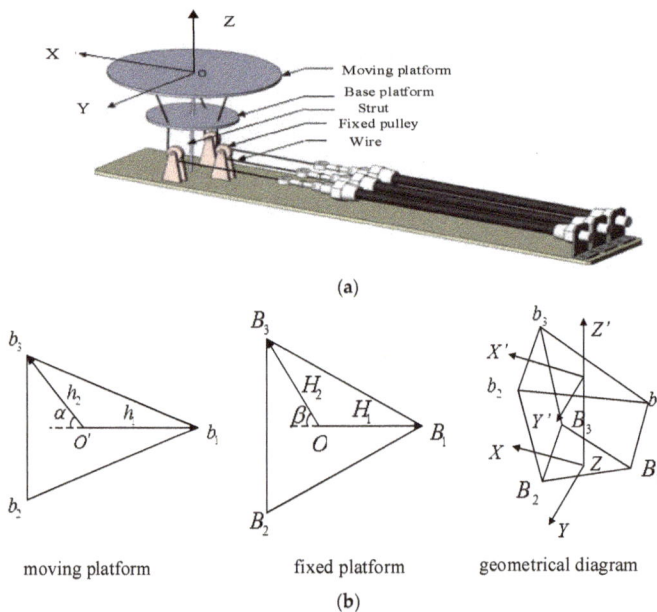

(a)

(b)

Figure 3. Kinematics of the designed 2-DOF ankle rehabilitation robot: (a) structure model, (b) geometrical diagram.

In order to control the robot end-effector to track a predefined trajectory for ankle movement training, the robot kinematic model must be studied [42], using which the joint space displacements can be determined from the end-effector orientation. As shown in Figure 3b, $b_1 b_2 b_3$ and $B_1 B_2 B_3$ represent the moving platform and the fixed platform, respectively. The vectors that connect the moving platform and the fixed platform can be written as $b_1 B_1$, $b_2 B_2$ and $b_3 B_3$. $O - X'Y'Z'$ and $O - XYZ$ are coordinate system of the moving platform and the fixed platform, respectively. A space vector in the moving coordinate can be transformed to the fixed via rotation matrix, which is widely used to establish inverse kinematics of the parallel rehabilitation robot [43]. Here $\alpha = 50°$, $\beta = 80°$, $h_1 = 0.07$ m, $h_2 = 0.08$ m, $H_1 = 0.05$ m, $H_2 = 0.06$ m. The rotation matrix can be expressed as:

$$T = T(y, \phi)T(x, \theta) = \begin{bmatrix} \cos\phi & \sin\phi\sin\theta & \sin\phi\cos\theta \\ 0 & \cos\theta & -\sin\theta \\ -\sin\phi & \cos\phi\sin\theta & \cos\phi\cos\theta \end{bmatrix}. \tag{1}$$

The solution of $b_1 B_1$, $b_2 B_2$ and $b_3 B_3$ is necessary for robot control and workspace analysis. It can be obtained by using the inverse kinematics. The link's length of this parallel robot is:

$$l_i = |L|_i = \left| Tr'_{b_i} + P - r_{B_i} \right| i = 1, 2, 3 \tag{2}$$

where L_i is the vector from B_i to b_i, P is the vector from O to O', r'_{b_i} is the vector from O' to $b_i (i = 1, 2, 3)$ and r'_{B_i} is the vector from O to $B_i (i = 1, 2, 3)$.

The dynamic model of the robot describes the relationship between the output torque and the desired angle as well as angular velocity [44]. The dynamics model is also the foundation of sliding mode control [45]. Define $q = \begin{bmatrix} \theta & \varphi & \phi \end{bmatrix}^T = \begin{bmatrix} \theta & \varphi & 0 \end{bmatrix}^T$ as the generalized coordinates of the robot's moving platform, thus the generalized speed of the moving platform is shown in Equation (3).

$$\omega = \tilde{E} \cdot \begin{bmatrix} \dot{\theta} \\ \dot{\varphi} \\ 0 \end{bmatrix} = \begin{bmatrix} \cos\varphi & 0 & 0 \\ 0 & 1 & 0 \\ -\sin\varphi & 0 & 1 \end{bmatrix} \begin{bmatrix} \dot{\theta} \\ \dot{\varphi} \\ 0 \end{bmatrix}. \tag{3}$$

Lagrange's equation is suitable for the complete system and it can solve the complex system dynamic equation in a simpler way [46]. So we use the Lagrange's equation to establish the dynamic equation of the moving platform:

$$M(q)\ddot{q} + C(q, \dot{q})\dot{q} + G(q) = \tau + \tau_d, \tag{4}$$

where $M(q)$, $C(q, \dot{q})$ and $G(q)$ represent the robot inertia matrix, the Coriolis centrifugal force matrix and the gravity matrix, τ is the robot torque and τ_d is the external disturbance torque. τ_d is mainly composed of human applied torque and the friction. The parameters in Equation (4):

$$\begin{aligned} M(q) &= TI_P T^T \\ C(q, \dot{q})\dot{q} &= \tilde{\omega} TTI_P T^T \\ G(q) &= -m\tilde{T}_{rm}g \end{aligned} \tag{5}$$

where m is the mass of the moving platform, I_p is the rotational inertia of the moving platform, r_m is the position vector of the moving platform centroid, $T_{rm} = Tr_m$ and $\tilde{T}_{rm}$ is the spiral matrix of T_{rm}. According to the formula, the driving force of each pneumatic muscle can be obtained, and finally to realize the accurate trajectory tracking of the robot platform.

3. Control Strategy

3.1. Backstepping Sliding Mode Control

The basic idea of backstepping design method is to decompose the complex nonlinear system into subsystems with lower orders, and then design Lyapunov function and intermediate virtual control for each subsystem [47]. Based on Equation (4), the controlled object model can be defined as

$$\begin{cases} \dot{q}_1 = q_2 \\ \dot{q}_2 = -M^{-1}Cq_2 + M^{-1}\tau - M^{-1}G + M^{-1}\tau_d \end{cases}, \tag{6}$$

where $q_1 = q$, q is the actual trajectory.

Assuming the desired position q_d, the controller can be designed by the following two steps:

Step 1: Define the tracking error $e_1 = q_1 - q_d$, then $\dot{e}_1 = \dot{q}_1 - \dot{q}_d = q_2 - \dot{q}_d$, and define the Lyapunov function as

$$V_1 = \frac{1}{2}e_1^T e_1. \tag{7}$$

So

$$\dot{V}_1 = e_1^T \dot{e}_1 = e_1^T (q_2 - \dot{q}_d). \tag{8}$$

Define

$$q_2 = e_2 + \dot{q}_d - c_1 e_1, \tag{9}$$

where $c_1 > 0$, e_2 is a virtual control law. From Equation (9), we can obtain

$$\begin{aligned} \dot{e}_1 &= \dot{q}_1 - \dot{q}_d \\ &= q_2 - \dot{q}_d + c_1 e - c_1 e_1 \\ &= e_2 - c_1 e_1 \end{aligned}. \tag{10}$$

From Equations (8) and (10) we can obtain

$$\dot{V}_1 = e_1^T \dot{e}_1 = e_1^T e_2 - c_1 e_1^T e_1. \tag{11}$$

If $e_2 = 0$, $\dot{V}_1 = -c_1 e_1^T e_1 = -c_1 (\|e_1\|_2)^2 \le 0$. So it is necessary to further design the control law.

Step 2: Define the switch function as

$$s = k_1 e_1 + e_2, \tag{12}$$

where $k_1 > 0$. Taking Equation (10) into (12), we can obtain

$$s = k_1 e_1 + \dot{e}_1 + c_1 e_1 = (k_1 + c_1)e_1 + \dot{e}_1. \tag{13}$$

The Lyapunov function is

$$V_2 = \frac{1}{2}e_1^T e_1 + \frac{1}{2}s^T s. \tag{14}$$

From Equation (14) we can obtain

$$\begin{aligned} \dot{V}_2 &= e_1^T \dot{e}_1 + s^T \dot{s} \\ &= e_1^T e_2 - c_1 e_1^T e_1 + s^T (k_1(e_2 - c_1 e_1) - M^{-1}C(e_2 + \dot{q}_d - c_1 e_1) \\ &\quad + M^{-1}\tau + M^{-1}\tau_d - M^{-1}G - \ddot{q}_d + c_1 \dot{e}_1) \end{aligned}. \tag{15}$$

So the control law can be written as

$$\tau_{BS-SMC} = \tau_{eq} + M\Delta\tau, \tag{16}$$

where

$$\tau_{eq} = M(-k_1(e_2 - c_1e_1) + M^{-1}C(e_2 + \dot{q}_d - c_1e_1) + M^{-1}G + \ddot{q}_d - c_1\dot{e}_1)$$
$$\Delta\tau = -h(s + \beta\mathrm{sgn}(s)) \tag{17}$$

where h and β are the parameters of exponential reaching law. They can determine the speed and time of the moving point approaching to the sliding surface.

3.2. Adaptive Backstepping Sliding Mode Control

The proposed ABS-SMC can estimate the external disturbance by establishing an disturbance observer [48]. Assuming that the external disturbance observer is $\hat{\tau}_d$.

Define

$$Q = \begin{bmatrix} q_1 \\ q_2 \end{bmatrix}. \tag{18}$$

So

$$\dot{Q} = \begin{bmatrix} \dot{q}_1 \\ \dot{q}_2 \end{bmatrix} = \begin{bmatrix} q_2 \\ \dot{q}_2 \end{bmatrix}$$
$$= \begin{bmatrix} q_2 \\ -M^{-1}Cq_2 - M^{-1}G + M^{-1}\tau + M^{-1}\tau_d \end{bmatrix}, \tag{19}$$

Equation (19) can be rewritten as:

$$\dot{Q} = \begin{bmatrix} q_2 \\ -M^{-1}Cq_2 - M^{-1}G \end{bmatrix} + \begin{bmatrix} 0 \\ M^{-1} \end{bmatrix}\tau + \begin{bmatrix} 0 \\ M^{-1} \end{bmatrix}\tau_d,$$
$$= f_1(Q) + f_2(Q)\tau + f_2(Q)\tau_d \tag{20}$$

where

$$f_1(Q) = \begin{bmatrix} q_2 \\ -M^{-1}Cq_2 - M^{-1}G \end{bmatrix}; f_2(Q) = \begin{bmatrix} 0 \\ M^{-1} \end{bmatrix}, \tag{21}$$

The disturbance observer is designed based on the difference between estimated output and actual output. Equation (20) can be rewritten as

$$f_2(Q)\tau_d = \dot{Q} - f_1(Q) - f_2(Q)\tau, \tag{22}$$

So the disturbance observer is designed:

$$\dot{\hat{\tau}}_d = \Gamma(\dot{Q} - f_1(Q) - f_2(Q)\tau - f_2(Q)\hat{\tau}_d), \tag{23}$$

Define vector $z = \hat{\tau}_d - p(Q)$. The observer gain can be expressed as $\Gamma = \partial p(Q)/\partial Q$. Let

$$\Gamma = \begin{bmatrix} \xi_2 & \xi_2 \end{bmatrix}, \xi_1 > 0, \xi_2 > 0 \tag{24}$$

$$p(Q) = \xi_1 q_1 + \xi_2 q_2 = \xi_1 q + \xi_2 \dot{q}. \tag{25}$$

$$\dot{z} = \dot{\hat{\tau}}_d - \dot{p}(Q). \tag{26}$$

Substituting Equations (23) and (25) into (26),

$$\dot{z} = \dot{\hat{\tau}}_d - \dot{p}(Q)$$
$$= \Gamma(-f_1(Q) - f_2(Q)\tau - f_2(Q)(z + p(Q)))$$
$$+ \begin{bmatrix} \xi_1 & \xi_2 \end{bmatrix}\begin{bmatrix} \dot{q} & \ddot{q} \end{bmatrix}^T - \xi_1\dot{q} - \xi_2\ddot{q} \tag{27}$$
$$= \Gamma(-f_1(Q) - f_2(Q)\tau - f_2(Q)(z + p(Q)))$$

Let $\tilde{\tau}_d = \tau_d - \hat{\tau}_d$. When the disturbance varies slowly relative to the observer dynamics, which is commonly assumed in observer design [48,49], it is reasonable that $\dot{\tau}_d = 0$, so we have

$$\dot{\tilde{\tau}}_d + \dot{\hat{\tau}}_d = 0. \tag{28}$$

Substituting Equations (24) and (25) into (28),

$$
\begin{aligned}
0 &= \dot{\tilde{\tau}}_d + \Gamma(\dot{Q} - f_1(Q) - f_2(Q)\tau - f_2(Q)\hat{\tau}_d) \\
&= \dot{\tilde{\tau}}_d + \Gamma(f_2(Q)\tau_d - f_2(Q)\hat{\tau}_d) = \dot{\tilde{\tau}}_d + \Gamma f_2(Q)\tilde{\tau}_d
\end{aligned} \tag{29}
$$

Substituting Equation (21) into (27), the disturbance observer can be written as

$$
\begin{aligned}
\hat{\tau}_d &= z + p(Q) \\
\dot{z} &= -(\xi_1\dot{q} + \xi_2 M^{-1}(-C\dot{q} - G + \tau) + \xi_2 M^{-1}(z + \xi_1 q + \xi_2 \dot{q}))
\end{aligned} \tag{30}
$$

Based on Equations (17) and (30), the adaptive control law can be written as

$$
\tau_{ABS-SMC} = \begin{aligned}
&M(-k_1(e_2 - c_1 e_1) + M^{-1}C(e_2 + \dot{q}_d - c_1 e_1) + M^{-1}G \\
&-M^{-1}\hat{\tau}_d + \ddot{q}_d - c_1 \dot{e}_1 - h(s + \beta\mathrm{sgn}(s)))
\end{aligned} \tag{31}
$$

According to these, the proposed ABS-SMC controller for the developed ankle rehabilitation robot with external disturbance in practice can be implemented based on the diagram in Figure 4, in which the controller observer can adaptively estimate the external disturbance.

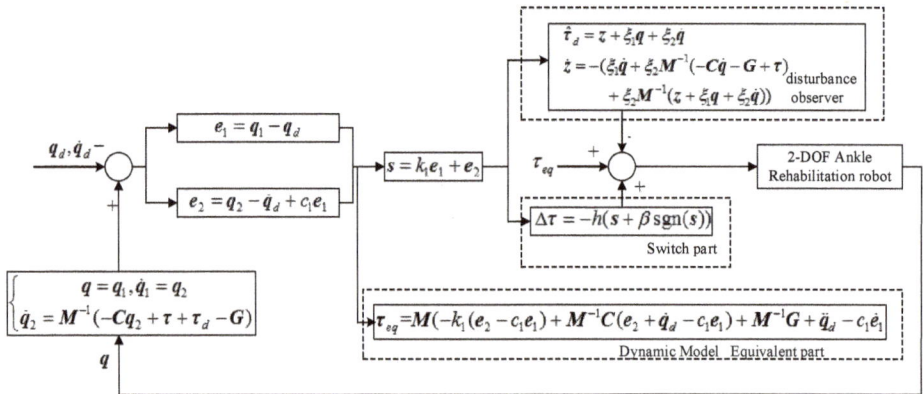

Figure 4. Implementation of ABS-SMC for the ankle rehabilitation robot.

3.3. Stability Analysis

To prove the stability of a closed-loop system, Lyapunov function is commonly used [28,29,50], through which we firstly prove that the estimation error of disturbance is bounded.

Remark 1. *For the dynamic model in (4) and the disturbance observer in (29), the estimation error $\tilde{\tau}_d$ is bounded.*

Proof. Define a Lyapunov function V_3 as follows:

$$V_3 = \frac{1}{2}\tilde{\tau}_d^T \tilde{\tau}_d. \tag{32}$$

Substituting Equations (21) and (24) into (23):

$$\dot{\hat{\tau}}_d = \begin{bmatrix} \xi_1 & \xi_2 \end{bmatrix} \left(\begin{bmatrix} \dot{q}_1 \\ \dot{q}_2 \end{bmatrix} - \begin{bmatrix} \dot{q}_2 \\ -M^{-1}Cq_2 - M^{-1}G \end{bmatrix} - \begin{bmatrix} 0 \\ M^{-1} \end{bmatrix} \tau - \begin{bmatrix} 0 \\ M^{-1} \end{bmatrix} \hat{\tau}_d \right).$$
$$= \xi_2(\ddot{q} + M^{-1}Cq_2 + M^{-1}G - M^{-1}\tau - M^{-1}\hat{\tau}_d) \tag{33}$$

Substituting Equation (6) into (33):

$$\dot{\hat{\tau}}_d = \xi_2(M^{-1}\tau_d - M^{-1}\hat{\tau}_d) = \xi_2 M^{-1}\tilde{\tau}_d. \tag{34}$$

So

$$\tilde{\tau}_d^T \dot{\tilde{\tau}}_d = \tilde{\tau}_d^T(\dot{\tau}_d - \dot{\hat{\tau}}_d) = -\tilde{\tau}_d^T \dot{\hat{\tau}}_d. \tag{35}$$

Substituting Equation (34) into (35), we have

$$\tilde{\tau}_d^T \dot{\tilde{\tau}}_d = -\xi_2 \tilde{\tau}_d^T M^{-1}\tilde{\tau}_d. \tag{36}$$

Because M^{-1} is a positive definite matrix and $\xi_2 > 0$, then

$$\dot{V}_3 = \tilde{\tau}_d^T \dot{\tilde{\tau}}_d \leq 0. \tag{37}$$

This indicates that the designed disturbance observer can track external disturbance, which means the estimation error $\tilde{\tau}_d$ is bounded, so Remark 1 is proved to be correct.

Then, we prove the stability of the combined system. As the robot moves within a confined space, the inertia matrix M is bounded and positive definite so M^{-1} exists and is bounded,

$$\|M^{-1}\tilde{\tau}_d\|_1 = \|M^{-1}\tau_d - M^{-1}\hat{\tau}_d\|_1 \leq \delta. \tag{38}$$

$\hat{\delta}$ is the estimated value of δ. Then define:

$$\dot{\hat{\delta}} = \gamma\|s\|_1. \tag{39}$$

where $\gamma > 0$ [51,52]. □

Remark 2. *As long as the parameters are appropriately set, the closed-loop system is stable for disturbance observer in (30) and control law in (31).*

Proof. The Lyapunov function is defined as

$$V = V_2 + \frac{1}{2\gamma}\tilde{\delta}^2 + \frac{1}{2}\tilde{\tau}_d^T \tilde{\tau}_d. \tag{40}$$

where $\tilde{\delta} = \delta - \hat{\delta}$.

From Equation (40), we can get

$$\dot{V} = \dot{V}_2 + \frac{1}{\gamma}\tilde{\delta}\dot{\tilde{\delta}} + \tilde{\tau}_d^T \dot{\tilde{\tau}}_d. \tag{41}$$

Substituting Equation (31) into (15), we can get

$$\begin{aligned} \dot{V}_2 + \frac{1}{\gamma}\tilde{\delta}\dot{\tilde{\delta}} &= e_1^T e_2 - c_1 e_1^T e_1 + s^T M^{-1}\tilde{\tau}_d - hs^T s - h\beta\|s\|_1 - \frac{1}{\gamma}\tilde{\delta}\dot{\hat{\delta}} \\ &\leq e_1^T e_2 - c_1 e_1^T e_1 + \delta\|s\|_1 - hs^T s - h\beta\|s\|_1 - \tilde{\delta}\|s\|_1 \\ &= e_1^T e_2 - c_1 e_1^T e_1 - hs^T s + (\delta - \tilde{\delta} - h\beta)\|s\|_1 \end{aligned} \tag{42}$$

Let $h\beta = \delta = \int \gamma \|s\|_1 dt$. Equation (42) can be rewritten as:

$$\dot{V}_2 + \frac{1}{\gamma}\tilde{\delta}\dot{\tilde{\delta}} \leq e_1^T e_2 - c_1 e_1^T e_1 - h s^T s + (\delta - \tilde{\delta} - h\beta)\|s\|_1$$
$$= e_1^T e_2 - c_1 e_1^T e_1 - h s^T s + (\delta - \tilde{\delta} - \delta)\|s\|_1 \qquad (43)$$
$$= e_1^T e_2 - c_1 e_1^T e_1 - h s^T s$$

Define $e = \begin{bmatrix} e_1^T & e_2^T \end{bmatrix}$, $e^T = \begin{bmatrix} e_1 \\ e_2 \end{bmatrix}$, and $= \begin{bmatrix} c_1 + hk_1^2 & hk_1 - \frac{1}{2} \\ hk_1 - \frac{1}{2} & h \end{bmatrix}$.

Then

$$eBe^T = \begin{bmatrix} e_1^T & e_2^T \end{bmatrix} \begin{bmatrix} c_1 + hk_1^2 & hk_1 - \frac{1}{2} \\ hk_1 - \frac{1}{2} & h \end{bmatrix} \begin{bmatrix} e_1 \\ e_2 \end{bmatrix}$$
$$= c_1 e_1^T e_1 - e_1^T e_2 + hk_1^2 e_1^T e_1 + hk_1 e_1^T e_2 + hk_1 e_2^T e_1 + h e_2^T e_2 \qquad (44)$$
$$= c_1 e_1^T e_1 - e_1^T e_2 + h s^T s$$

Substituting Equation (44) into (43):

$$\dot{V}_2 + \frac{1}{\gamma}\tilde{\delta}\dot{\tilde{\delta}} \leq -eBe^T. \qquad (45)$$

If we make be a positive definite matrix, then

$$\dot{V}_2 + \frac{1}{\gamma}\tilde{\delta}\dot{\tilde{\delta}} \leq -e^T Be \leq 0. \qquad (46)$$

Because

$$|B| = h(c_1 + hk_1^2) - (hk_1 - \tfrac{1}{2})^2$$
$$= h(c_1 + k_1) - \tfrac{1}{4} \qquad (47)$$

By appropriately setting h, c_1, k_1, we can make $|B| > 0$, so that B is a positive definite matrix and guarantee $\dot{V}_2 + \frac{1}{\gamma}\tilde{\delta}\dot{\tilde{\delta}} \leq 0$.

From Equation (37), we can get

$$\tilde{\tau}_d^T \dot{\tilde{\tau}}_d \leq 0. \qquad (48)$$

Substituting Equations (46) and (48) into (41):

$$\dot{V} = \dot{V}_2 + \frac{1}{\gamma}\tilde{\delta}\dot{\tilde{\delta}} + \tilde{\tau}_d^T \dot{\tilde{\tau}}_d \leq 0. \qquad (49)$$

Therefore, as long as the mentioned parameters are appropriately set, we can ensure the system be stable. In this way, Remark 2 is proved to be correct. □

4. Experimental and Results Discussion

In order to confirm the performance of the proposed control method, experiments were carried out on the 2-DOF ankle rehabilitation robot. The experiments can be divided into four groups: (1) step response experiment; (2) sine trajectory tracking experiment (without subject); (3) robustness test with human subjects; and (4) sudden external disturbance experiment. BS-SMC has been widely used in recent years and achieved good control performance [38–40], so we conduct the experiments to compare the proposed control method with BS-SMC to verify its control capacity and advantages.

4.1. Step Response

To simulate step response, the moving platform was firstly set to its initial pose ($\theta = 0°$, $\varphi = 0°$). Then, at t = 10 s, the expected position of the moving platform was set as $\theta = 10°$ and $\varphi = 10°$. The experimental results of both BS-SMC and ABS-SMC are shown in Figure 5.

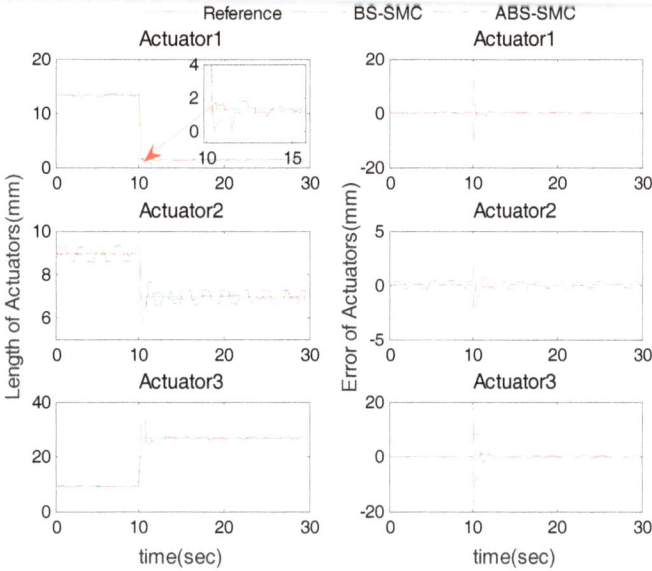

Figure 5. Actuator position tracking results and errors in step response experiment with robot controlled by BS-SMC and ABS-SMC respectively.

Figure 5 shows the step response of three PMs under different control methods. It can be seen that both the proposed ABS-SMC and BS-SMC were able to generate delay less than 0.5 s, but the ABS-SMC reached the desired trajectory more quickly after a short shock. The response time of the proposed control method was 1 s while that of the BS-SMC was about 1.5 s. In addition, there was always vibration existing near the desired trajectory in the BS-SMC experiment, while the proposed ABS-SMC could effectively reduce chattering and guaranteed the operation safety. Moreover, the overshoot of ABS-SMC was significantly smaller than that of BS-SMC. For example, the tracking overshoot of Actuator 3 was about 5 mm when controlled by ABS-SMC. If the overshoot is too large, the patient's foot may have to rotate at a large angle in a short time, which may cause the secondary injury to the patient. On the other hand, after the system reached the steady state, the error of the ABS-SMC was smaller than 0.5 mm while the maximum error of the BS-SMC was 2 mm.

4.2. Sine Trajectory Tracking Experiment (without Subject)

The desired trajectory was set $\theta = 10\sin(2\pi ft)(\deg)$, $\varphi = 10\cos(2\pi ft)(\deg)$, $f = 10\,\text{Hz}$. The results of sine trajectory tracking with no subject involved (load = 0) are shown in Figures 6 and 7. From Figure 6, we can see that the proposed method had higher control accuracy and smaller chattering than BS-SMC, due to its ability to compensate the external disturbance, which can effectively guarantee the safety and stability of the rehabilitation operations. In order to further quantitatively compare the performance between ABS-SMC and BS-SMC, maximum error (ME) and average error (AE) of the robot control results were calculated for statistical evaluation. Table 1 shows the position tracking errors of the two control methods. Taking Actuator 1 as an example, for the proposed control method, the ME and AE were 0.84 mm and 0.39 mm respectively, while the ME and AE of BS-SMC were

1.48 mm and 0.64 mm. Compared with BS-SMC, the ME and AE of ABS-SMC were reduced by about 43% and 40% respectively. In Table 2, the ME (0.69°) and AE (0.19°) of the rotation angle around X-axis were reduced by 53% and 70%, compared with BS-SMC (1.48° and 0.57°). Compared with BS-SMC, the proposed ABS-SMC cannot only improve the position control accuracy, but also has a lower chattering level attributing to its ability of disturbance estimation.

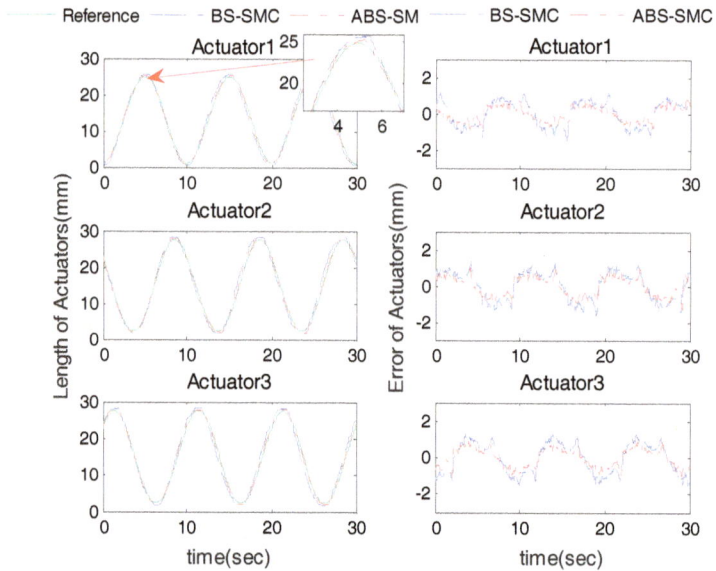

Figure 6. Actuator position tracking results (without subject).

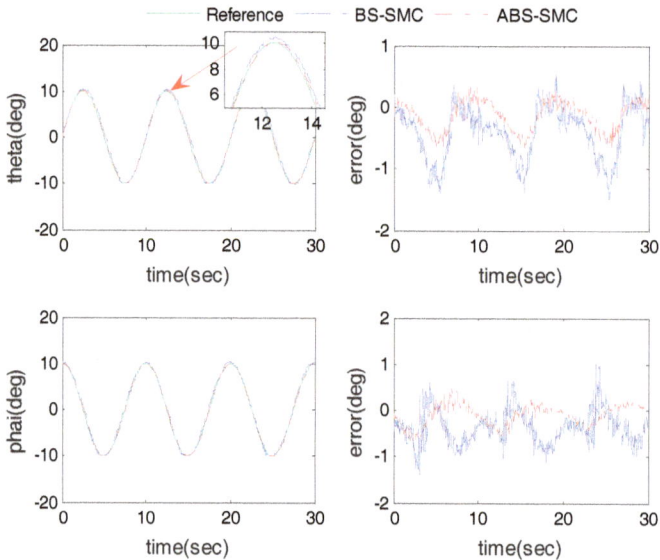

Figure 7. Robot end-effector angle tracking results (without subject).

Table 1. Statistical analysis of actuator position tracking errors under different control methods (without subject).

Methods		Maximum Error (mm)			Average Error (mm)		
		A1	A2	A3	A1	A2	A3
Position tracking results	ABS-SMC	0.84	1.05	0.93	0.39	0.47	0.46
	BS-SMC	1.48	1.64	1.55	0.64	0.72	0.75

Table 2. Statistical analysis of end-effector angle tracking errors under different control methods (without subject).

Methods		Maximum Error (°)		Average Error (°)	
		θ	φ	θ	φ
Angle tracking results	ABS-SMC	0.69	0.68	0.19	0.20
	BS-SMC	1.48	1.41	0.44	0.44

4.3. Robustness Test with Human Subjects

In order to verify the robustness of the proposed controller, especially when interacting with human users, five healthy subjects were involved in the experiment. The information of all subjects is shown in Table 3. The participants were instructed to fix their right foot on the robot moving platform so that they can follow the moving platform for passive training. This trial has been approved by the Human Participants Ethics Committees from Wuhan University of Technology, China and written informed consent was obtained from each participant. The experimental results were compared with BS-SMC to verify its superior ability by taking advantage of external disturbance estimation. We take Subject 1 as an example with results shown in Figures 8 and 9.

Table 3. Information of all involved subject.

Participants	Gender	Age	Height (cm)	Weight (kg)
Subject 1	male	23	175	65
Subject 2	male	22	178	64
Subject 3	female	23	160	49
Subject 4	female	24	165	50
Subject 5	male	25	180	70

The results of the sine wave tracking with Subject 1 are shown in Figures 8 and 9. Compared with BS-SMC, we can see that proposed control method has smaller tracking errors. In the case of Actuator 1, as shown in Tables 1 and 4, when the ABS-SMC was applied to the robot, compared to the experiment without subject, the ME and AE of position tracking result increased by about 0.26 mm and 0.04 mm only. However, when BS-SMC was used, the ME and AE increased by 1.23 mm and 0.66 mm. Comparing Tables 2 and 5, taking the rotation angle around X axis as an example, in the use of ABS-SMC and when subject participated, the ME and AE only increased by about 0.21° and 0.01°, but the ME and AE increased by 0.56° and 0.10° when using BS-SMC.

In Figure 8c, the desired trajectory was sinusoidal, so the torque applied by the subject to the moving platform showed a similar pattern. ABS-SMC regarded the exerted force as an external disturbance, thus the estimated external disturbance torque also revealed similar sine changes. On the other hand, it can be seen from Figure 8d that the control law of the proposed ABS-SMC was quite different from that of the BS-SMC, especially when it reached the extreme point. This is because the external disturbance reached the maximum at the extreme point of the control law. It can also be noticed that the estimated external disturbance of Z-axis was much smaller than X and Y axes. This is because the designed robot cannot rotate around the Z-axis. The ideal Z-axis torque should be zero, but in practice the moving platform still has a slight rotation in the Z-axis.

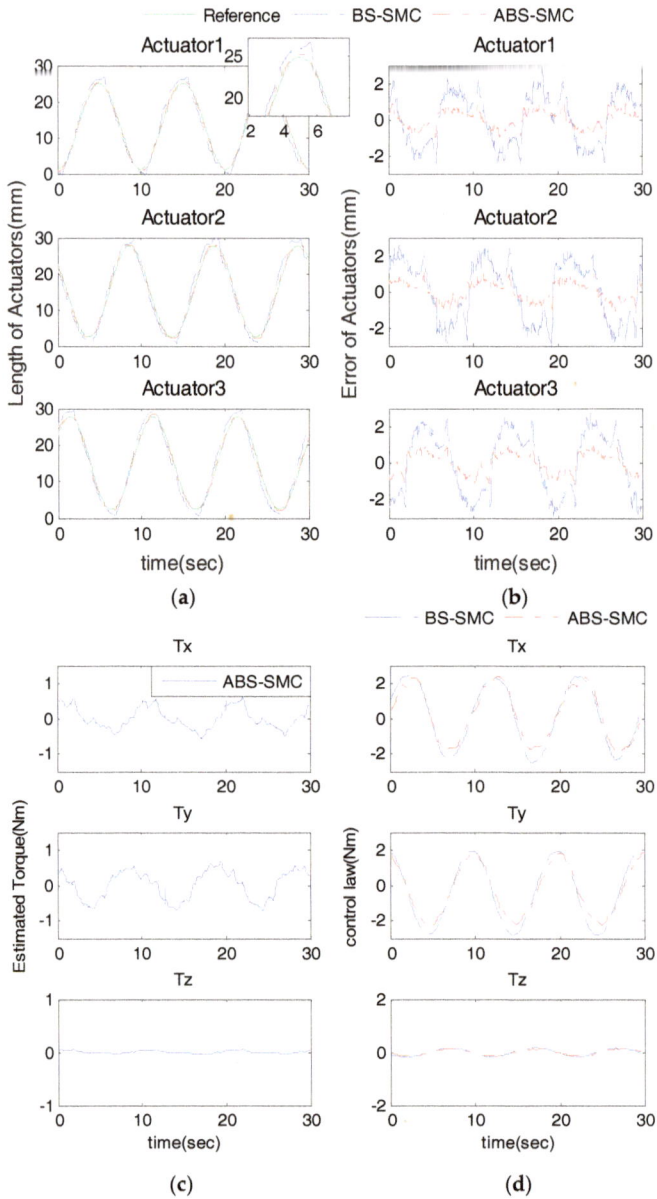

Figure 8. Actuator position tracking results with subject 1: (**a**) actuator position tracking results; (**b**) the actuator tracking errors; (**c**) the estimated external torque (using ABS-SMC) and (**d**) the control output tuning processing via ABS-SMC disturbance estimation.

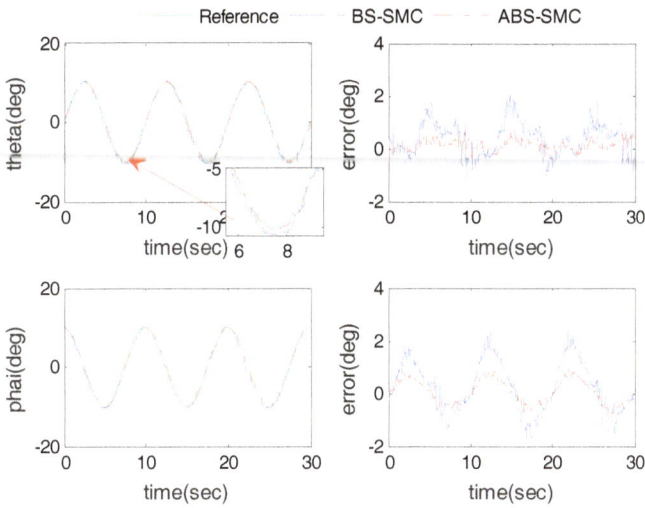

Figure 9. End-effector angle tracking results with subject 1.

Figure 10 further shows the errors of three actuators with all five participants. We can see that the proposed ABS-SMC is able to obtain smaller errors which also changed more smoothly. It can be again validated that the proposed ABS-SMC is able to obtain better robustness. The statistical details in Tables 4 and 5 indicate the robustness of the ABS-SMC scheme for its adaptability to different subjects with varying capabilities. When different subjects involved, the actuators' ME changed very slightly. The minimum ME was 1.10 mm and the maximum 2.07 mm. The change of AE was also small (0.37~0.49 mm). When using BS-SMC to control the robot, the ME ranged 2.71~5.30 mm, and the AE ranged 1.14~1.56 mm; therefore, the stability and control accuracy of ABS-SMC were better than BS-SMC, which could adapt to different people's rehabilitation training. Therefore, we can conclude that the ABS-SMC has a better robustness as it estimates the exerted disturbance and adjusts the control law in real time, resulting in higher control accuracy and reduced chattering.

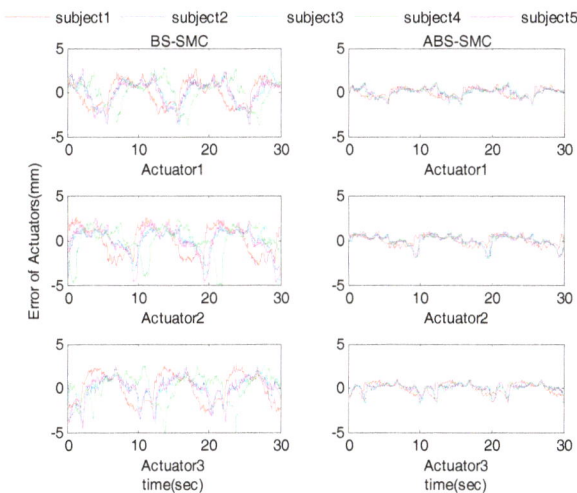

Figure 10. Actuator tracking error results with five subjects.

Table 4. Statistical analysis of actuator position tracking errors under different control methods (with five subjects).

	Participants	Methods	Maximum Error (mm)			Average Error (mm)		
			A1	A2	A3	A1	A2	A3
	Subject 1	ABS-SMC	1.10	1.13	1.33	0.43	0.47	0.49
		BS-SMC	2.71	3.60	3.24	1.30	1.48	1.56
	Subject 2	ABS-SMC	1.52	2.07	1.76	0.39	0.47	0.37
Position		BS-SMC	3.71	4.67	4.20	1.19	1.43	1.07
tracking	Subject 3	ABS-SMC	1.53	2.02	1.81	0.40	0.47	0.37
results		BS-SMC	3.90	5.01	4.19	1.17	1.46	1.10
	Subject 4	ABS-SMC	1.77	2.07	1.88	0.39	0.48	0.38
		BS-SMC	3.86	5.22	5.30	1.22	1.27	1.29
	Subject 5	ABS-SMC	1.39	1.97	1.66	0.39	0.47	0.37
		BS-SMC	3.74	4.96	4.63	1.14	1.34	1.09

Table 5. End-effector angle tracking errors under different control methods (with five subjects).

	Participants	Methods	Maximum Error (°)		Average Error (°)	
			θ	φ	θ	φ
	Subject 1	ABS-SMC	0.90	0.99	0.20	0.39
		BS-SMC	2.04	2.50	0.54	0.75
	Subject 2	ABS-SMC	1.12	0.99	0.29	0.28
Angle		BS-SMC	2.25	2.18	0.50	0.78
tracking	Subject 3	ABS-SMC	1.21	1.18	0.29	0.34
results		BS-SMC	2.91	2.36	0.67	0.78
	Subject 4	ABS-SMC	1.41	1.13	0.43	0.34
		BS-SMC	3.32	2.75	0.63	0.66
	Subject 5	ABS-SMC	1.14	0.89	0.27	0.28
		BS-SMC	2.97	2.17	0.92	0.94

4.4. Sudden External Disturbance

To further confirm the anti-interference ability of the proposed ABS-SMC, a certain resistance was applied on the 2-DOFs ankle rehabilitation robot. During different training cycles, the strength and duration of the resistance are shown in Table 6 and the experimental results are compared with BS-SMC. It can be seen that the trajectories of the actuator 2 and 3 were exactly the same when the trajectory of the moving platform is $\theta = 0°$, $\varphi = 10\cos(2\pi ft)°$. In order to ensure the applied force consistent for the two control methods comparison, the ABS-SMC was used to control the actuator 1 and actuator 2, while the BS-SMC was used to control the actuator 3 of the rehabilitation robot. The experimental results are shown in Figure 11.

Table 6. Resistance force and duration of four phases in the experiment.

	Man-Made Resistance	Size (N)	Duration (s)
Phase i (P i)	None	0	0
Phase ii (P ii)	Applied	10	2
Phase iii (P iii)	Applied	30	2
Phase iv (P iv)	Applied	30	3

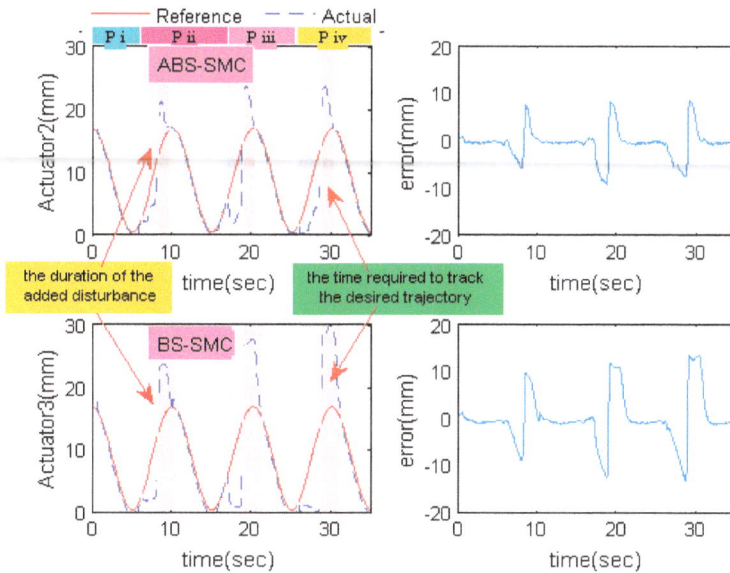

Figure 11. Actuator trajectory tracking results with abrupt disturbances.

Figure 11 shows the trajectory tracking curve after applying sudden disturbances. It can be seen that the time required for ABS-SMC to track the desired trajectory was about 1.53 s and the maximum error was about 7.50 mm in phase ii, while the time required for the BS-SMC was 2.39 s and the maximum error 9.61 mm. In phase iii, compared with phase ii, the time that the proposed ABS-SMC required to tracks the desired trajectory only increased by 3.92% and the maximum error increased by 7.60%. However, the required time and maximum error increased by 7.60% and 20.9% respectively in BS-SMC. Similar patterns were found in phase iv, the data increased by 4.81% and 4.96% respectively in ABS-SMC, while in BS-SMC increased by 8.40% and 17.53% respectively. So we can conclude that the time required to re-follow the desired trajectory by using ABS-SMC was reduced and the maximum error also remained small under uncertain resistances. The ABS-SMC can achieve a better anti-interference performance compared with the BS-SMC, attributing to its ability of estimating external disturbance and adjusting the control output accordingly.

To further verify superior ability of the proposed method, we also compared our results with other recently published works, which also aimed to control the PMs-driven rehabilitation robot. As summarized in Table 7, the proposed control method shows a better performance. Zhang et al. [53] used adaptive patient-cooperative control method to control a compliant ankle rehabilitation robot driven by PMs. They conducted the experiments with the subject, and the root mean square deviation (RMSD) was 2.34°. Jamwal et al. [18] used a fuzzy-based disturbance observer (FBDO) to control a 3-DOF ankle rehabilitation robot driven by PMs. The maximum error (ME) and average error (AE) of end-effector were 22.93% and 6.43%. The team also designed a robust iterative feedback tuning control scheme to improve the performance, and the ME and RMSD of trajectory tracking of the robot were about 12.48% and 1.40° [9]. In addition, Su at al. [54] proposed a model-based chattering mitigation robust variable control (CRVC) method and applied this method to control a lower limb rehabilitation robot driven by PMs. The ME of the end-effector was 15.00% and The RMSD was 2.34°. In this paper, when there was a participant, the ME, AE and RMSD were 7.05%, 2.15% and 0.78°, respectively. It can be seen from the above analysis that the control performance of the proposed method is obviously better than that of the above methods.

Table 7. Comparison of existing control methods and the proposed method for PMs-driven parallel rehabilitation robot. (*, unknown).

Literature	End-Effector Tracking Error					
	Without Human Participant			With Human Participant		
	ME (%)	AE (%)	RMSD	ME (%)	AE (%)	RMSD
[9]	11.18	*	1.35	12.48	*	1.40
[18]	*	*	*	22.93	6.43	*
[53]	*	*	*	*	*	2.34
[54]	*	*	*	15.00	*	*
Current study	3.45	1.00	0.44	7.05	2.15	0.78

From the experimental results analysis, it can be concluded that the ABS-SMC estimates external disturbance and adaptively adjusts the control law so the performance is obviously better than that of the BS-SMC and the recent published control schemes in [15,23,43,44] in terms of response speed, control accuracy, robustness and ability to resist external disturbance. This controller can meet the rehabilitation demands of patients under dynamic conditions.

5. Conclusions

In this paper, a 2-DOF parallel robot was developed for ankle rehabilitation and the inverse kinematics model as well as the dynamics model of the robot were constructed. This paper proposed an ABS-SMC for PMs by introducing a disturbance observer, so the external disturbances can be estimated and the control output can be adjusted in real time. Experimental results show that the ABS-SMC had better trajectory tracking performance compared with the conventional method. The proposed method can greatly reduce chattering, which may reduce secondary damage to the patient. When participants were involved, the tracking error of traditional method obviously increased while the error of the proposed method remained small. In addition, the ABS-SMC has a better anti-interference ability. When the ankle rehabilitation robot was applied with greater resistance, the proposed method could quickly track the desired trajectory after removing the resistance. How the control would perform under uncertainties in the model and the applied torque is also need to be studied in the future. Because of the complexity of the ankle rehabilitation robot, it is difficult to establish a precise dynamic model. Our model here can match the real system to a large extent, which can also be reflected from the experimental results. However, the model uncertainties should be optimized further and the applied torque can be measured in real time by using a force/torque sensor to reach a more accurate model that will in turn improve the control performance. To improve the patient's participation in the future work, patient force feedback must be considered. In this case, the performance of current position/force hybrid control and impedance control can be improved by incorporating the proposed ABS-SMC method. Furthermore, functional electrical stimulation, and biological signals should also be applied to the control of the robot to improve the patient's voluntary participation and rehabilitation training performance.

Acknowledgments: This work was supported by the National Natural Science Foundation of China under Grants No. 51675389, 51705381, and 51475342.

Author Contributions: C.Z. conducted research for the related works; J.Z. and W.M. conceived research subject and designed the experiments; C.Z. performed the experiments; Q.A. and W.M. analyzed the data; Q.A. and J.Z. wrote the paper; C.Z., W.M. and M.Y. revised the manuscript; Q.L., S.X. and M.Y. supervised the paperwork and provided review, comments, assessment, etc. All authors have read and approved the final manuscript.

Conflicts of Interest: The authors declare no conflict of interest.

References

1. Zhang, M.; Davies, T.C.; Xie, S. Effectiveness of robot-assisted therapy on ankle rehabilitation—A systematic review. *J. Neuroeng. Rehabil.* **2013**, *10*, 30. [CrossRef] [PubMed]
2. Staniszewski, M.; Zybko, P.; Wiszomirska, I. Influence of a nine-day alpine ski training programme on the postural stability of people with different levels of skills. *Biomed. Hum. Kinet.* **2016**, *8*, 24–31. [CrossRef]
3. Omar, S.M.M.H.; El-Kalaa, F.A.; Ali, E.S.F.; El-Karim, A.A.A.; Sekily, N.M.E. Anatomical and magnetic resonance imaging study of the medial collateral ligament of the ankle joint. *Alex. J. Med.* **2016**, *52*, 73–81. [CrossRef]
4. Hertel, J. Functional anatomy, pathomechanics, and pathophysiology of lateral ankle instability. *J. Athl. Train.* **2002**, *37*, 364. [PubMed]
5. Farulla, G.A.; Pianu, D.; Cempini, M.; Cortese, M.; Russo, L.O.; Indaco, M.; Nerino, R.; Chimienti, A.; Oddo, C.M.; Vitiello, N. Vision-based pose estimation for robot-mediated hand telerehabilitation. *Sensors* **2016**, *16*, 208. [CrossRef] [PubMed]
6. Saadat, M.; Rastegarpanah, A.; Abdullah, C.Z.; Rakhodaei, H.; Borboni, A.; Maddalena, M. Path's slicing analysis as a therapist's intervention tool for robotic rehabilitation. In *Advances in Service and Industrial Robotics, Proceedings of the 26th International Conference on Robotics in Alpe-Adria-Danube Region (RAAD 2017), Turin, Italy, 21–23 June 2017*; Ferraresi, C., Quaglia, G., Eds.; Springer International Publishing: Cham, Switzerland, 2018; pp. 901–910.
7. Borboni, A.; Maddalena, M.; Rastegarpanah, A.; Saadat, M.; Aggogeri, F. Kinematic performance enhancement of wheelchair-mounted robotic arm by adding a linear drive. In Proceedings of the IEEE International Symposium on Medical Measurements and Applications, Benevento, Italy, 15–18 May 2016; pp. 1–6.
8. Ai, Q.; Liu, Q.; Yuan, T.; Lu, Y. Gestures recognition based on wavelet and LLE. *Australas. Phys. Eng. Sci. Med.* **2013**, *36*, 167–176. [CrossRef] [PubMed]
9. Meng, W.; Xie, S.Q.; Liu, Q.; Lu, C.Z.; Ai, Q. Robust iterative feedback tuning control of a compliant rehabilitation robot for repetitive ankle training. *IEEE/ASME Trans. Mechatron.* **2017**, *22*, 173–184. [CrossRef]
10. Meng, W.; Liu, Q.; Zhou, Z.; Ai, Q.; Sheng, B.; Xie, S.S. Recent development of mechanisms and control strategies for robot-assisted lower limb rehabilitation. *Mechatronics* **2015**, *31*, 132–145. [CrossRef]
11. Liu, Q.; Liu, D.; Meng, W.; Zhou, Z.; Ai, Q. Fuzzy sliding mode control of a multi-DOF parallel robot in rehabilitation environment. *Int. J. Humanoid Robot.* **2014**, *11*, 1450004. [CrossRef]
12. Rastegarpanah, A.; Saadat, M.; Borboni, A. Parallel robot for lower limb rehabilitation exercises. *Appl. Bionics Biomech.* **2016**, *2016*. [CrossRef] [PubMed]
13. Rastegarpanah, A.; Saadat, M.; Borboni, A.; Stolkin, R. Application of a parallel robot in lower limb rehabilitation: A brief capability study. In Proceedings of the 1st International Conference on Robotics and Automation for Humanitarian Applications (RAHA 2016), Kerala, India, 18–20 December 2016.
14. Pehlivan, A.U.; Sergi, F.; O'Malley, M.K. A subject-adaptive controller for wrist robotic rehabilitation. *IEEE/ASME Trans. Mechatron.* **2015**, *20*, 1338–1350. [CrossRef]
15. Repperger, D.W.; Phillips, C.A.; Neidhard-Doll, A.; Reynolds, D.B.; Berlin, J. Actuator design using biomimicry methods and a pneumatic muscle system. *Control Eng. Pract.* **2006**, *14*, 999–1009. [CrossRef]
16. Ba, D.X.; Ahn, K.K. Indirect sliding mode control based on gray-box identification method for pneumatic artificial muscle. *Mechatronics* **2015**, *32*, 1–11. [CrossRef]
17. Xie, S.Q.; Jamwal, P.K. An iterative fuzzy controller for pneumatic muscle driven rehabilitation robot. *Expert Syst. Appl.* **2011**, *38*, 8128–8137. [CrossRef]
18. Jamwal, P.K.; Xie, S.Q.; Hussain, S.; Parsons, J.G. An adaptive wearable parallel robot for the treatment of ankle injuries. *IEEE/ASME Trans. Mechatron.* **2014**, *19*, 64–75. [CrossRef]
19. Park, Y.-L.; Chen, B.-R.; Young, D.; Stirling, L.; Wood, R.J.; Goldfield, E.; Nagpal, R. Bio-inspired active soft orthotic device for ankle foot pathologies. In Proceedings of the 2011 IEEE/RSJ International Conference on Intelligent Robots and Systems: Celebrating 50 Years of Robotics (IROS'11), San Francisco, CA, USA, 25–30 September 2011; Institute of Electrical and Electronics Engineers Inc.: San Francisco, CA, USA; pp. 4488–4495.
20. Sawicki, G.S.; Ferris, D.P. A pneumatically powered knee-ankle-foot orthosis (kafo) with myoelectric activation and inhibition. *J. Neuroeng. Rehabil.* **2009**, *6*, 23. [CrossRef] [PubMed]

21. Murphy, P.; Adolf, G.; Daly, S.; Bolton, M.; Maurice, O.; Bonia, T.; Mavroidis, C.; Yen, S.-C. Test of a customized compliant ankle rehabilitation device in unpowered mode. In Proceedings of the 36th Annual International Conference of the IEEE Engineering in Medicine and Biology Society, Chicago, IL, USA, 26–30 August 2014; Volume 2014, pp. 3057–3060.

22. Lin, L.-H.; Yen, J.-Y.; Wang, F.-C. System identification and robust control of a pneumatic muscle actuator system. In Proceedings of the 2nd International Conference on Engineering and Technology Innovation 2012 (ICETI 2012), Kaohsiung, Taiwan, 2–6 November 2012; Trans Tech Publications: Kaohsiung, Taiwan, 2013; pp. 1936–1940.

23. Chang, M.-K.; Liou, J.-J.; Chen, M.-L. T-s fuzzy model-based tracking control of a one-dimensional manipulator actuated by pneumatic artificial muscles. *Control Eng. Pract.* **2011**, *19*, 1442–1449. [CrossRef]

24. Zhao, J.; Zhong, J.; Fan, J. Position control of a pneumatic muscle actuator using RBF neural network tuned PID controller. *Math. Prob. Eng.* **2015**, *2015*. [CrossRef]

25. Zhang, J.-F.; Yang, C.-J.; Chen, Y.; Zhang, Y.; Dong, Y.-M. Modeling and control of a curved pneumatic muscle actuator for wearable elbow exoskeleton. *Mechatronics* **2008**, *18*, 448–457. [CrossRef]

26. Choi, T.-Y.; Choi, B.-S.; Seo, K.-H. Position and compliance control of a pneumatic muscle actuated manipulator for enhanced safety. *IEEE Trans. Control Syst. Technol.* **2011**, *19*, 832–842. [CrossRef]

27. Jiang, F.; Tao, G.; Li, Q. Analysis and control of a parallel lower limb based on pneumatic artificial muscles. *Adv. Mech. Eng.* **2016**, *9*, 1–14. [CrossRef]

28. Fan, Q.Y.; Yang, G.H. Adaptive actor-critic design-based integral sliding-mode control for partially unknown nonlinear systems with input disturbances. *IEEE Trans. Neural Netw. Learn. Syst.* **2016**, *27*, 165–177. [CrossRef] [PubMed]

29. Liu, S.Y.; Liu, Y.C.; Wang, N. Nonlinear disturbance observer-based backstepping finite-time sliding mode tracking control of underwater vehicles with system uncertainties and external disturbances. *Nonlinear Dyn.* **2017**, *88*, 465–476. [CrossRef]

30. Mohammadi, A.; Tavakoli, M.; Marquez, H.J.; Hashemzadeh, F. Nonlinear disturbance observer design for robotic manipulators. *Control Eng. Pract.* **2013**, *21*, 253–267. [CrossRef]

31. Yang, H.J.; Yu, Y.; Qiu, J.; Hua, C.C. Active disturbance rejection tracking control for a nonlinear pneumatic muscle system. *Int. J. Control Autom. Syst.* **2017**, *15*, 2376–2384. [CrossRef]

32. Zhu, X.C.; Tao, G.L.; Cao, J. Pressure observer-based adaptive robust trajectory tracking control of a parallel manipulator driven by pneumatic muscles. *J. Zhejiang Univ.-SCI A* **2007**, *8*, 1928–1937. [CrossRef]

33. Wu, J.; Huang, J.; Wang, Y.J.; Xing, K.X. Nonlinear disturbance observer-based dynamic surface control for trajectory tracking of pneumatic muscle system. *IEEE Trans. Control Syst. Technol.* **2014**, *22*, 440–455. [CrossRef]

34. Elobaid, Y.M.T.; Huang, J.; Wang, Y.J. Nonlinear disturbance observer-based robust tracking control of pneumatic muscle. *Math. Prob. Eng.* **2014**, *2014*. [CrossRef]

35. Yang, H.J.; Yu, Y.; Zhang, J.H. Angle tracking of a pneumatic muscle actuator mechanism under varying load conditions. *Control Eng. Pract.* **2017**, *61*, 1–10. [CrossRef]

36. Jia, F.; Hou, L.; Wei, Y.; You, Y.; Yan, L. Adaptive fuzzy sliding mode control for hydraulic servo system of parallel robot. *Indones. J. Electr. Eng.* **2014**, *12*, 4125–4133. [CrossRef]

37. Kanellakopoulos, I.; Kokotovic, P.V.; Morse, A.S. Systematic design of adaptive controllers for feedback linearizable systems. In Proceedings of the 1991 American Control Conference, Boston, MA, USA, 26–28 June 1991; pp. 649–654.

38. Petit, F.; Daasch, A.; Albu-Schaffer, A. Backstepping control of variable stiffness robots. *IEEE Trans. Control Syst. Technol.* **2015**, *23*, 2195–2202. [CrossRef]

39. Taheri, B.; Case, D.; Richer, E. Force and stiffness backstepping-sliding mode controller for pneumatic cylinders. *IEEE-ASME Trans. Mechatron.* **2014**, *19*, 1799–1809. [CrossRef]

40. Esmaeili, N.; Alfi, A.; Khosravi, H. Balancing and trajectory tracking of two-wheeled mobile robot using backstepping sliding mode control: Design and experiments. *J. Intell. Robot. Syst.* **2017**, *87*, 601–613. [CrossRef]

41. Pusey, J.; Fattah, A.; Agrawal, S.; Messina, E. Design and workspace analysis of a 6–6 cable-suspended parallel robot. *Mech. Mach. Theory* **2004**, *39*, 761–778. [CrossRef]

42. Gao, G.; Lu, J.; Zhou, J. Kinematic modeling for a 6-DOF industrial robot. In Proceedings of the 2012 International Conference on Mechatronic Systems and Materials Application (ICMSMA 2012), Qingdao, China, 8–9 September 2012; pp. 471–474.

43. Ayas, M.S.; Altas, I.H. Fuzzy logic-based adaptive admittance control of a redundantly actuated ankle rehabilitation robot. *Control Eng. Pract.* **2017**, *59*, 44–54. [CrossRef]

44. Yu, Y. Q.; Du, Z. C.; Yang, J.-X.; Li, Y. An experimental study on the dynamics of a 3-RRR flexible parallel robot. *IEEE Trans. Robot.* **2011**, *27*, 992–997. [CrossRef]

45. Hosseini, A.; Karimi, H.; Zarafshan, P.; Massah, J.; Parandian, Y. Modeling and control of an octorotor flying robot using the software in a loop. In Proceedings of the 4th International Conference on Control, Instrumentation, and Automation (ICCIA 2016), Qazvin, Iran, 27–28 January 2016; pp. 52–57.

46. Li, X.; Wang, X.F.; Wang, J.H. A kind of Lagrange dynamic simplified modeling method for multi-DOF robot. *J. Intell. Fuzzy Syst.* **2016**, *31*, 2393–2401. [CrossRef]

47. Shao, K.; Ma, Q. Global fuzzy sliding mode control for multi-joint robot manipulators based on backstepping. In Proceedings of the 8th International Conference on Intelligent Systems and Knowledge Engineering (ISKE 2013), Shenzhen, China, 20–23 November 2013; pp. 995–1004.

48. Xing, K.; Huang, J.; Wang, Y.; Wu, J.; Xu, Q.; He, J. Tracking control of pneumatic artificial muscle actuators based on sliding mode and non-linear disturbance observer. *IET Control Theory Appl.* **2010**, *4*, 2058–2070. [CrossRef]

49. Niu, J.; Yang, Q.Q.; Wang, X.Y.; Song, R. Sliding mode tracking control of a wire-driven upper-limb rehabilitation robot with nonlinear disturbance observer. *Front. Neurol.* **2017**, *8*, 646. [CrossRef] [PubMed]

50. Chen, M.; Yu, J. Disturbance observer-based adaptive sliding mode control for near-space vehicles. *Nonlinear Dyn.* **2015**, *82*, 1671–1682. [CrossRef]

51. Zhang, Y.M.; Yan, P. Sliding mode disturbance observer-based adaptive integral backstepping control of a piezoelectric nano-manipulator. *Smart Mater. Struct.* **2016**, *25*, 125011. [CrossRef]

52. Gao, H.; Lv, Y.; Ma, G.; Li, C. Backstepping sliding mode control for combined spacecraft with nonlinear disturbance observer. In Proceedings of the 2016 UKACC 11th International Conference on Control, Belfast, UK, 31 August–2 September 2016.

53. Zhang, M.; Xie, S.Q.; Li, X.; Zhu, G.; Meng, W.; Huang, X.; Veale, A. Adaptive patient-cooperative control of a compliant ankle rehabilitation robot (CARR) with enhanced training safety. *IEEE Trans. Ind. Electron.* **2017**, *65*, 1398–1407. [CrossRef]

54. Su, C.; Chai, A.; Tu, X.K.; Zhou, H.Y.; Wang, H.Q.; Zheng, Z.F.; Cao, J.Y.; He, J.P. Passive and active control strategies of a leg rehabilitation exoskeleton powered by pneumatic artificial muscles. *Int. J. Pattern Recognit. Artif. Intell.* **2017**, *31*, 1759021. [CrossRef]

sensors

MDPI

Article

Knee Impedance Modulation to Control an Active Orthosis Using Insole Sensors

Ana Cecilia Villa-Parra [1,2,*,†], Denis Delisle-Rodriguez [1,3], Jessica Souza Lima [4], Anselmo Frizera-Neto [1] and Teodiano Bastos [1,*]

[1] Postgraduate Program in Electrical Engineering, Federal University of Espirito Santo, Vitoria 29075-910, Brazil; delisle05@gmail.com (D.D.-R.); frizera@ieee.org (A.F.-N.)

[2] Biomedical Engineering Research Group GIIB, Universidad Politécnica Salesiana, Cuenca 010105, Ecuador

[3] Center of Medical Biophysics, University of Oriente, Santiago de Cuba 90500, Cuba

[4] Postgraduate Program in Biotechnology, Universidade Federal do Espirito Santo, Vitoria 29043-900, Brazil; jpaola.fisio@gmail.com

* Correspondence: acvillap@ieee.org (A.C.V.-P.); teodiano.bastos@ufes.br (T.B.); Tel.: +55-27-4009-2077

† Current address: Av. Fernando Ferrari, 514, Goiabeiras, Vitória CEP 29075-910, Brazil.

Received: 10 October 2017; Accepted: 22 November 2017; Published: 28 November 2017

Abstract: Robotic devices for rehabilitation and gait assistance have greatly advanced with the objective of improving both the mobility and quality of life of people with motion impairments. To encourage active participation of the user, the use of admittance control strategy is one of the most appropriate approaches, which requires methods for online adjustment of impedance components. Such approach is cited by the literature as a challenge to guaranteeing a suitable dynamic performance. This work proposes a method for online knee impedance modulation, which generates variable gains through the gait cycle according to the users' anthropometric data and gait sub-phases recognized with footswitch signals. This approach was evaluated in an active knee orthosis with three variable gain patterns to obtain a suitable condition to implement a stance controller: two different gain patterns to support the knee in stance phase, and a third pattern for gait without knee support. The knee angle and torque were measured during the experimental protocol to compare both temporospatial parameters and kinematics data with other studies of gait with knee exoskeletons. The users rated scores related to their satisfaction with both the device and controller through QUEST questionnaires. Experimental results showed that the admittance controller proposed here offered knee support in 50% of the gait cycle, and the walking speed was not significantly different between the three gain patterns ($p = 0.067$). A positive effect of the controller on users regarding safety during gait was found with a score of 4 in a scale of 5. Therefore, the approach demonstrates good performance to adjust impedance components providing knee support in stance phase.

Keywords: active knee orthosis; admitance control; footswitch; gait cycle; knee impedance

1. Introduction

Walking is more difficult for persons that suffer gait impairments due to age, stroke, paralysis or spinal cord injury [1,2]. They usually present muscles weakness, knee instability, gait asymmetry and reduction of gait velocity [3,4], which may produce alterations in sensory or motor systems, leading to injury, disability, risk of falls, loss of independence and reduction in the quality of life [5].

Robotic assisted systems can provide functional compensation for lower-limbs during gait, making possible to improve the human locomotion assistance and the gait rehabilitation through powered exoskeletons and active orthoses [6–9]. The objective of these devices is to help lower-limb impaired people to make their joints move through external movement compensation, using suitable mechanical structures, actuators and control systems. Preliminary findings report promising results, as the fact of sub-acute stroke patients experimenting added benefit from exoskeletal gait training [10], and powered exoskeletons providing individuals with thoracic-level motor-complete spinal cord injury the ability to walk [11].

For the implementation of proper gait training and rehabilitation plans, control strategies that consider both the ability and impairment of the user are required [12]. In this sense, an impedance controller offers the possibility of regulating the mechanical impedance at joints according to the user's disability level and their voluntary participation to promote a compliant human–robot interaction [13–15]. Here, the impedance is regulated through the relation between force, position and its time-derivate, which is given by three components: stiffness, damping and inertia. Thus, a robotic-assisted system can provide interactive gait training adjusting the amount of support to be assisted [8]. In fact, some reviews report that the use of an adaptive impedance control strategy provides a gait motion training that is comparable to the one provided by physical therapists [6].

In this context, some robotic devices use variable impedance, such as the mechanism reported in [16], which employs a variable damping to substitute the stabilizing effect of eccentric quadriceps' contractions during stance flexion in walking. In addition, there is the robotic orthosis reported in [14], which uses an adaptive impedance control to provide assistance at low compliance level to severely impaired subjects adapting the compliance to an increased level for subjects with less severe impairments. In [17], an impedance control is used as method to effectively transfer the task-oriented impedance profile from the human master to the robotic slave device.

Due to the fact that humans change their joint impedances during gait by regulating the postures and the muscle-contraction levels to maintain the stability, robotic devices must integrate methods for a suitable impedance modulation to assist the movement through the gait cycle. Impedance modulation allows promoting a compliant human-robot interaction to provide an effective human support through assisting the limited motor capability of the user [12,13]. Despite this, few studies have explored suitable and reliable methods to execute this modulation in gait applications, which are necessary in rehabilitation robots to guarantee a dynamic performance [18]. The literature provides information about impedance modulation for assist-as-needed control strategies based on interaction torque estimation methods and trajectory references [8,12,18,19]. In this case, a common limitation is the discontinuous model, just like to turn on or off the robotic assistance, rather than offering a seamless impedance tuning process [18]. Robots that use manual impedance level adjustment to adapt the support to patient's capabilities or training progress have also been reported [12]. Some methods try to estimate the joint stiffness using electromyography signals combined with kinetic and kinematic measurements to estimate muscle force, together with models that relate muscle force to stiffness [20,21], which would be of great interest for control strategies. However, these methods have still not been applied in control systems for robotic devices to assist gait.

Furthermore, strategies such as stance control (SC) using impedance control have been little explored, although it is reported as a strategy that can be used to increase walking speed, reduce energy expenditure and gait asymmetry (for both affected and unaffected legs), allowing less stress for paretic musculature in patients with muscular weakness [22–24]. A stance control strategy provides knee stability and protects the joint from collapsing during the standing and stance phase of walking, releasing the knee to allow free motion during the swing phase [25].

A study about the mechanics of the knee during the stance phase of the gait, reported in [26], suggests that, ideally, the mechanism that adjusts impedance at the knee should be based on the gait speed and weight in order to mimic the behavior of the human knee joint. In this sense, a suitable impedance modulation can allow a smooth switching between the stance phase and swing phase to apply impedance compensation during gait under the SC principle, which is a remarkable challenge to warranty a suitable response of robotic orthosis [25,27]. Thus, in contrast with mechanical knee orthoses, a stance controller implemented with a variable impedance controller can be considered as a promising orthotic intervention for assistive devices, in order to provide the patient with adequate knee stability and allow a more normal gait.

The objective of this work is to propose a new method for online impedance modulation to switch the knee impedance throughout the gait cycle in order to implement a stance controller with an admittance controller (one of the variations of impedance controllers). Our impedance modulation method uses the gait velocity, height and weight of the user to generate a gain variable pattern to increase or decrease impedance parameters during gait. Information about the gait phases obtained from an instrumented insole composed of force sensors is used. To validate the approach, the controller was implemented and tested, with different subjects, in an active knee orthosis.

The novelty of the proposed control scheme relies on the use of the footswitch data of the instrumented insole to regulate the knee impedance of the user without additional sensors, through the generation of gain patterns that adjust the impedance components. Previous studies show that instrumented insoles provide information about plantar pressure that can be used to implement strategies for human motion recognition [28] and detect gait sub-phases [29]. In addition, due to the fact that it is a method based on direct measurement of ground reactions having high accuracy [30], several analyses of walking strategies in stroke survivors and older adults are being developed based on data gathered from these instrumented insoles [31,32].

The method proposed here uses two gain variable patterns, which are based on knee torque and knee velocity during gait, in order to evaluate the suitable condition to implement the SC controller. This control strategy also provides the possibility of investigating knee impedance variations in humans, such as done by other studies focused on upper limbs [13], which is of vital interest to researchers involved with the design and control of variable impedance prosthetic and orthotic devices [33]. This paper is organized as follows. Section 2 presents the admittance control strategy and the knee impedance adjustment method, the gait phase detection system composed of an instrumented insole, the description of the active knee orthosis and the experimental protocol used to validate the controller. Section 3 shows experimental results to evaluate the method, and Section 4 presents the discussion.

2. Materials and Methods

2.1. Admittance Controller

According to the description of the SC principle, during gait, it is necessary that the knee impedance variation allows both body support and free movement of the leg. This dynamic requires high resistance at the movement, which can be defined using a system with force feedback. In this sense, admittance controllers are stable in high stiffness conditions; therefore, they are more suitable for implementation of an SC, due to the high and stable stiffness needed to avoid knee collapse during stance phase [34]. An admittance controller is one variation of impedance controller and its performance is determined by both the precision of force sensor and actuator position. Compared with impedance control, admittance behavior is often more easily implemented in hardware [35]. Thus, a proper measure of the effectiveness of a system, which is meant to produce a rapid motion response to external forces, is the mechanical admittance Y [36], defined as:

$$Y = v/F, \tag{1}$$

where v is the velocity of the controlled system at the point of interaction, and F is the contact force at that point. A large admittance corresponds to a rapid velocity induced by applied forces. The dynamic behavior for the interaction between the actuator and the environment (in this case, the user during gait) can be expressed by the model shown in Figure 1.

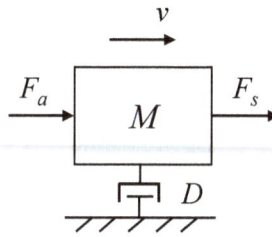

Figure 1. Schematic of one-mass dynamic system.

In this model, the plant parameters are assumed to have values M and D for the mass and damping, respectively, in which an actuator exerts a force F_a, and the environment a force F_s. Then, the equation of motion for the system velocity is

$$M\dot{v} + Dv = F_a + Fs. \tag{2}$$

F_a and F_s can be measured with a force sensor in order to obtain an interaction force F, hence it can be considered $F_a + F_s = F$.

In the Laplace domain, (1) can be expressed as

$$v(s) = F(Ms + D)^{-1}. \tag{3}$$

For the implementation, the use of a velocity controller in the active knee orthosis is assumed. Based on Equations (1) and (3), the desired admittance can be expressed as:

$$Y(s) = (Ms + D)^{-1}. \tag{4}$$

The gain pattern to modulate the inertia and damping is applied to the relation of M and D, maintaining a ratio $r = 0.2$ without considering units, where r was experimentally obtained here and expressed as:

$$r = M/D, \tag{5}$$

with $M > 0$ and $D > 0$.

2.2. Knee Impedance Modulation

In order to implement the SC control strategy with the admittance controller, a modulation through a variable gain G to increase or decrease the impedance components (damping and inertia) is required. The modulation must be according to the gait sub-phases to adapt the knee joint impedance during gait. Usually, most of the gait cycles are divided in the sequence of the following sub-phases [37]: (1) Initial contact (IC), defined by the heel contact; (2) Loading response and mid-stance (MS), defined by a flat foot contact; (3) Terminal stance (TS), defined by the heel off; (4) Swing (SW), defined by the foot off, as shown in Figure 2a.

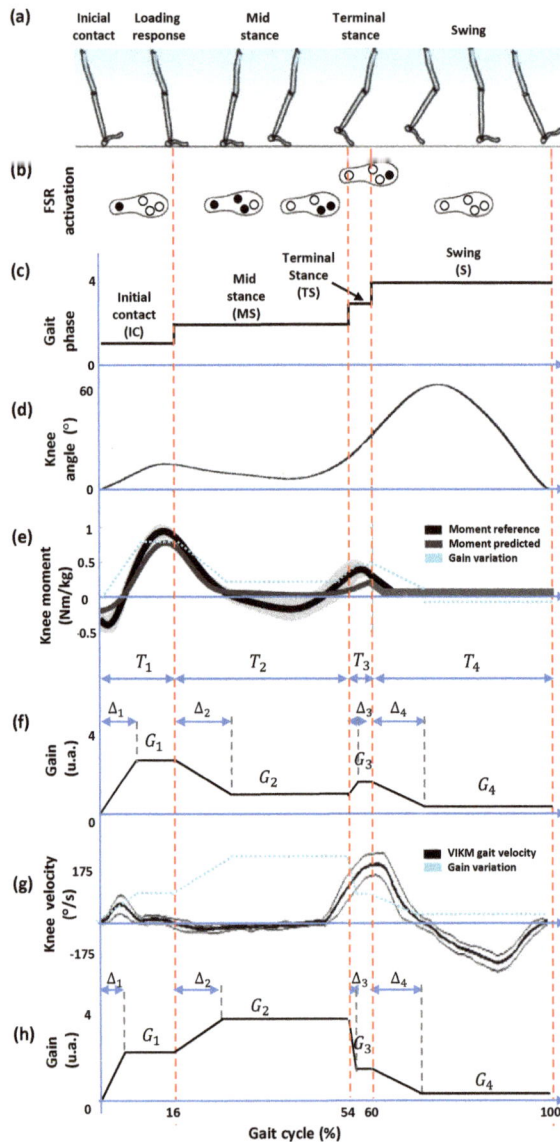

Figure 2. Events related to gait phases. (**a**) sub-phases of the gait cycle; (**b**) on-off sequence of force sensing resistors (FSR) throughout the gait cycle; (**c**) footswitch signal generated by the instrumented insole to identify gait phases; (**d**) knee angle throughout the gait cycle; (**e**) knee moment during gait, correspondent to the reference and predicted values of [21]; the gain variation was considered to define the gait pattern to knee impedance modulation during gait; (**f**) gain pattern P_1 based on the knee moment to decrease/increase gain values during gait phases for stance control; (**g**) knee velocity during gait using the variable impedance knee mechanism (VIKM) [38]; (**h**) gain pattern P_2 based on the the knee velocity to decrease/increase gain values during gait phases for a stance control.

This sequence offers information to develop an impedance modulation for an online variation of the knee joint impedance. The objective is to block the knee joint only in the stance phase to resist the knee flexion and allow free knee extension and free knee motion in the swing phase [25,39], in order to achieve, during gait, the knee angle, moment and velocity, as shown in Figure 2d–f.

For that, a different value of G for each sub-phase must be defined and vary smoothly. Figure 2f,h shows two examples of variation of G in a gait cycle. In both cases, the value for each sub-phase is G_1 for IC, G_2 for MS, G_3 for TS and G_4 for SW, which requires suitable times to increase/decrease G during the gait cycle, defined as: Δ_{t1}, Δ_{t2}, Δ_{t3} and Δ_{t4}. Considering that the weight and the gait velocity are the two major parameters that affect the mechanical parameters of the knee [26], both weight and the gait velocity are considered here to define the corresponding G and Δ_t.

The first example of variation of G, known as *pattern* 1 (P_1), shown in Figure 2f, corresponds to a pattern based on the knee moment variation shown in Figure 2e, which is the knee moment reported in a study of a model of a neuromuscular mechanism to regulate knee joint impedance during human locomotion [21]. Here, P_1 is adapted at the knee moment tendency throughout the gait sub-phases, in which G_1 has the highest values in the IC phase when the knee generates the first flexion. In sub-phase MS, G_2 decreases with a little increment in TS. The second example of variation of G shown in Figure 2h, termed *pattern* 2 (P_2), is a pattern obtained from a tendency marked in Figure 2g, which shows the knee velocity during walking using the variable impedance knee mechanism of an SC orthosis [38]. In this case, the highest value of G_2 is generated in the sub-phase MS, when the knee maintains the angle but the knee torque decreases. In both cases, an impedance modulation using P_1 and P_2 can generate a knee impedance that allows a shock damping during the weight acceptance stage (sub-phases IC and MS) where the knee applies a large moment.

For both patterns, the increase/decrease of G can be executed in times Δ_1, Δ_2, Δ_3 and Δ_4 for IC, MS, TS and SW, respectively. Hence, values of Δ depend on the period of duration of each sub-phase of the gait cycle. Then, considering i as the phase number assigned as follows: $i = 1$ for IC, $i = 2$ for MS, $i = 3$ for TS and $i = 4$ for SW, the duration of each sub-phase can be expressed as

$$T_i = t_{GC}(Q_i/100)f_s, \tag{6}$$

where T_i is the duration of each sub-phase in seconds, t_{GC} is the time of the gait cycle in seconds, Q_i is the percentage of each phase with respect to the gait cycle, and f_s is the sampling frequency in samples per second. As shown in Figure 2f, a suitable Δ_i does not have to exceed the corresponding T_i.

According to gait studies [40], t_{GC} can be estimated through Equation 7:

$$t_{GC} = SL/v_u, \tag{7}$$

where SL is the stride length in meters, and v_u is the user velocity in meters per second. SL can be estimated from the users height H in meters multiplied by the constant 0.826 [40]. Hence, T_i can be expressed as

$$T_i = 0.826(HQ_if_s)/100v_u. \tag{8}$$

Experimental tests to validate Q_i with the instrumented insole were conducted, obtaining the following percentages for each phase: $16 \pm 4\%$, $38 \pm 6\%$, $6 \pm 0.8\%$ and $40 \pm 4\%$ for IC, MS, TS and SW phases, respectively. Based on the knee moment and velocity shown in Figure 2e,g, IC and MS are the more critical phases, which occur when a knee support is required. In this case, Δ_t should allow a time of stabilization in order to sustain the knee with a G constant for each phase.

Therefore, for this method, values of Q_i were defined as: $Q_1 = 10\%$; $Q_2 = 20\%$ and $Q_4 = 40\%$, as shown in Figure 3a,b. This consideration allows having a minimum period of time to increase or decrease the corresponding G and applies to patterns P_1 and P_2 (knee moment and velocity).

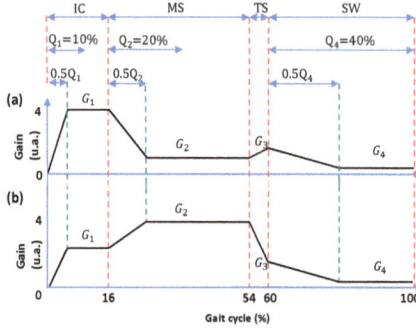

Figure 3. Percentage of each phase with respect to the gait cycle (Q_i) taken into account in this approach for (a) gain pattern P_1; (b) gain pattern P_2.

In relation to the sub-phase *TS*, it can be seen in Figure 3 that it has short duration with respect to other phases, and does not allow a suitable time for stabilization of *G*. For that reason, in order to simplify the method, 30% was chosen as the percentage for Q_3.

Considering that Δ represents 50% of its corresponding *T*, to obtain a smooth switching between the levels of *G*, Equation (9) is used:

$$\Delta_i = (0.0413i/v_u)Hf_s. \tag{9}$$

Figure 4 shows the flowchart of the algorithm implemented in Simulink/Matlab (2014b, The MathWorks Inc.) for online gain pattern generation, where *Phd* is the default phase from which the pattern *G* begins to be generated; *Phs* is the current phase recognized through the insole, and δG is the gain increment for each phase.

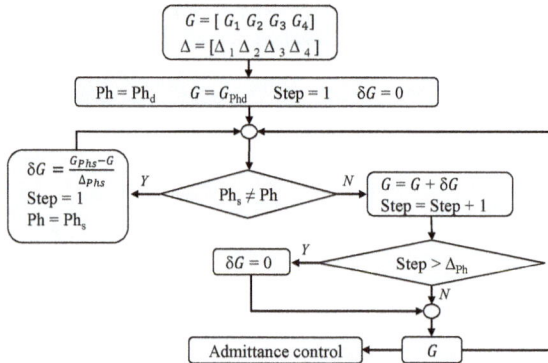

Figure 4. Flowchart of the algorithm used to generate the pattern *G*, where *Phs* is the output of the gait phase detector, and *Phd* is the default phase (recommended sub-phase mid-stance).

Using the aforementioned gain patterns P_1 and P_2, the modulation of *M* and *D* in each gait sub-phase during the gait cycle can be expressed as

$$M_i = M_d G_i, \tag{10}$$

$$D_i = D_d G_i, \tag{11}$$

where M_d and D_d are the inertia and damping default values, respectively, according to Equation (5).

2.3. Gait Phase Detection

The gait phase detection is required to implement the impedance modulation method proposed here, which is done by the instrumented insole built with force sensing resistors (FSRs) shown in Figure 5a. Four FSRs are placed on the plantar surface of the foot. Figure 5b shows the sensor locations, which are defined in the function of the peaks of the plantar pressure data reported in [41,42], corresponding to hallux bone (FSR1), 1st metatarsal (FSR2), 5th metatarsal (FSR3) and calcaneus (FSR4). These locations allow for acquiring more relevant ground reaction forces generated during gait to recognize stance sub-phases, which are suitable to use in feet with normal arch, high arch and flat foot Figure 5c.

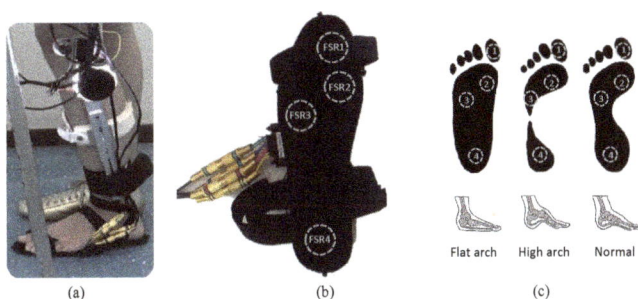

(a)　　　　　　　　　(b)　　　　　　　　　(c)

Figure 5. (**a**) instrumented insole implemented at the active knee orthosis; (**b**) FSR locations; (**c**) FSR locations at flat arch, high arch and normal foot.

The sensors employed are FlexiForce A401 (Tekscan Inc, Boston, MA, USA), which are force-sensing resistors with a sensing area of 25.4 mm and standard force range of 111 N. An electronic circuit was implemented to obtain output voltages proportional to the plantar pressure. To validate the insole data, a pressure sensitive gait mat GAITRite Electronic Walkway Platinum (CIR Systems Inc., Peekskill, NY, USA), 9 m long was employed. The signals of the insole were acquired with a DAQ USB-6009 (sampling frequency of 120 Hz) using the DAQ Express^{TM} driver of National Instruments © (Austin, TX, USA) and Matlab software. The mat data were acquired at 1 kHz using the PKMAS (ProtoKinetics Movement Analysis) software (Franklin, NJ, USA) [43]. The acquisition data were synchronized throughout an external pulse. Two subjects (man: 35 years; 1.72 m; 70 kg and woman: 78 years, 1.75 m, 80 kg) walked at a comfortable velocity on the mat using the insole. Each subject completed six trials (each trial with six steps) completing 36 gait cycles. A concordance correlation analysis was performed to estimate the reliability of the insole pressure signals in relation to the foot pressure measured by the mat.Then, a gait phase detection algorithm based on a truth table from the combinations of the sensors during gait was programed in Matlab Simulink. The signals were acquired through an analog to digital acquisition card, model Diamond-MM-32DX-AT (32 inputs of 16 bits, 4 outputs of 12 bits, with maximum sampling frequency of 250 kHz) of a PC-104 computer, sampled at a frequency of 1 kHz, and conditioned through a low-pass filter Butterworth of 5th-order, with cutoff frequency of 10 Hz. Afterwards, the signals were compared to a threshold of 0.5 V in order to obtain contact information (on-off) from the footswitch. In order to recognize the gait sub-phases *IC*, *MS*, *TS* and *SW*, the combinations shown in Figure 2b were considered. Then, a truth table implemented in Simulink/Matlab, which includes these combinations, was used to obtain a logic scheme to generate the footswitch signal shown in Figure 2c.

2.4. Active Knee Orthosis

Figure 6 shows the active knee orthosis developed at the Federal University of Espirito Santo (UFES/Brazil) known as ALLOR (Advance Lower Limb Orthosis for Rehabilitation), which was used to test the admittance controller with the modulation method proposed here.

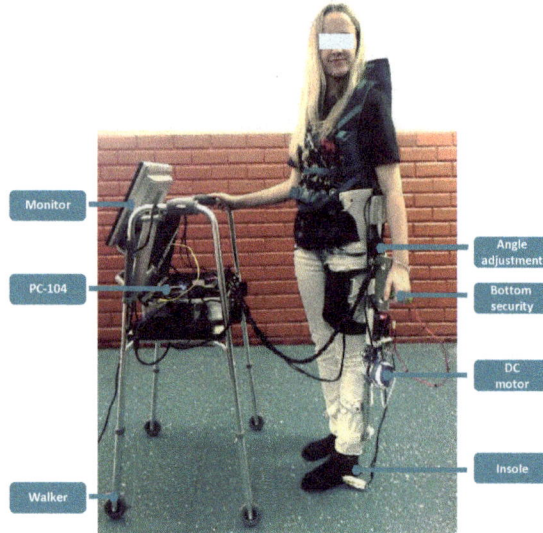

Figure 6. Advance Lower Limb Orthosis for Rehabilitation (ALLOR) built for this research.

ALLOR is a two degree of freedom orthosis composed of an active knee joint and a passive hip, which moves in the sagittal plane during the walking. The hip joint has a manual flexion and extension angle regulator from 0 to 80 degrees. Although this joint is not active, the regulation, according to the user requirements, allows for establishing a safe range of motion. During gait, the physiological range of motion (flexion and extension) must be adjusted to $\pm20°$ for hip, while the movements of the frontal plane are restricted. ALLOR is mounted on the left leg of the user with the axis of rotation of the orthosis joint aligned with the axis of the user knee and hip joints. To ensure a correct alignment during operation, a backpack and rigid braces at the thigh and shank with velcro straps are used. ALLOR weighs 3.4 kg (including 0.8 kg of the backpack) and is adaptable to different anthropometric setups, which include heights of 1.5 to 1.85 m and weights from 50 to 95 kg. It provides both mechanical power to the knee joint and feedback information related to knee angle, interaction torque and gait phases. It was developed for knee rehabilitation in both sit position and during gait. In this last case, the user must use the walker shown in Figure 6.

The components of the active knee joint are a brushless flat motor (model 408057), a Harmonic Drive gearbox (model CSD-20-160-2A-GR) and an analog pulse-width modulation (PWM) servo drive (model AZBH12A8). Additionally, ALLOR is equipped with a strain gauge arrangement (Wheatstone bridge configuration), which measures the torque produced by its interaction with the user. A precision potentiometer model 157S103MX) is used as an angular position sensor to measure knee angles. ALLOR also uses Hall Effect sensors inside the motor to compute angular speeds of the actuator. The computer used to implement the control software is a PC/104, which is a standard for embebbed computers, in which the architecture is built by adding interconnected modules through an industry standard architecture (ISA) data bus. The modules are a motherboard, power source, ethernet communication and an analog to digital (A/D) acquisition card, model Diamond-MM-32DX-AT (32 inputs of 16 bits, four outputs of 12 bits, with maximum sampling frequency of 250 kHz). All sensors, acquisition and velocity driver are connected through the A/D card. The whole system requires 24 V/12A DC power supply and uses a controller area network (CAN) bus running at 1 Mbps. The control software was developed in Simulink/Matlab and uses a real-time target library. Safety conditions are incorporated at the ALLOR control system along with mechanical stops, which ensure that the actuator operates within the normal range of motion of the knee, allowing safe use.

Figure 7 shows the admittance controller implemented in Simulink/Matlab, which is based on Equation (2). *Ph(t)* is the phase, which is recognized online by the gait phase detector through the use of the instrumented insole. *G(t)* is a variable gain for the impedance modulation. The controller also includes an outer force control loop implemented over a inner velocity control loop, in which the motor controller performs the velocity closed-loop control with information feed from Hall sensors on the motor structure. In this controller, the axis of the subject knee joint is considered to be aligned to the axis of the knee joint of the active orthosis.

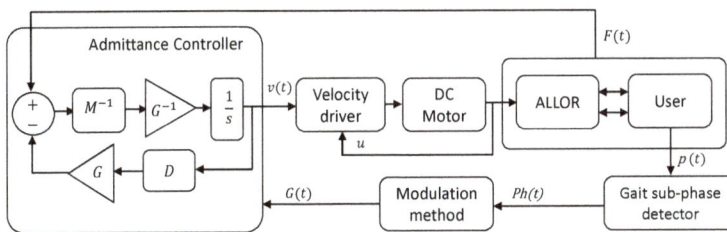

Figure 7. Admittance controller implemented at the active knee orthosis.

2.5. Experimental Protocol

In order to evaluate the proposed method, the following protocol was conducted with ALLOR. Three healthy subjects, female (26 $\pm$ 5.13 years; height 1.62 $\pm$ 0.03 m; weight 56 $\pm$ 8.75 kg) without lower-limb injury or locomotion deficits, participated in the tests. Written, informed consent was obtained from each subject before participation. The Ethics Committee of the Federal University of Espirito Santo approved this protocol, with number: 64801316.5.0000.5542. At the beginning of the test, the subjects were asked to perform a trial with the walker and without ALLOR, walking a distance of 10 m at a comfortable speed for each one. Then, the gait velocity was calculated to obtain the *v* reference value needed to adjust *G*. Then, ALLOR was mounted on the subject to perform three level-ground walking trials in a distance of 10 m with the following patterns for *G*: (1) knee moment based-pattern shown in Figure 3a, termed P_1, with $G_1 = 0.7$ W, $G_2 = 0.2$ W, $G_3 = 0.3$ W and $G_4 = 0.1$ W; (2) knee velocity based-pattern shown in Figure 3b, termed P_2, with $G_1 = 0.4$ W, $G_2 = 0.7$ W, $G_3 = 0.2$ W and $G_4 = 0.1$ W; (3) pattern termed P_3 to perform a gait without knee modulation, maintaining G_4 corresponding to SW phase in all the gait cycle, hence $G_1 = G_2 = G_3 = G_4 = 0.1$ W, where *W* is the user's weight.

For the three patterns of *G*, the impedance parameters *M* and *D* are set as 0.5 kg and 2.5 N/(m/s), respectively, which are obtained experimentally from gait tests with ALLOR. The trials were carried out at slow speed, determined by the subject, and were performed with the acquisition hardware attached to a four wheel walker as shown in Figure 6, in order to have a mobile platform during the tests. Each trial had an average of seven steps, and three trials with each pattern *G* were performed. The patterns (P_1, P_2 and P_3) were randomly applied at the controller in order to not influence their perception regarding the effects introduced by each modulation pattern.

During these experiments, the subjects were asked to accomplish their normal gait patterns, considering the imposed system (ALLOR and walker) and a slow speed. The use of a walker in this study was with the goal of emulating the same conditions of patients or subjects with disabilities, which will need the walker to improve their stability and ambulatory ability themselves in order to feel safety during gait. A sequence of a subject performing this protocol is illustrated in Figure 8.

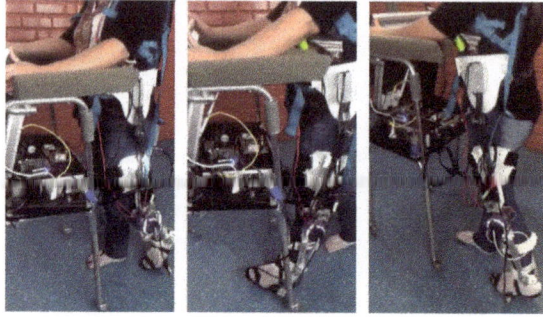

Figure 8. Sequence of an experiment conducted at 0.2 m/s by a subject wearing ALLOR with the stance control using the knee impedance modulation based on the knee moment during gait.

2.6. Statistical and User's Satisfaction Analysis

For statistical analysis, data from the three subjects that participated in the test, related to speed of walking, cadence, stance phase percent and maximum knee flexion in swing phase, were used. Friedman test (non-parametric statistical test) was used to compare the three gain modulation patterns. The level of significance was set at $p < 0.05$.

Finally, a survey to measure satisfaction with the use of assistive technology, the adapted Quebec User Evaluation of Satisfaction with Assistive Technology (QUEST 2.0) was used [44]. QUEST 2.0 may be used to evaluate the user satisfaction through 12 questions separated in two items: assistive technology and services. In this study, only the issues related to assistive technology (dimensions, weight, adjustments, safety, durability, simplicity of use, comfort, and effectiveness) were evaluated, since it is a non-commercial product in the phase of controlled tests. The score for each question ranges from 1 to 5 (1 "not satisfied at all"; 2 "not very satisfied"; 3 "more or less satisfied"; 4 "quite satisfied"; and 5 "very satisfied"), and, finally, an average score is taken for the number of valid questions answered. The subjects were asked to select the three most important items.

3. Results

3.1. Instrumented Insole

Figure 9 shows the mat and insole pressure data during the test with subject 1. For both subjects, 144 steps were collected and the pressure data of the insole presented acceptable values of precision ($r > 0.92 \pm 0.02$) with respect to the mat pressure, which has an accuracy of ($Cb > 0.82 \pm 0.02$), and acceptable reproducibility ($pc > 0.89 \pm 0.03$).

Although the pressure data presented differences in the first sub-phases of stance phase, as shown in Figure 9, the insole showed a good response to recognize stance and swing phase. Hence, the algorithm used in this research to determine the gait phases uses data from the insole.

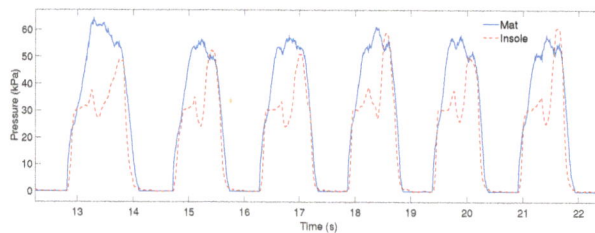

Figure 9. Mat and insole plantar pressure data during six steps.

3.2. Knee Impedance Modulation

Two gain patterns for knee impedance modulation were used to control the knee orthosis: the first based on the knee moment during gait P_1, the second based on knee velocity P_2. For purposes of comparison with both knee modulation patterns, a third pattern P_3 was also used to develop a free gait without knee impedance modulation. Results are shown in Figure 10 to demonstrate the efficiency of the knee modulation proposed for SC assistance, where it is possible to see the variation of the knee angle during gait using patterns P_1 and P_2. Here, the variation of the knee angle in time shows that the subject walked approximately with the same velocity. The knee angle showed similar amplitudes for both patterns. Even though the footswitch signals presented a false negative in sub-phase TS as shown in Figure 10a, the method may adapt the modulation of the gain G in this period and maintain the expected value of G.

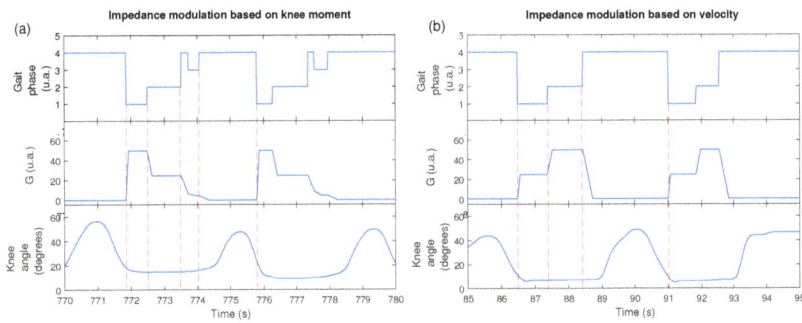

Figure 10. Footswitch signal, gain variation and knee angle during gait with impedance modulation. (**a**) modulation obtained from the pattern P_1 based on normal velocity; (**b**) modulation obtained from the pattern P_2 based on the knee moment.

For the pattern G based on velocity P_2, Figure 11a–c shows the variation of G during the gait, which generates different footswitch signals at different walking speeds.

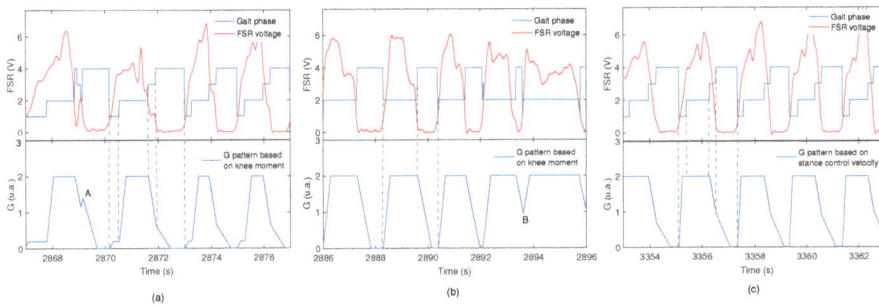

Figure 11. M variation during gait cycle. (**a**) variation of M during a gait test, which generates four sub-phases: initial contact (IC), mid-stance (MS), terminal stance (TS) and swing (SW); (**b**) example with three sub-phases; (**c**) example that shows the variation of M during a gait cycle with noise.

The footswitch signal of the instrumented insole showed good performance to measure the four gait sub-phases considered for the impedance modulation. The gait pattern also was different in some cases, as shown in Figure 10, which is considered common due to the gait dynamic. These examples demonstrate that a specific subject does not present a single characteristic gait cycle. It is reported in literature that the percentage of atypical cycles in the healthy adults is from 1% to 3% [29]. Despite this, the modulation method proposed here was able to generate pattern G to obtain an SC performance even with a non ideal footswitch signal, as shown in Figure 11.

Figure 12 shows the knee torque and knee angle of subject 1, with impedance modulation patterns P_1, P_2, and P_3 during the gait. The maximum torque is presented at the beginning of the knee flexion (marked with red dashed line) for P_1 and P_2. Another increment of knee torque was obtained at the beginning of the stance phase (marked with gray dashed line) for the three patterns.

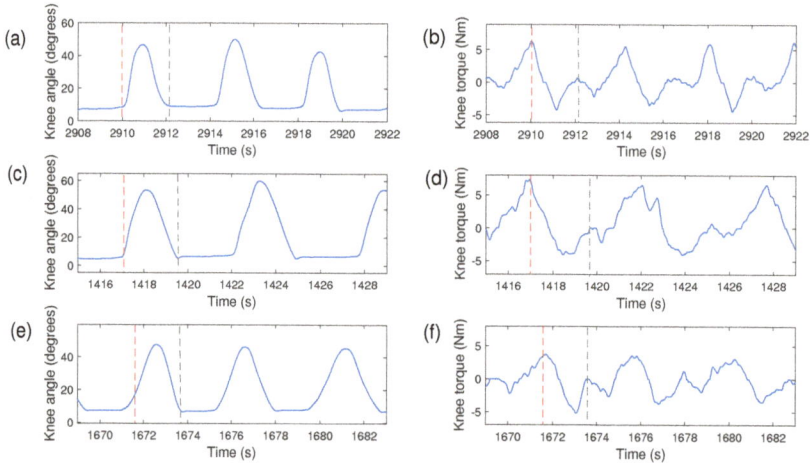

Figure 12. Knee angle and knee torque during knee impedance modulation with patterns G: P_1 (**a,b**); P_2 (**c,d**); P_3 (**e,f**).

In [19], a gait analysis with an active knee orthosis without a walker was conducted, in which the torque with a position control at 0.28 m/s was approximately ±5 Nm. In addition, this study reported that the torque with an adaptive impedance control at 0.28 m/s and 0.44 m/s may have values of ±10 Nm. In our work, walking with ALLOR and the walker at 0.2 m/s implies an interaction torque of ±5 Nm, as shown in Figure 12. In this sense, the new method for knee impedance modulation proposed here presents less knee torque than the torque presented in [19], with both position control and knee impedance modulation. Hence, the method based on FSR sensors for gait phase segmentation may be used to modulate knee impedance without demanding additional knee torque from the user during walking.

Table 1 shows the temporospatial and kinematic information of the subjects when walking using the three G patterns P_1, P_2 and P_3.

Table 1. Mean and standard deviation for temporospatial parameters, and maximum flexion during swing phase for subjects wearing the active knee orthosis controlled by the stance control strategy using three types of impedance modulation patterns.

	Gait Velocity (m/s)	Cadence (steps/min)	Stance Phase (%gait cycle)	Maximum Flexion in Swing Phase (°)
P1	0.18 (0.07)	26.76 (6.87)	49.73 (7.87)	36.44 (9.56)
P2	0.14 (0.05)	22.41 (3.25)	46.19 (9.04)	39.92 (12.40)
P3	0.18 (0.04)	24.02 (6.22)	44.51 (7.75)	39.0 (10.61)
p-value	0.0670	0.0032 *	0.4493	0.1534

P1, Gain pattern based on knee moment; P2, Gain pattern based on knee velocity; P3, Gain pattern for gait without knee support in the stance phase. * significant difference

For the three subjects, the results shown in Table 1 demonstrate no significant difference in gait velocity ($p = 0.067$), percentage of ST phase of the gait cycle ($p = 0.44$) and maximum flexion in SW ($p = 0.153$) between P_1, P_2 and P_3. However, a significant difference in walking with P_2 resulted in a slower cadence in 16% and 10% compared with P_1 and P_3, respectively ($p = 0.0032$).

Considering both modulation patterns, pattern G that presents better temporospatial parameters was the knee moment based on pattern P_1, which reported highest walking speed, cadence and stance phase percentage of the gait cycle compared with P_2. However, regarding the kinematics, the maximum knee flexion was increased ($39.92 \pm 12.40°$) using the pattern P_2 based on knee velocity.

In addition, the duration of the swing phase during walking generally represented 50% to 56% of the gait cycle, as shown in Table 1. A study that describes gait analysis using an exoskeleton with walker [45] reports a swing phase around 37% of the gait cycle with healthy subjects. On the other hand, in [46], a gait analysis with an hybrid neuroprosthesis for SC is performed, which reports that the swing phase during evaluation with nondisabled subjects represents 36% to 51% of the gait cycle. Then, the swing phase percentage obtained during gait with the proposed method agrees with a gait analysis that considers SC. Furthermore, a study of gait analysis with assistive devices tested with pathological cases, such as spinal cord injury [46], reports that the swing phase represents 25% of the gait cycle. In [47], other research of a gait analysis using a knee-ankle-foot orthoses with a powered knee joint is reported, whose swing phase during evaluation with poliomyelitis subjects represents 36% to 51% of the gait cycle. Then, in this sense, an important future task is to analyze the real-time adjustment of knee impedance in pathological gait.

Walker assisted gait with healthy subjects has been reported values between 0.17 m/s and 0.29 m/s depending on the body weight bearing patterns of the leg [48]. In [49], a gait assisted by a smart walker without orthosis in post-stroke subjects showed gait velocity values between 0.23 m/s and 0.44 m/s. Then, for the three patterns used here, the walking velocity is within that range of the first case, which means that the effect of ALLOR with knee impedance modulation does not produce a significant speed reduction in walker-assisted gait. In addition, this value indicates that the incorporation of a smart walker can be considered for test with post-stroke subjects.

Regarding the maximum knee flexion during swing phase, the three patterns present similar values, agreeing with [45], where the walking speed using a powered gait orthosis with a walker has been reported as being $48 \pm 10°$. Here, it should be made clear that, during walker assisted gait, the gait velocity and knee flexion in phase SW are lower than a normal gait.

Regarding the QUEST survey, the user satisfaction with ALLOR controlled by the proposed approach was scored as: dimensions: 4.00 ± 0.00, weight: 3.67 ± 0.58, adjustment: 3.33 ± 0.58, safety: 4.00 ± 0.00, durability: 3.00 ± 1.00, ease of use: 3.33 ± 1.15, comfort: 3.33 ± 0.58 and effectiveness: 3.33 ± 0.58, in a range of 0 to 5.

Based on the experience after and during this study, it was verified that the use of our system requires a therapist or assistant to mount the orthosis on the user. The total time required for this task is approximately 8 min with subjects familiarized with ALLOR. When it is being used for the first time, more minutes are required, in order to adjust the length of leg and thigh segments along with hip angle adjustment. In this case, the total amount of time is from be 20 to 25 min.

4. Discussion

This work evaluated the effect on walking with an active knee orthosis (ALLOR) while using two knee impedance modulation patterns: P_1 (based on knee moment) and P_2 (based on knee velocity), which incorporates an SC strategy. The main functional purpose of the SC strategy is to provide free movements in swing phase and provide a support to knee joints during stance phase.

The three patterns present no significant difference in walking speed, stance phase percentage of the gait cycle and maximum flexion during swing phase, as indicated in Table 1. The results of our study demonstrated that the proposed patterns P_1 and P_2 could be used to improve knee support in stance phase. Hence, both gain patterns are suitable to modulate the knee impedance and assist the knee joint under the SC strategy using an admittance controller.

The variation of the knee impedance was performed considering two implications for the design of stance control orthoses: walking speed and weight. In fact, literature shows that the following parameters: stiffness, knee flexion and extension, and maximum moment change with gait speed [26]. On the other hand, it is also reported that the stiffness of the parallel assistive device should be modified as the load or pilot weight changes [26]. In this sense, both patterns G change the impedance at knee during gait cycle, increasing or decreasing it, according to the both weight and velocity of the subject. Then, the proposed approach may also be considered to evaluate knee impedance variation to design efficient assistive devices. The weight percentage considered for each gait phase was due to the characteristics of the subjects during tests, which are healthy adults that have normal gait development. For this study, gait assisted by a walker was chosen due to it allowing offering safety to the users. It is worth mentioning that parallel bars, crutches or canes can be also used as support elements instead of the walker. For future tests with subjects with disabilities, the total weight of the user will be considered, according to recommendations for the design of parallel assistive devices [26].

In addition, for gait phase recognition in pathological cases, such as stroke survivors with foot drop problems, which present different footswitch signals (percentage of atypical cycles from 11% to 100% in pathological subjects) [29], an alternative for gait phase detection might be necessary. Therefore, an individual study to define it is recommended to design an insole with additional sensors or programming a gait-phase detector for each case must be conducted. In addition, data fusion techniques may be used, taking into account the knee angles acquired by goniometers or inertial measurement units (IMUs). Based on the experience and participants' comments, the instrumented insole was comfortable to use. In future works, the insole will be used to study plantar pressure in order to detect alterations in gait, and allow comparing stroke survivors with healthy people. Stroke survivors usually adopt walking strategies, such as heel walking, planar stride or low heel pressure. These gait alterations can evolve to more complex musculoskeletal disorders, which influence functional activities. Plantar pressure can inform about these alterations, calculating the gait variability over time [31], and therapists can use this data as feedback to help strategies for rehabilitation avoid the evolution of gait disorders.

In relation to user satisfaction, results show that the lowest score (3.00) was related to "durability", while questions on "adjustment", "ease of use", "comfort" and "effectiveness" received a mean value (3.33) on the QUEST score. In this sense, some hardware adjustments are needed to obtain a more robust system and improve the "adjustment", "ease of use" and "comfort" items, such as new materials to adjust the exoskeleton and to decrease structure and the hardware weight. The comfort is associated with adjustment of actuators and biomechanics of human movement. Some factors such as sensors, straps and weight affect the gait of healthy people, causing more energy costs [15]. Physiological theories have been developed to address these limitations in wearable robots [50], but more clinical trials are necessary to determine how these adjustments influence normal and pathological gaits, making these exoskeletons more easy to use in daily activities. It is worth noting that participants in this study this system for the first time. Based on their comments after the experimental protocol, the time required to adjust the device will be improved. We considered that offering unilateral knee assistance for healthy subjects can influence this discomfort.

Regarding the "effectiveness", a clinical protocol with a therapist is needed to address this issue in practice, in order to evaluate ALLOR with knee impedance modulation and its effect on patients. With this purpose, a graphical user interface will be adapted for the therapist who accompanies the rehabilitation, in order to facilitate the programming and monitoring of variables, such as: knee angle and torque, plantar pressure, number of steps and choosing pattern G for knee impedance modulation.

To conclude, our control method constitutes an approach to assist knee movement in stance phase. Future works will focus on implementing a position controller for swing phase or functional electrical stimulation FES, in order to apply force to the advance leg. In addition, future efforts will investigate correlations between the FSR activation and the knee joint impedance during walking on treadmills and stair climbing.

Acknowledgments: This work was supported by CNPq (304192/2016-3), CAPES (88887.095626/2015-01), FAPES (72982608) (Brazil), and SENESCYT (Convocatoria Abierta 2012) (Ecuador). We thank the members of the ALLOR team for supporting the research.

Author Contributions: A.C.V.-P. developed the knee exoskeleton and controller, performed the experiments, analyzed the data and drafted the manuscript. D.D.-R. contributed to designing the experimental protocol, drafting the manuscript and analyzed the data. J.S.L. designed the experimental protocol, contributed to performing the experiments, and analyzed the data and interpretation of findings. A.F.-N. and T.B. contributed with materials and analysis tools, and contributed to the data analysis and interpretation of the findings. All authors read, edited, and approved the final manuscript.

Conflicts of Interest: The authors declare no conflict of interest.

Abbreviations

The following abbreviations are used in this manuscript:

ALLOR	Advance Lower Limb Orthosis for Rehabilitation
D	Damping
F	Force
FSR	Force sensing resistor
G	Gain for impedance modulation
H	ALLOR user's height
IC	Initial contact phase
M	Mass
p	Plantar pressure
P_1	Gain pattern for impedance modulation based on knee moment during gait
P_2	Gain pattern for impedance modulation based on knee velocity during gait
P_3	Gain pattern for impedance modulation to obtain a free movement
Ph	Phase
Q	Percentage
QUEST	Quebec User Evaluation of Satisfaction with Assistive Technology
SC	Stance control
ST	Stance phase
SW	Swing phase
t	Time
W	ALLOR user's weight

References

1. Mahlknecht, P.; Kiechl, S.; Bloem, B.R.; Willeit, J.; Scherfler, C.; Gasperi, A.; Rungger, G.; Poewe, W.; Seppi, K. Prevalence and Burden of Gait Disorders in Elderly Men and Women Aged 60–97 Years: A Population-Based Study. *PLoS ONE* **2013**, *8*, e69627.
2. Balaban, B.; Tok, F. Gait Disturbances in Patients With Stroke. *PM&R* **2014**, *6*, 635–642.
3. Salzman, B. Gait and balance disorders in older adults. *Am. Fam. Physician* **2010**, *82*, 61–68.
4. Hendrickson, J.; Patterson, K.K.; Inness, E.L.; McIlroy, W.E.; Mansfield, A. Relationship between asymmetry of quiet standing balance control and walking post-stroke. *Gait Posture* **2014**, *39*, 177–181.
5. Weerdesteyn, V.; de Niet, M.; van Duijnhoven, H.J.R.; Geurts, A.C.H. Falls in individuals with stroke. *J. Rehabil. Res. Dev.* **2008**, *45*, 1195–1213.
6. Dzahir, M.A.M.; Yamamoto, S.I. Recent trends in lower-limb robotic rehabilitation orthosis: Control scheme and strategy for pneumatic muscle actuated gait trainers. *Robotics* **2014**, *3*, 120–148.
7. Tucker, M.R.; Olivier, J.; Pagel, A.; Bleuler, H.; Bouri, M.; Lambercy, O.; del R Millán, J.; Riener, R.; Vallery, H.; Gassert, R. Control strategies for active lower extremity prosthetics and orthotics: A review. *J. Neuroeng. Rehabilit.* **2015**, *12*, 1.

8. Chen, B.; Ma, H.; Qin, L.Y.; Gao, F.; Chan, K.M.; Law, S.W.; Qin, L.; Liao, W.H. Recent developments and challenges of lower extremity exoskeletons. *J. Orthop. Transl.* **2016**, *5*, 26–37.

9. Villa-Parra, A.; Broche, L.; Delisle-Rodríguez, D.; Sagaró, R.; Bastos, T.; Frizera-Neto, A. Design of active orthoses for a robotic gait rehabilitation system. *Front. Mech. Eng.* **2015**, *10*, 242–254.

10. Louie, D.R.; Eng, J.J. Powered robotic exoskeletons in post-stroke rehabilitation of gait: A scoping review. *J. Neuroeng. Rehabilit.* **2016**, *13*, 53.

11. Louie, D.R.; Eng, J.J.; Lam, T. Gait speed using powered robotic exoskeletons after spinal cord injury: A systematic review and correlational study. *J. Neuroeng. Rehabilit.* **2015**, *12*, 82.

12. Cao, J.; Xie, S.Q.; Das, R.; Zhu, G.L. Control strategies for effective robot assisted gait rehabilitation: The state of art and future prospects. *Med. Eng. Phys.* **2014**, *36*, 1555–1566.

13. Tsuji, T.; Tanaka, Y. Tracking control properties of human-robotic systems based on impedance control. *IEEE Trans. Syst. Man Cybern. Part A Syst. Hum.* **2005**, *35*, 523–535.

14. Hussain, S.; Xie, S.Q.; Jamwal, P.K. Adaptive impedance control of a robotic orthosis for gait rehabilitation. *IEEE Trans. Cybern.* **2013**, *43*, 1025–1034.

15. Huo, W.; Mohammed, S.; Moreno, J.C.; Amirat, Y. Lower limb wearable robots for assistance and rehabilitation: A state of the art. *IEEE Syst. J.* **2016**, *10*, 1068–1081.

16. Bulea, T.C.; Kobetic, R.; To, C.S.; Audu, M.L.; Schnellenberger, J.R.; Triolo, R.J. A variable impedance knee mechanism for controlled stance flexion during pathological gait. *IEEE/ASME Trans. Mechatron.* **2012**, *17*, 822–832.

17. Ajoudani, A. *Transferring Human Impedance Regulation Skills to Robots*; Springer: Berlin, Heidelberg, 2016.

18. Meng, W.; Liu, Q.; Zhou, Z.; Ai, Q.; Sheng, B.; Xie, S.S. Recent development of mechanisms and control strategies for robot-assisted lower limb rehabilitation. *Mechatronics* **2015**, *31*, 132–145.

19. Figueiredo, J.; Félix, P.; Santos, C.P.; Moreno, J.C. Towards human-knee orthosis interaction based on adaptive impedance control through stiffness adjustment. *IEEE Int. Conf. Rehabil. Robot.* **2017**, 406–411. doi:10.1109/ICORR.2017.8009281.

20. Pfeifer, S.; Vallery, H.; Hardegger, M.; Riener, R.; Perreault, E.J. Model-based estimation of knee stiffness. *IEEE Trans. Biomed. Eng.* **2012**, *59*, 2604–2612.

21. Sartori, M.; Maculan, M.; Pizzolato, C.; Reggiani, M.; Farina, D. Modeling and simulating the neuromuscular mechanisms regulating ankle and knee joint stiffness during human locomotion. *J. Neurophysiol.* **2015**, *114*, 2509–2527.

22. Rafiaei, M.; Bahramizadeh, M.; Arazpour, M.; Samadian, M.; Hutchins, S.; Farahmand, F.; Mardani, M. The gait and energy efficiency of stance control knee-ankle-foot orthoses: A literature review. *Prosthet. Orthot. Int.* **2016**, *40*, 202–214.

23. Zissimopoulos, A.; Fatone, S.; Gard, S.A. Biomechanical and energetic effects of a stance-control orthotic knee joint. *J. Rehabilitat. Res. Dev.* **2007**, *44*, 503–513.

24. Zacharias, B.; Kannenberg, A. Clinical Benefits of Stance Control Orthosis Systems: An Analysis of the Scientific Literature. *J. Prosthet. Orthot.* **2012**, *24*, 2–7.

25. Ir, M.; Azuan, N. Stance-Control-Orthoses with Electromechanical Actuation Mechanism: Usefulness, Design Analysis and Directions to Overcome Challenges. *J. Neurol. Neurosci.* **2015**, doi:10.21767/2171-6625.100049.

26. Shamaei, K.; Dollar, A.M. On the mechanics of the knee during the stance phase of the gait. *IEEE Int. Conf. Rehabil. Robot.* **2011**, *2011*, 5975478, doi:10.1109/ICORR.2011.5975478.

27. Yakimovich, T.; Lemaire, E.D.; Kofman, J. Engineering design review of stance-control knee-ankle-foot orthoses. *J. Rehabilitat. Res. Dev.* **2009**, *46*, 257.

28. Han, Y.; Cao, Y.; Zhao, J.; Yin, Y.; Ye, L.; Wang, X.; You, Z. A self-powered insole for human motion recognition. *Sensors* **2016**, *16*, 1502.

29. Agostini, V.; Balestra, G.; Knaflitz, M. Segmentation and classification of gait cycles. *IEEE Trans. Neural Syst. Rehabilit. Eng.* **2014**, *22*, 946–952.

30. Shahabpoor, E.; Pavic, A. Measurement of walking ground reactions in real-life environments: A systematic review of techniques and technologies. *Sensors* **2017**, *17*, 2085.

31. Munoz-Organero, M.; Parker, J.; Powell, L.; Mawson, S. Assessing walking strategies using insole pressure sensors for stroke survivors. *Sensors* **2016**, *16*, 1631.

32. Moufawad El Achkar, C.; Lenoble-Hoskovec, C.; Paraschiv-Ionescu, A.; Major, K.; Büla, C.; Aminian, K. Physical behavior in older persons during daily life: Insights from instrumented shoes. *Sensors* **2016**, *16*, 1225.

33. Tucker, M.R.; Moser, A.; Lambercy, O.; Sulzer, J.; Gassert, R. Design of a wearable perturbator for human knee impedance estimation during gait. *IEEE Int. Conf. Rehabil. Robot.* **2013**, *2013*, 6650372, doi:10.1109/ICORR.2013.6650372.

34. Espinosa, D.A. Hybrid Walking Therapy with Fatigue Management for Spinal Cord Injured Individuals. Ph.D. Thesis, Carlos III University, Madrid, Spain, 2013.

35. Buerger, S.P.; Hogan, N. Impedance and Interaction Control. In *Robotics and Automation Handbook*; CRC Press: Boca Raton, FL, USA, 2005.

36. Newman, W.S. Stability and Performance Limits of Interaction Controllers. *J. Dyn. Syst. Meas. Control* **1992**, *114*, 563–570.

37. Chen, B.; Zheng, E.; Wang, Q.; Wang, L. A new strategy for parameter optimization to improve phase-dependent locomotion mode recognition. *Neurocomputing* **2015**, *149*, 585–593.

38. Bulea, T.C.; Kobetic, R.; Triolo, R.J. Restoration of stance phase knee flexion during walking after spinal cord injury using a variable impedance orthosis. *Conf. Proc. IEEE Eng. Med. Biol. Soc.* **2011**, *2011*, 608–611.

39. Yan, T.; Cempini, M.; Oddo, C.M.; Vitiello, N. Review of assistive strategies in powered lower-limb orthoses and exoskeletons. *Robot. Auton. Syst.* **2015**, *64*, 120–136.

40. Arnos, P. Age-Related Changes in Gait: Influence of Upper-Body Posture. Ph.D. Thesis, University of Toledo, Madrid, Spain, 2007.

41. Wafai, L.; Zayegh, A.; Woulfe, J.; Aziz, S.M.; Begg, R. Identification of Foot Pathologies Based on Plantar Pressure Asymmetry. *Sensors* **2015**, *15*, 20392–20408.

42. Callaghan, M.; Whitehouse, S.; Baltzopoulos, V.; Samarji, R. A Comparison of the Effects of First Metatarsophalangeal Joint Arthrodesis and Hemiarthroplasty on Function of Foot Forces using Gait Analysis. *Foot Ankle Online J.* **2011**, *4*, 1.

43. Lynall, R.C.; Zukowski, L.A.; Plummer, P.; Mihalik, J.P. Reliability and validity of the protokinetics movement analysis software in measuring center of pressure during walking. *Gait Posture* **2017**, *52*, 308–311.

44. De Carvalho, K.E.C.; Júnior, G.; Bolívar, M.; Sá, K.N. Translation and validation of the Quebec User Evaluation of Satisfaction with Assistive Technology (QUEST 2.0) into Portuguese. *Rev. Bras. Reumatol.* **2014**, *54*, 260–267.

45. Kang, S.J.; Ryu, J.C.; Moon, I.H.; Kim, K.H.; Mun, M.S. Walker gait analysis of powered gait orthosis for paraplegic. In *World Congress on Medical Physics and Biomedical Engineering 2006*; Springer: Berlin, Heidelberg, 2007; pp. 2889–2891.

46. To, C.S.; Kobetic, R.; Bulea, T.C.; Audu, M.L.; Schnellenberger, J.R.; Pinault, G.; Triolo, R.J. Stance control knee mechanism for lower-limb support in hybrid neuroprosthesis. *J. Rehabil. Res. Dev.* **2011**, *48*, 839.

47. Arazpour, M.; Moradi, A.; Samadian, M.; Bahramizadeh, M.; Joghtaei, M.; Ahmadi Bani, M.; Hutchins, S.W.; Mardani, M.A. The influence of a powered knee–ankle–foot orthosis on walking in poliomyelitis subjects: A pilot study. *Prosthet. Orthot. Int.* **2016**, *40*, 377–383.

48. Bachschmidt, R.A.; Harris, G.F.; Simoneau, G.G. Walker-assisted gait in rehabilitation: A study of biomechanics and instrumentation. *IEEE Trans. Neural Syst. Rehabil. Eng.* **2001**, *9*, 96–105.

49. Loterio, F.A.; Valadão, C.T.; Cardoso, V.F.; Pomer-Escher, A.; Bastos, T.F.; Frizera-Neto, A. Adaptation of a smart walker for stroke individuals: A study on sEMG and accelerometer signals. *Res. Biomed. Eng.* **2017**, doi:10.1590/2446-4740.01717.

50. Li, J.; Zhang, Z.; Tao, C.; Ji, R. Structure design of lower limb exoskeletons for gait training. *Chin. J. Mech. Eng.* **2015**, *28*, 878–887.

MDPI

St. Alban-Anlage 66

4052 Basel

Switzerland

Tel. +41 61 683 77 34

Fax +41 61 302 89 18

www.mdpi.com

Sensors Editorial Office

E-mail: sensors@mdpi.com

www.mdpi.com/journal/sensors

www.ingramcontent.com/pod-product-compliance
Lightning Source LLC
Chambersburg PA
CBHW051849210326
41597CB00033B/5834